Climate Change, Human Rights, and Adaptive Mobility

Climate Change, Human Rights, and Adaptive Mobility

LAUREN SAKAE NISHIMURA
University of Melbourne

OXFORD
UNIVERSITY PRESS

OXFORD
UNIVERSITY PRESS

Great Clarendon Street, Oxford, OX2 6DP,
United Kingdom

Oxford University Press is a department of the University of Oxford.
It furthers the University's objective of excellence in research, scholarship,
and education by publishing worldwide. Oxford is a registered trade mark of
Oxford University Press in the UK and in certain other countries.

Published in the United States of America by Oxford University Press
198 Madison Avenue, New York, NY 10016, United States of America.

British Library Cataloguing in Publication Data
Data available

Library of Congress Control Number: 2025938669

ISBN 9780198930037

DOI: 10.1093/9780198930068.001.0001

Printed and bound by
CPI Group (UK) Ltd., Croydon, CR0 4YY

The manufacturer's authorised representative in the EU for product safety is
Oxford University Press España S.A., Parque Empresarial San Fernando de Henares,
Avenida de Castilla, 2 – 28830 Madrid (www.oup.es/en or product.safety@oup.com).
OUP España S.A. also acts as importer into Spain of products made by the manufacturer.

MIX
Paper | Supporting responsible forestry
FSC® C013604

For Matt, Maya, and Jackson, my greatest loves

Acknowledgements

This book is the result of work that began over a decade ago and benefited from many colleagues and mentors along the way. Foremost, I would like to thank my doctoral supervisors at the University of Oxford, Catherine Redgwell and Cathryn Costello, for their years of generous feedback, support, and guidance. Their brilliance and scholarship continue to serve as an inspiration as I chart my own academic path.

The research for this book occurred in the various places I have been fortunate enough to call home. In Oxford, in addition to my doctoral supervisors, I am particularly thankful for the feedback received from Liz Fisher, Nazila Ghanea, and Miles Jackson, as well as for the conversations and camaraderie with Erica Bower, Yulia Ioffe, Max Harris, Fernanda Farina, Melita Leoussis, Adora Elisapeta Lovestrand, Lena Kolb, Courtney Wittekind, Matt Walton, and Jill Kavanagh. In addition, I am indebted to Walter Kälin, David Cantor, and Ilan Kelman for the feedback they provided on an early draft of portions of my thesis at an environmental displacement conference at the University of Sheffield. I am also enormously grateful to all involved in the master's in International Human Rights Law programme for setting me on this course. Special thanks are owed to Nazila Ghanea and to Andrew Shacknove, whose continued advice and friendship have been invaluable. In addition, I want to thank Rowan Ryrie for her friendship, solidarity, and making Oxford feel like home.

I am also thankful for my time in Michigan, as a Grotius Fellow and research scholar at the University of Michigan Law School. I benefited greatly from discussions with Yahli Shereshevsky, Tami Groswald Ozery, Salvo Nicolosi, Konstantin Chatziathanasiou, David Hughes, Jing Geng, and Ulf Stückemann. I am particularly grateful to James Hathaway for his insights and thoughtful critiques. Our conversations and work on the Michigan Guidelines on Refugee Freedom of Movement gave me a richer appreciation for the role of refugee law and the resolve to move farther afield in my analysis.

The book developed significantly during a McKenzie postdoctoral fellowship in Melbourne. I am indebted to the University of Melbourne and Melbourne Law School for providing both the support and time needed for further research and writing. I want to thank Margaret Young for her mentorship, advice, and thought-provoking conversations. Special thanks are due to Sundhya Pahuja for her wisdom and generosity, and for making me feel welcome from the moment we met. Thank you to the colleagues that have taken the time to offer guidance, stimulating conversations, and encouragement, including Mark Deng, Michelle Foster, Alison

Duxbury, Jenny Morgan, Mas Generis, Elly Harrould-Kolieb, Marika Sosnowski, Jayani Nadarajalingam, Lulu Weis, Adil Hasan Khan, André Dao, Jordy Silverstein, Bec Nelson, Tomás Daly, Farrah Ahmed, Cait Storr, Christine Parker, Andrew Kenyon, Adrienne Stone, Robi Rado, John Tobin, Alice Palmer, Jackie Peel, Arj Dibley, Radha Govil, Lucy O'Brien, Averyl Gaylor, Jason Odering, Zoe Nay, Roanna McClelland, Ella Vines, Bek Markey-Towler, and Kate Jama. Julia Dehm and Adil Hasan Khan made the transition to Melbourne for my family a seamless one, and I will always be thankful for the serendipity that initially brought us together. My thanks also for discussions in the climate reparations reading group, with organizers Julia Dehm and Erin Fitz-Henry, Dylan Asafo, Elise Klein, and fellow participants.

I would like to express additional gratitude for feedback on drafts of articles and chapters that became this book. Insights and encouragement from my thesis examiners Jane McAdam and Lavanya Rajamani provided the impetus to undertake further research and bring this book into being. I am incredibly grateful to Erica Bower and Zoe Nay for thoughtful comments and suggestions, as well as their friendship and much-needed levity during difficult times. My thanks as well to Fay Gibbons, Paulina dos Santos Major, Merel Alstein, and the team at Oxford University Press for their support and patience, and to the anonymous reviewers who provided comments on my proposal.

Several sections in chapters of the book build on arguments previously published in 'Adaptation and Anticipatory Action: Integrating Human Rights Duties into the Climate Change Regime' (2022) 12(2) *Climate Law* 99–127 and 'The Paris Agreement's Contribution to the Development of the Law on Adaptation' in Alexander Zahar (ed), *Research Handbook on the Law of the Paris Agreement* (Edward Elgar 2024). I appreciated the feedback and editorial expertise of Alexander Zahar on both pieces. Portions of the case studies in my doctoral thesis, which are updated in Chapters 7 through 9, also draw on a study I undertook on behalf of the Office for the UN High Commissioner for Human Rights (OHCHR). Along with the support of the Platform on Disaster Displacement (PDD), I drafted a report that was submitted to the Human Rights Council and published by the PDD, 'The slow onset effects of climate change and human rights protection for cross border migrants' OHCHR (22 March 2018) UN Doc A/HRC/37/CRP.4. I want to recognize Ben Schachter, Pia Oberoi, Atle Solberg, and Sarah Koeltzow, in particular, at OHCHR and PDD.

The focus of my research is further informed by other professional experiences and education. I am grateful for my time working with asylum seekers in the United States and with EarthRights International in Chiang Mai and Yangon. These experiences were both inspiring and challenged me to think hard about the way people move and the conditions that they face. It also made clear that what matters most is the ability to have input and control in the processes that affect our homes and environment. Thank you to Ernesto, Frank, Jose and Luis, and community members and activists in Thilawa and Dawei for sharing your stories. I would also like to thank my mentors and teachers at Georgetown University Law Centre, particularly Richard Lazarus for helping to foster my interest in environmental law, and Philip Schrag and Jaya Ramji-Nogales in the Center for Applied Legal Studies, a clinical programme where I represented my first client seeking asylum.

Finally, I offer my deepest gratitude to my family and friends. Thank you to my parents, Leslie and Rik Mayfield and Wayne and Liane Nishimura, for unwavering love and support. Thank you for the different perspectives provided, for instilling values that are part of my being, and for showing me the awe and beauty of the natural world. I also wish to recognize the generations that came before me. My grandmothers—Sakae Nishimura, Dolores Vrooman, Donna Smith, and Margaret Bauer—provided an example of how to navigate adversity and criticism as fierce, independent women. Thank you to Gerry and Michelle Smith and Teresa Brandon and Bruce Weitzel for their positivity, support, and care. I am blessed with incredible siblings and in-laws to whom I owe much gratitude. I am thankful for the encouragement from and adventures with Parrish Nishimura and Tiffany Lin, Nikki Nishimura, Luke and Katie Mayfield, Jenny Willis and Kimon de Greef, and Sunny Nakae. I am enormously thankful to Sue Willis and Paul Schissler for welcoming me into their family and for the crucial support during pregnancies and the pandemic that made finishing my doctorate possible. I am grateful to my dear friends, old and new, for their support and encouragement across continents.

And the most heartfelt thank you to the three loves of my life. To Matt Schissler, thank you for being the person I start and end my days with, for being the calm in the chaos, and for the laughter, wisdom, and nourishment. Thank you also for ensuring I have had the time and space to think and write, and for your feedback throughout. To my children, Maya Sakae and Jackson Parrish, I began this journey before you existed and I continue, every day, because of you both. May I, may we all, do more to care for the world of which we are a part.

Contents

Table of Cases

INTERNATIONAL COURTS AND TRIBUNALS

African Commission on Human and Peoples' Rights

Committee on the Rights of the Child

European Court of Human Rights

Inter-American Court of Human Rights

International Court of Justice

International Tribunal for the Law of the Sea

Permanent Court of Arbitration

United Nations Human Rights Committee

World Trade Organization

NATIONAL COURTS

Brazil

Netherlands

New Zealand

Table of Treaties and Other Instruments, Decisions, and Reports

TREATIES

OTHER INTERNATIONAL INSTRUMENTS

NATIONAL LEGISLATION

Bangladesh

Fiji

DECISIONS

African Commission on Human and Peoples' Rights (ACHPR)

United Nations Convention to Combat Desertification (UNCCD)

United Nations Framework Convention on Climate Change (UNFCCC)

UNFCCC Conference of the Parties (COP)

Kyoto Protocol COP Serving as the Meeting of the Parties (CMP)

POLICIES, GUIDANCE, AND REPORTS

Adaptation Fund

African Union

Bangladesh Government

Burkina Faso Government

Committee on Economic, Social and Cultural Rights (CESCR)

Intergovernmental Panel on Climate Change (IPCC)

International Labour Organization

International Law Commission (ILC)

Office of the High Commissioner for Human Rights (OHCHR)

Organisation for Economic Co-operation and Development (OECD)

Pacific Island Forum

Solomon Islands Government

Sudan Government

Tuvalu Government

United Kingdom Government Office for Science

Abbreviations

ACHPR	African Commission for Human and Peoples' Rights
ARSDRR	Africa Regional Strategy for Disaster Risk Reduction
BCCRF	Bangladesh Climate Change Resilient Fund
BCCSAP	Bangladesh Climate Change Strategy and Action Plan
BCCTF	Bangladesh Climate Change Trust Fund
CAF	Cancun Adaptation Framework
CBDR-RC	common but differentiated responsibilities and respective capabilities
CEDAW	Convention on the Elimination of All Forms of Discrimination Against Women
CESCR	Committee on Economic, Social and Cultural Rights
CIESIN	Center for International Earth Science and Information Network
CMA	Conference of the Parties serving as the meeting of the Parties to the Paris Agreement
COP	Conference of the Parties
COSIS	Commission of Small Island States on Climate Change and International Law
CRC	Convention on the Rights of the Child
DRR	disaster risk reduction
ECOWAS	Economic Community of West African States
ECtHR	European Court of Human Rights
ExCom	Executive Committee of the WIM
FAO	Food and Agriculture Organization
FbF	forecast-based-financing
FPIC	Free, prior and informed consent
FRDP	Framework for Resilient Development in the Pacific
GCF	Green Climate Fund
GCM	Global Compact for Safe, Orderly and Regular Migration
GEF	Global Environment Facility
GGA	global goal on adaptation
GHGs	greenhouse gases
GIZ	Gesellschaft für Internationale Zusammenarbeit
GST	Global Stocktake
HRC	Human Rights Committee
IACtHR	Inter-American Court of Human Rights
ICCPR	International Covenant on Civil and Political Rights
ICERD	International Convention on the Elimination of All Forms of Racial Discrimination
ICESCR	International Covenant on Economic, Social and Cultural Rights
ICJ	International Court of Justice

ICRMW	International Convention on the Protection of the Rights of All Migrant Workers and Members of Their Families
IDMC	Internal Displacement Monitoring Centre
IDP	internally displaced person
IFRC	International Federation of Red Cross and Red Crescent Societies
IGAD	Intergovernmental Authority on Development
IIED	International Institute for Environment and Development
ILA	International Law Association
ILC	International Law Commission
ILO	International Labour Organization
IOM	International Organization for Migration
IPCC	Intergovernmental Panel on Climate Change
ITLOS	International Tribunal for the Law of the Sea
IUCN	International Union for Conservation of Nature
KJIP	Kiribati Joint Implementation Plan for Climate Change and Disaster Risk Management
LDCF	LDC Fund
LDCs	least developed countries
LECZ	low-elevation coastal zones
LEG	LDC Expert Group
MiGOF	Migration Governance Framework
MoDMR	Ministry of Disaster Management and Relief
MoEFCC	Ministry of Environment, Forest and Climate Change
MPI	Migration Policy Institute
NAP	National Adaptation Plan
NAPA	National Adaptation Programme of Action
NCs	National Communications
NDC	nationally determined contribution
NOAA	National Oceanic and Atmospheric Administration
NRC	Norwegian Refugee Council
NSMDCIID	National Strategy on the Management of Disaster and Climate Induced Internal Displacement
NWP	Nairobi Work Programme
NZHC	High Court of New Zealand
NZIPT	New Zealand Immigration and Protection Tribunal
NZSC	Supreme Court of New Zealand
OAU	Organization of African Unity
OCHA	UN Office for the Coordination of Humanitarian Affairs
OECD	Organisation for Economic Co-operation and Development
OHC	ocean heat content
OHCHR	Office of the UN High Commissioner for Human Rights
PAICC	Paris Agreement Implementation and Compliance Committee
PCA	Permanent Court of Arbitration
PDD	Platform on Disaster Displacement
RMI	Republic of the Marshall Islands

SCCF	Special Climate Change Fund
SIDS	small island developing states
SOP	Standard Operating Procedures for Planned Relocation
TFD	WIM Task Force on Displacement
UDHR	Universal Declaration of Human Rights
UN	United Nations
UNCCD	UN Convention to Combat Desertification
UNCHR	UN Commission on Human Rights
UNCLOS	UN Convention on the Law of the Sea
UNDP	UN Development Programme
UNDRIP	UN Declaration on the Rights of Indigenous Peoples
UNDRR	UN Office for Disaster Risk Reduction
UNEP	UN Environment Programme
UNFCCC	UN Framework Convention on Climate Change
UNGA	UN General Assembly
UNHCR	UN High Commissioner for Refugees
UNHRC	UN Human Rights Council
UNICEF	UN Children's Fund
USAID	United States Agency for International Development
VCLT	Vienna Convention on the Law of Treaties
WAP	W-Arly-Pendjari
WHO	World Health Organization
WIM	Warsaw International Mechanism for Loss and Damage
WMO	World Meteorological Organization
WTO	World Trade Organization
YSPA	Young Power in Social Action

Introduction

In February 2023, the United Nations (UN) Security Council gathered to discuss the threats posed by sea level rise. Among the warnings of vanishing coastlands, flooding, saltwater intrusion, and food and water insecurity, speakers highlighted the risk of vast displacement. The UN Secretary-General, Antonio Guterres, put such a risk in stark terms. He explained that many communities and entire low-lying countries could disappear, and that this would cause the world to 'witness a mass exodus of entire populations on a biblical scale'.[1] While the language of his statement is remarkable, this kind of prediction is not new. As early as its first report in 1990, the Intergovernmental Panel on Climate Change (IPCC) declared that millions could be displaced by shoreline erosion, coastal flooding, and severe drought caused by climate change.[2] Over the decades that followed, researchers' estimates of displacement ranged from tens of millions to over a billion by the middle of this century.[3] In 2018 and 2021, a set of World Bank reports titled *Groundswell* estimated that, by 2050, in a worst-case scenario climate change could lead to the internal displacement of 216 million people across six regions.[4]

Climate change is and will continue to be a driver of human mobility. There is clear scientific consensus that the adverse impacts of climate change are widespread and will only worsen as the world continues to warm. Across countries and communities, those most vulnerable are disproportionately affected, with climate impacts exacerbating vulnerabilities and putting peoples' livelihoods and lives at risk. These risks, in turn, can lead people to leave their homes.[5] Yet such movement will not be monolithic or happen all at once, despite the images conjured by

[1] 'Climate Change-Induced Sea-Level Rise Direct Threat to Millions around World, Secretary-General Tells Security Council' *UN Meetings Coverage and Press Releases* (14 February 2023).

[2] IPCC, 'Climate Change: The 1990 and 1992 IPCC Assessments, IPCC First Assessment Report Overview and Policymaker Summaries and 1992 IPCC Supplement' (IPCC 1992) 103.

[3] Estimates of large numbers of migrants have been questioned for their methodological basis and utility. Discussed in Chapter 1, III; see eg David Durand-Delacre and others, 'Climate Migration Is About People, Not Numbers' in Steffen Böhm and Sian Sullivan (eds), *Negotiating Climate Change in Crisis* (Open Book Publishers 2021).

[4] Kanta Kumari Rigaud and others, 'Groundswell: Preparing for Internal Climate Migration' (World Bank 2018); Viviane Clement and others, 'Groundswell Part 2: Acting on Internal Climate Migration' (World Bank 2021).

[5] See IPCC, 'Summary for Policymakers' in HO Pörtner and others (eds), *Climate Change 2022: Impacts, Adaptation and Vulnerability. Contribution of Working Group II to the Sixth Assessment Report of the Intergovernmental Panel on Climate Change* (Cambridge University Press 2022) 11, 15–16.

Climate Change, Human Rights, and Adaptive Mobility. Lauren Sakae Nishimura, Oxford University Press.
 DOI: 10.1093/9780198930068.003.0001

references to a 'biblical scale' mass exodus. People will move in a variety of ways and with varying levels of resources and support. Movement can occur rapidly or over time, and can affect individuals, groups, or communities. Movement can be temporary, but it will become more permanent as climate impacts begin to render places uninhabitable. It can cross borders, but it is expected to occur more frequently internally, or within countries. And climate impacts can contribute to immobility, with some people unwilling or unable to move.[6]

For all their rhetorical power, such warnings of an impending exodus thus actually *underestimate* the scope and complexity of the problem, thereby distorting the range of potential solutions. These warnings leave out much of the mobility, and immobility, that will occur—and is already occurring—in the context of climate change. Moreover, emphasis on the negative consequences and security threats posed by such displacement can limit the types of responses considered and the legal protections available to migrants. New approaches are needed. This book offers new legal arguments based on an alternative way to conceive of climate-related human mobility. It seeks to ground proactive, adaptive action in obligations of international law.

I. Current approaches to climate change and human mobility

Many of the legal and policy approaches to addressing climate change and human mobility take mobility, particularly displacement, as their conceptual starting point. They therefore explore legal frameworks that address human mobility, including existing international legal regimes and proposals for new frameworks that would create a category of climate-related migrant entitled to legal protection. As Chapter 1 details, these international frameworks share key features that can be limiting in the context of climate change.

The first feature is their concern with specific forms of movement, often forced movement or displacement. They generally do not apply to anticipatory or voluntary movement, which includes the movement certain climate impacts—particularly slower processes of environmental change—are thought to trigger.[7] Different kinds of climate impacts and their interaction contribute to multiple forms of mobility. Sudden-onset events tend to be discrete, last a matter of days, and have immediate effects. They include storms, hurricanes, and flooding.

[6] G Cissé and others, 'Health, Wellbeing and the Changing Structure of Communities' in HO Pörtner and others (eds), *Climate Change 2022: Impacts, Adaptation and Vulnerability* (n 5) 1080–83; Foresight, 'Migration and Global Environmental Change' (UK Government Office for Science 2011) Final Project Report 37.

[7] See Sumudu Atapattu, 'Climate Change, Human Rights, and Forced Migration: Implications for International Law' (2009) 27 Wisconsin International Law Journal 607; Daniel Bodansky, Jutta Brunnée, and Lavanya Rajamani, *International Climate Change Law* (Oxford University Press 2017) 316.

Sudden-onset events frequently lead to displacement, or movement that is predominately forced or compelled. In contrast, slow-onset events are usually gradual and transform the environment over a period of months or years. They include sea level rise, land and forest degradation, desertification, ocean acidification, rising temperatures, and loss of biodiversity. Slow-onset events can contribute to displacement, but they are often associated with migration, or movement that involves more choice and that is undertaken before the worst harm from climate impacts occurs. Climate change is increasing the frequency and intensity of both kinds of events, and adverse effects will continue to accelerate and compound over time.[8] However, the different experiences and types of movement that can occur in the context of climate change are not equally accounted for under current mobility-related international law frameworks. This precludes access to legal status and assistance for many migrants that are not forced to move due to extreme conditions or events.

A second feature is that legal protection under these frameworks tends to be tied to a category of migrant—a 'refugee' or 'stateless person', for example. Yet many people who move in the wake of climate impacts do not fit neatly into these categories, leaving a gap in international legal protection. New categories such as 'climate migrant' attempt to fill this gap, but they consistently founder due to difficulty establishing causal links between specific impacts of climate change and an individual's movement.[9] This is due, in part, to the interaction of the impacts of climate change and the context in which they occur. Climate change is rarely the sole driver of movement. Climate events interact with a wide range of socio-economic, political, cultural, and geographic factors that influence the ability and decision to move. As a result, most people move for a confluence of reasons, and different people confronted with the same conditions may make different decisions about whether to stay or to move. Qualifying for protection based on one of these categories of migrants generally requires individual status determinations, which are subject to jurisdictional constraints and can lead to inconsistent outcomes. Furthermore, for those forced to flee their country, legal status is meant to confer a form of surrogate protection by creating obligations under international law for another country to step in when individuals are unable or unwilling to obtain such protection from their own countries. Many countries that face significant risks from climate change, however, want to continue to provide protection to

[8] IPCC, *Climate Change 2021: The Physical Science Basis. Contribution of Working Group I to the Sixth Assessment Report of the Intergovernmental Panel on Climate Change* (V Masson-Delmotte and others (eds), Cambridge University Press 2021); J Hansen and others, 'Ice Melt, Sea Level Rise and Superstorms: Evidence from Paleoclimate Data, Climate Modeling, and Modern Observations That 2°C Global Warming Could Be Dangerous' (2016) 16 Atmospheric Chemistry and Physics 3761.

[9] See Jane McAdam, 'Swimming against the Tide: Why a Climate Change Displacement Treaty Is Not the Answer' (2011) 23 International Journal of Refugee Law 2, 14–15; Walter Kälin and Nina Schrepfer, 'Protecting People Crossing Borders in the Context of Climate Change Normative Gaps and Possible Approaches' (UN High Commissioner for Refugees (UNHCR) 2012) 28–29.

their people, and people may also want to continue to be considered nationals and do not want to lose connection to their homeland.[10]

Thirdly, there are limitations imposed both temporally and by a focus on severe risks. Current frameworks tend to focus on protection *after* people have moved or crossed borders and do not apply until people experience or face serious harm. For example, refugees and those granted protection from *refoulement* must generally be outside their country. Consequently, such protections do not apply until people have left their country, while most mobility associated with environmental change occurs internally.[11] For these migrants, home countries should ensure access to fundamental human rights, but it is often an inability to do so that prompts people to flee. The severity of the risk required further limits the timing of protection under current international frameworks. Complementary protection offered through the principle of *non-refoulement*—which protects people from being returned to a country—illustrates this point. It applies in cases where people are at risk of irreparable harm, or when they face life-threatening situations; torture; or cruel, inhuman or degrading treatment upon return. While these kinds of claims are possible, there has yet to to be a case in which the effects of climate change were found to meet the required threshold of harm. This is true even where people are already facing significant restrictions to potable water and difficulty growing crops due to saltwater intrusion caused by sea level rise,[12] which raises questions about how dire a situation must be before relief is granted. Extreme weather events like hurricanes, heavy rains, and flooding can quickly displace people. But many of the risks from climate change accrue over time or as impacts increasingly worsen, which makes them a poor fit for protection frameworks that apply in extreme circumstances or after their occurrence.

Finally, and most fundamentally, these approaches individualize what should be a collective problem. Making legal status and protection contingent on meeting specific criteria or qualifying as a certain category of migrant constructs the problem as one to be solved on an individual basis. This asks individuals to navigate bureaucratic systems for themselves, rather than treating the role of climate change in mobility as an issue demanding collective solutions. Some individuals will succeed and qualify for legal protection under existing frameworks while others in similar situations will not. Moreover, when entire geographic areas are at risk and whole communities may need to relocate, solutions must be envisioned that do more than require people to act exclusively as individuals in search of protection that is often ad hoc and too late. Collective action will be essential, and international law can and should play a part in the concerted efforts needed to

[10] Discussed further in Chapter 8, exploring the Pacific Islands.

[11] See D Dodman and others, 'Cities, Settlements and Key Infrastructure' in *Climate Change 2022: Impacts, Adaptation and Vulnerability* (n 5) 930; Cissé and others (n 6) 1045, 1080.

[12] As was the case in *Teitiota v New Zealand*, discussed in Chapters 1 and 4.

provide support, adequate legal protection, and safety and dignity for those facing significant harm from climate change.

II. Reversing the approach

This book offers a new approach to climate change and human mobility. Rather than beginning with frameworks on mobility and asking how they address the challenges created by climate change, it makes the international regime on climate change the starting point. More specifically, it explores adaptation within this international regime and how the obligations related to adaptation that it contains can be turned towards issues of human mobility. A focus on the climate regime enables action that can help prevent displacement, through both measures that mitigate the causes of climate change and also those that adjust to its actual or expected impacts. The former, mitigation, can reduce overall warming and severity of adverse effects and thereby decrease the number of people likely to be displaced. The latter is adaptation, which can include measures that enable people to stay in place as long as possible and to move safely elsewhere.

Adaptation is a process of adjustment and it can thus include measures that are both responsive and, crucially, anticipatory. Adaptation is therefore not necessarily limited temporally, nor are obligations on adaptation focused on specific types of movement or categories of individual migrant. As a result, mobility can be understood as part of the process of adaptation, to respond to the risks posed by climate change. The book refers to this as adaptive mobility. As conceived of in the chapters that follow, adaptive mobility is an anticipatory, proactive approach to movement that is rooted in human rights. It involves planning for and facilitating mobility to address reasonably foreseeable risks and to help ensure access to critical rights and resources. Adaptive mobility includes migration and planned relocation. Planned relocation is the coordinated movement of people away from disaster or risk-prone areas.[13] Unlike temporary evacuation, it is considered a permanent solution. Planned relocation often involves action by governments, and it is more likely to succeed when it includes the participation of affected communities for resettlement elsewhere, in conditions that ensure human rights. Adaptive mobility should be informed by the choices and preferences of people, and it must be guided by the international legal obligations of countries. In providing safe, rights-protective

[13] This book adopts language used in the international regime on climate change, which has referred to three forms of 'climate change induced' movement: displacement, migration, and planned relocation. Decision 1/CP.16, 'The Cancun Agreements: Outcome of the work of the Ad Hoc Working Group on Long-term Cooperative Action under the Convention' (15 March 2011) FCCC/CP/2010/7/Add.1, para 14(f); see also The Nansen Initiative, 'Agenda for the Protection of Cross-Border Displaced Persons in the Context of Disasters and Climate Change' vol 1 (2015) paras 8, 16–22 (echoing the Cancun Agreements' language).

means to move elsewhere, adaptive mobility can also help prevent forced immobility or precarious migration.

This book aims to provide a legal basis to shape adaptation, and it argues that international law can provide concrete, actionable grounds to prompt anticipatory and proactive measures that protect people in advance of serious harm. Anchoring an approach to mobility in adaptation within the international climate change regime takes advantage of the regime's extensive reach. The international climate change regime includes three climate treaties—the 1992 United Nations Framework Convention on Climate Change (UNFCCC), the 1999 Kyoto Protocol, and the 2015 Paris Agreement[14]—the Conference of the Parties to these treaties, the bodies created by the parties, and the rules and decisions they make. This book focuses on the UNFCCC and the Paris Agreement, as the Kyoto Protocol is primarily aimed at mitigation and did not create any new legal obligations on adaptation. These treaties are a product of negotiation and compromise, and do not go far enough in efforts mitigate the causes of climate change.[15] Yet both the UNFCCC and the Paris Agreement enjoy near-universal participation and do not allow for reservations, which means that almost every country is subject to their terms.[16] As such, the obligations they create have the potential to be a powerful legal tool to guide climate action, particularly on adaptation.

There are three kinds of adaptation obligations identified in this book: to plan for and implement adaptation, to assist, and to cooperate. These obligations do not mandate any specific action, enabling a wider focus on measures that can be taken before, during, and after people move. Such a focus can take advantage of the knowledge and foresight that the science on climate change affords to argue for action on adaptation now, rather than only after people are displaced. But the breadth of adaptation obligations can create ambiguity about their content. Addressing this ambiguity and making adaptation obligations more concrete is thus one of the core tasks of this book. Its chapters set out how human rights law can help clarify the content of these obligations, and they explain how integrating positive duties from human rights can lead to key requirements related to human mobility.

[14] UNFCCC (adopted 9 May 1992, entered into force 21 March 1994) 1771 UNTS 107; Kyoto Protocol to the UNFCCC (adopted 11 December 1997, entered into force 16 February 2005) 2303 UNTS 162 (Kyoto Protocol); Paris Agreement (adopted 12 December 2015, entered into force 4 November 2016) 3156 UNTS 79.

[15] The combined emission reductions pledged by parties to the Paris Agreement could still lead to a 2.5–2.9 degree Celsius (°C) warming by the end of the century, well short of the Agreement's goal to hold the global average temperature 'to well below 2°C' while pursuing a 1.5°C limit. See Paris Agreement, art 2(1)(a); Conference of the Parties serving as the meeting of the Parties to the Paris Agreement (CMA), 'Nationally Determined Contributions under the Paris Agreement: Synthesis Report by the Secretariat' (2022) FCCC/PA/CMA/2022/4.

[16] There are 198 parties to the UNFCCC and 195 to the Paris Agreement. UN Treaty Collection, 'Status of UNFCCC' ch XXVII: Environment <https://treaties.un.org/Pages/ViewDetailsIII.aspx?src=TREATY&mtdsg_no=XXVII-7&chapter=27&Temp=mtdsg3&clang=_en> accessed 9 June 2023; UN Treaty Collection, 'Status of Paris Agreement' <https://treaties.un.org/Pages/ViewDetails.aspx?src=IND&mtdsg_no=XXVII-7-d&chapter=27&clang=_en> accessed 9 June 2023.

These include the need for countries to plan for and facilitate mobility as a part of a process of adaptation and, as part of this process, to ensure the participation of affected persons and prioritize those most vulnerable. This approach to mobility requires action *before* people are unable to access their rights and resources and *before* they are either rendered immobile or forced into precarious, irregular movement. It aims to enable a proactive, anticipatory form of adaptive mobility.

The climate regime also provides a collective dimension to adaptive mobility, including through its adaptation obligations. Obligations to assist and cooperate are the foundation for the book's argument for adequate support from developed to developing countries in measures to plan for and facilitate adaptive mobility.[17] Such a requirement is bolstered by the operative principles, collective goals, and differentiated responsibilities on adaptation set out in climate treaties, which are described further in Chapter 2. When implemented with obligations to plan for and facilitate adaptive mobility, support obligations can help shift the burden away from countries that are facing some of the greatest harm from climate change but that have contributed least to its causes. They put the onus on developed countries due to their capacity and contribution to climate change.

III. Migration as adaptation and adaptive mobility

As a necessary complement to providing a new legal approach, this book aims to contribute to a specific framing of climate change and human mobility, one that emphasizes the needs, rights, and agency of people. In doing so, it moves away from warnings that emphasize an exodus and analyses premised on climate change as a driver of displacement, which can justify a focus on the national security risks and negative consequences of such displacement. Discussions of mass movement at the Security Council, conflict and the political tensions this could create, and disruptions to stability and peace are examples of the security concerns raised in current debates. These concerns are valid, particularly if more concerted, urgent climate action is not taken. Yet making them the primary framing for climate-related movement can constrain understanding of both the challenges posed by climate change and the kinds of solutions available and necessary. This framing can, for example, deflect the blame for movement from those causing climate change to those suffering its consequences, while glossing over action or inaction by states.[18] It can reinforce marginalization, racialized distinctions, and

[17] The distinction between developing and developed countries is one made in climate treaties, as discussed in Chapter 2.

[18] Like the climate change regime, this book uses the term 'state' synonymously with country. There is no formal definition of state under international law, and the term broadly refers to an independent country with standing in international law. See Montevideo Convention on the Rights and Duties of States (adopted 26 December 1963, entered into force 26 December 1934) 165 LNTS 19, art 1 (setting out criteria of statehood; treaty widely considered customary law).

the othering of migrants, and it can also enable responses that restrict mobility and create barriers to migration.[19] Restrictive policies and practices that create onerous processes, close borders, or criminalize entry or stay can push people into irregular, precarious journeys. They could lead to an increase in the kind of displacement that is the focus of security concerns, exacerbating risks to people and the places to which they migrate.

While acknowledging that most movement involves some degree of choice and compulsion,[20] this book contrasts displacement with adaptive mobility. The connection between human mobility and adaptation might seem intuitive. People move as an adaptation strategy, in the face of climate risk or other changing conditions. They will continue to do so even without access to external support from governments or the international community. The ability to move freely and the challenges migrants encounter in their journeys, however, will depend largely on the resources they can access and the legal pathways and protections that are available. Recognition of the connection between adequate and timely planning, resources, and protection and the experience of mobility is foundational to the arguments that will follow, as is the belief that law should not be static. While decisions and plans can be put into place now to improve the journeys of migrants in the future, legal approaches must be adaptable and evolve alongside the challenges they address.

Leaving one's home is rarely an easy prospect. But safe, dignified movement is possible, and mobility has the potential to be a beneficial strategy rather than a source of further suffering. The adaptive mobility advocated for in this book offers an approach grounded in legal obligations to ensure anticipatory, planned, and rights-protective pathways elsewhere. It focuses on risks to people and their rights, which enables discussion of measures that prioritize human rights and needs. These measures may therefore consider access to food, water, health services, housing, and livelihoods, as well as the conditions for migrants in transit and at destination. This can contribute to better outcomes for people. It can help avoid the kind of displacement national security narratives are built upon, particularly when impacts to host communities are considered and mobility involves more choice than compulsion. While such an approach does not protect against all forms of

[19] For more, see eg Betsy Hartmann, 'Rethinking Climate Refugees and Climate Conflict: Rhetoric, Reality and the Politics of Policy Discourse' (2010) 22 Journal of International Development 233; Lauren Nishimura, '"Climate Change Migrants": Impediments to a Protection Framework and the Need to Incorporate Migration into Climate Change Adaptation Strategies' (2015) 27 International Journal of Refugee Law 107, 120–21; Carmen G Gonzalez, 'Migration as Reparation: Climate Change and the Disruption of Borders' (2020) 66 Loyola Law Review 401, 425–26.

[20] Compulsion can arise from several sources that include the contextual risks people face, the exercise of power by a relevant authority, or conditions after people move that make it difficult to return. David James Cantor, 'Conceptualising "Relocation" Across Displacement Contexts' (2024) 15 Journal of International Humanitarian Legal Studies 23, 36. The book seeks to offer an approach that enables mobility before people are compelled to move due to contextual drivers, and it advocates for the role of authority to be one that facilitates rather than forces movement.

harm associated with climate impacts or mobility, it does provide affected communities with avenues to proactively move with more dignity and agency. When mobility is forced, more stress and greater disruptions to lives and livelihoods ensue.[21] Furthermore, a focus on people and their rights translates into action by states at different points in the process of adaptation. Such a process can include measures to reduce vulnerability, enable people to stay, provide participatory processes, and plan for and facilitate adaptive mobility, all of which can help prevent precarious movement. This approach to adaptation and mobility is not meant to dictate outcomes, as the range of possible measures evinces. Instead, it underscores the importance of putting people at the centre of climate action. Doing so is necessary both as a matter of international law (as Chapters 3 to 6 argue) and for vulnerable countries and communities to successfully adapt, which in some circumstances requires undertaking adaptive mobility (as illustrated in Chapters 7 to 9). Such an approach also focuses on collective action and support for adaptation—and measures to enable adaptive mobility specifically—both so that individuals do not carry the burden alone and so that responsibility is shared amongst countries.

IV. Structure of the book

This book uses legal analysis and case studies to build arguments for a new approach to climate change and human mobility, one that leads to adaptive mobility and support for developing countries in efforts to enable such mobility. It looks to international law to provide a legal basis for this approach, arguing that adaptation obligations can be interpreted and applied to compel and shape state planning for, facilitation of, and support for adaptive mobility. To accomplish this, the book identifies adaptation obligations in the climate regime, uses treaty interpretation to clarify these obligations, and closes with case studies to illustrate how legal arguments can work in action.

Chapter 1 sets the stage for the rest of the book. It discusses the connection between climate change and human mobility and the limitations of current international law and policy approaches to climate-related mobility. The chapter establishes that gaps in legal protection exist for migrants, gaps this book speaks to through arguments for adaptive mobility. It also explains several concepts and conclusions that are necessary for subsequent arguments, including the evidentiary basis for predicted climate impacts, the disproportionate way impacts affect many developing countries, and the importance of and approach to vulnerability in the book. This knowledge and the ability to plan for climate impacts underlie

[21] See eg IPCC, 'Managing the Risks of Extreme Events and Disasters to Advance Climate Change Adaptation', A Special Report of Working Groups I and II of the IPCC (Field CB and others (eds), Cambridge University Press 2012) 300; The Nansen Initiative (n 13) 38–39; Cissé and others (n 6) 1080.

the arguments in this book, which relies on the foreseeability of the risks they pose to trigger proactive adaptation that prioritizes some of the most vulnerable people, as well as support for adaptive efforts in vulnerable developing countries.

Chapter 2 focuses on adaptation in the climate change regime, which provides the foundation for the book's approach to climate change and human mobility. It first analyses the meaning of climate change adaptation and emphasizes that it is a process which evolves with changing conditions and which can lead to actions in anticipation of, or reaction to, these conditions. It also argues that in addition to environmental conditions, social, economic, political, and cultural conditions must be understood and accounted for in adaptation processes, so that they may adequately protect human interests and rights. Chapter 2 explains that adaptation can include a range of measures, some of which can prevent displacement and allow people to stay in place for as long as possible and others that are necessary to plan for or facilitate adaptive mobility. The chapter next describes the evolution of adaptation in the climate change regime, from the initial perception of adaptation as a secondary response—one expected to occur naturally or if mitigation efforts failed—to its increasing importance as knowledge about the impacts of climate change grows. It then discusses the relationship between adaptation and loss and damage, as well as the history of human mobility in the regime. The bulk of the chapter analyses adaptation obligations. It discusses the nature of these obligations and their scope, content, and relationship to other provisions in climate treaties. It identifies three types of adaptation obligations: to act on adaptation, including by planning for and implementing measures; for certain developed countries to assist developing country parties with adaptation, primarily through financial support; and to cooperate in adaptation efforts.[22] These obligations provide the legal foundation for arguments to compel and shape state action on adaptive mobility.

Adaptation obligations are broad, but they can be made more concrete. Chapters 3 to 6 concretize them through the integration of international human rights law and the incorporation of the climate regime's operative principles. The latter include the principles of precaution, equity, and common but differentiated responsibilities and respective capabilities (CBDR-RC). These chapters argue that duties associated with human rights can help clarify what adaptation obligations require and shape adaptive action, including on human mobility. Chapter 3 builds on the analysis of the climate change regime and provides the central methodology of the book. It starts from the premise that international law is a system that should aim for coherence and consistency. To that end, Chapter 3 argues that the tools of treaty interpretation provide a means to understand adaptation obligations in a way that accords with the broader international legal system. It then applies the interpretive process to focus on climate treaties' terms and operative principles,

[22] See UNFCCC, art 4(1)(b), (e), (3), (4); Paris Agreement, arts 7(1), (13), 9(1).

context, object and purpose, and the systemic integration of relevant rules of international law.[23] While these are all part of a single interpretive process, the analysis examines how each element contributes to the interpretation of adaptation obligations. It demonstrates that the integration of international human rights law is necessary, setting out the steps that lead to such a conclusion: establishing the general relevance of human rights law, arguing for an evolutive interpretation in line with the nature of adaptation and the climate regime, and discussing the broad applicability of human rights. It also links treaty interpretation to adaptive mobility as a form of adaptation, highlighting how the elements of interpretation contribute to legal arguments for more proactive, anticipatory mobility.

Chapter 4 takes the general argument for integration and demonstrates how it can clarify the first type of adaptation obligation—obligations to act on adaptation. It focuses on two specific substantive human rights: the rights to life and to an adequate standard of living. The latter is analysed through its component rights, the rights to adequate food, water, housing, and health. While the interdependence and indivisibility of human rights are not disputed,[24] these substantive rights are critical to survival and subsistence, and they play a significant role in whether a person is able to stay or move elsewhere. Integration of the positive duties associated with these rights leads to an interpretation that requires countries to act in the face of foreseeable risks from climate impacts to ensure access to essential resources and a life with dignity. In certain circumstances, these duties will necessitate that countries undertake adaptive mobility measures, or measures to plan for and facilitate adaptive mobility within borders and, where possible, across borders.

Chapter 5 explores how procedural rights can help shape adaptive mobility. Procedural rights include rights of participation in decision-making and access to information, and this chapter argues that their integration leads to an interpretation of adaptation obligations that requires countries to provide information and opportunities to participate and provide input in the process of adaptive mobility. The chapter underscores the role of affected persons' agency, participation, and perception plays in mobility outcomes. It closes by applying the climate regime's principle of precaution, which supports adaptive action sooner than reliance on human rights alone to address the foreseeable risks posed by climate change and to incorporate human rights benchmarks and safeguards into planning and anticipatory action. This step in the interpretive process is important, because when positive obligations must be undertaken under human rights law is not always clearly specified or tends to be clarified after a violation occurs.

[23] See Vienna Convention on the Law of Treaties (adopted 23 May 1969, entered into force 27 January 1980) 1155 UNTS 331 (VCLT) arts 31(1), 31(2), 31(3)(c).

[24] For more on environmental harms and the indivisibility of human rights, see Erin Daly and James R May, 'Indivisibility of Human and Environmental Rights' in James R May and Erin Daly (eds), *Human Rights and the Environment: Legality, Indivisibility, Dignity and Geography*, Elgar Encyclopedia of Environmental Law Series, vol VII (2019) 171–82.

Chapter 6 turns to adaptation obligations of assistance and cooperation. It analyses duties from human rights law that can be integrated into the interpretation and implementation of these obligations, to direct them towards support for adaptive mobility measures. It also argues that these adaptation obligations add a collective dimension to support, which can shift or share the burden of adaptation efforts and adaptive mobility. As with the principle of precaution, the principles of equity and CBDR-RC help guide adaptation obligations and clarify who is due support and who must provide it. The chapter explains how these principles, bolstered by human rights duties, translate into adaptation obligations that require developed countries to support developing countries in their adaptation efforts, including adaptive mobility measures. Such support is crucial when developing countries lack capacity to undertake adaptation or provide essential resources. It is also a key addition to this book's approach to mobility. At a minimum, support must include financial and technical assistance for internal adaptive mobility measures. As this book argues, in some contexts it can and should also include cooperative efforts to facilitate mobility across borders. However, there are challenges to assistance and cooperation that must be overcome. The chapter discusses these obstacles and raises potential ways forward, as well as open questions that remain. These questions provide fertile ground for future research and areas of advocacy, to fill gaps in financing and support for climate adaptation and adaptive mobility.

Chapters 7 to 9 explore the book's legal arguments through three case studies: Bangladesh, the Pacific Islands, and the Sahel. These case studies are emblematic of the general need for a more proactive, anticipatory approach to mobility. However, what that mobility can and should look like differs depending on the context. Drawing on existing empirical studies and relevant legal and policy analysis, the case studies highlight a range of climate change impacts and forms of mobility, while also discussing different aspects of the legal arguments raised in prior chapters. Each explores areas that are highly vulnerable and expected to face significant and foreseeable climate impacts. They also focus on vulnerable developing countries, as these countries are owed assistance and cooperation with adaptation in the climate regime. Each chapter shows how the integration of human rights, knowledge about climate impacts, and precaution can shape adaptation and lead to adaptive mobility. Yet the different locations are highlighted to raise a range of questions about internal and international mobility, planned relocation, and the tension between mobility and immobility. Bangladesh is a low-lying country, which is expected to face significant risks as a result. It is already preparing for increased internal displacement and provides a context to explore issues of internal adaptive mobility. The Pacific Islands raise issues of immobility and adaptive mobility as planned relocation, highlighting the tension between choosing to stay and planning to relocate and the need to consider and facilitate both. The Sahel offers the opportunity to focus on the role of adaptive mobility and international movement. The case study of the region explores impacts to drylands, testing the limits of the argument for adaptation obligations by considering whether and how these obligations can lead to facilitation of and support for international migration. Together, these case studies offer insight into the ambition and limitations of arguments for adaptive mobility. While not dictating action, they argue that the integration of human rights and key guiding principles like precaution throughout the process of adaptation leads to improved outcomes for those most vulnerable to climate change. They make the case for adaptive mobility specifically, to help avoid displacement, forced immobility, and precarious journeys by migrants.

The book concludes by returning to the problem that motivated developing a new legal approach—the limitations of current international law and policy in addressing climate change and human mobility. Predicted widespread displacement need not become our reality, even in circumstances where movement may be unavoidable. The approach to adaptive mobility laid out in this book aims to compel action before people are displaced, undertake precarious journeys, or must seek legal status after they move—as refugees or otherwise. And it aims to intervene in ways that mobility-focused frameworks cannot, to enable people to stay as long as possible and make migration part of a process of adaptation. Measures to facilitate adaptive mobility, with their inclusion of longer-term planning, anticipatory action, and human rights, can better account for and assist those vulnerable to climate impacts to ensure safer journeys and conditions upon arrival. Obligations of developed countries to support developing countries are also critical to the implementation of such measures and the success of adaptive mobility.

Ultimately, the book contributes to wider discussions about climate change and international law. It uses international law to approach an issue that intersects and interacts across global, regional, and local governance spaces. Its arguments, particularly in addressing areas of potential legal conflict, can therefore apply where climate change confronts other areas of law. Yet the book focuses on climate change and human mobility, in part because their connection tests the limits of international law and raises larger questions about equity and justice. It makes a scholarly, practical, and normative argument for the value of action using existing frameworks, not as a solution by itself but as something that can be useful to the broader, collective struggles that are also necessary. Climate change poses a massive and foreseeable global challenge that puts at risk the survival of people and the planet. Creative and innovative approaches—across disciplines and sectors—are urgently needed. This book looks to the current tools of international law and the ongoing evolution of the climate change regime to propose one such approach.

1

Gaps in Protection: Climate Change, Human Mobility, and Current Approaches

I. Introduction

This chapter analyzes the limits of current approaches to climate-related human mobility under international frameworks, to contrast them with the book's approach based on adaptation. The chapter first discusses how the impacts of climate change interact with contextual conditions and an individual's circumstances, and how this interaction exacerbates vulnerability and exposure to risks. The chapter then builds on this understanding of differential impacts to discuss the influence of climate change on human mobility. It explains that while displacement is a long-standing concern, a more nuanced perspective is emerging. Two conclusions arise from this perspective that are of continued importance for the book. First, that people are rarely forced to move because of the impacts of climate change alone. These impacts instead operate alongside a range of other factors to influence mobility in a variety of ways. This does not mean climate change is not a critical factor. Climate change creates foreseeable risks that will result in more people on the move, with increasing displacement likely absent concerted climate action. Second, there is increasing emphasis put on human mobility as a form of adaptation. Expecting the individual and their network to marshal resources, however, can undermine the benefits and success of such mobility. The risks migrants face will persist without an expansion in the legal pathways or protections offered, which is one of the reasons why the legal arguments developed in the book offer a much-needed intervention. The chapter next explains why anticipatory, adaptive movement does not fit neatly into existing approaches to human mobility under international law. It outlines the limitations of international law frameworks, the legal gaps they create, and the need for a new approach.

II. The interaction between climate change, vulnerability, and risk

A. Risk and the impacts of climate change

Since the Intergovernmental Panel on Climate Change (IPCC) issued its first assessment report in 1990, evidence of the impacts of climate change and the threats

Climate Change, Human Rights, and Adaptive Mobility. Lauren Sakae Nishimura, Oxford University Press.
 DOI: 10.1093/9780198930068.003.0002

they pose have continued to grow more certain and alarming. The IPCC is an intergovernmental body assembled under the United Nations (UN) tasked with assessing the science related to climate change. Its sixth assessment report makes clear that human-caused global warming is contributing to 'widespread and rapid changes' to the atmosphere, ecosystems, and oceans, which lead to far-reaching adverse impacts on nature and people.[1] Efforts to prevent or reduce the emissions of greenhouse gases (GHGs) that cause climate change, through mitigation measures, are critical to reducing the harm from impacts going forward. Yet current mitigation commitments are insufficient and will likely lead to an increase of global temperatures of more than 1.5°C by the end of the century.[2] Even considerably more mitigation, however, cannot stop all the changes underway. The last decade ushered in the hottest temperatures on record. From 2015 through 2022, annual global temperatures reached at least 1°C above pre-industrial levels, and each year represented one of the warmest recorded.[3] The World Meteorological Organization (WMO) expects temperatures to surge, setting new average global temperature records in the next five years (2023 through 2027).[4]

Increasing global temperatures will further result in changes to oceans. As the largest heat sink on the planet, oceans play a critical role in regulating the climate, absorbing 90 per cent of the excess heat from rising temperatures. Ocean temperatures and ocean heat content (OHC) are therefore also rising, with the pace of warming accelerating. The OHC is the amount of heat stored by the oceans and measured by water temperatures, and it reached record highs from 2019 through 2022.[5] These records are expected to be repeatedly broken going forward. July 2023, for example, set both global surface and ocean temperature records.[6] Ocean warming contributes to projected 'unprecedented conditions' of surface acidification, oxygen decline, and marine heatwaves.[7] Ocean warming can also fuel more intense weather systems, including heavier rainfall and extreme heat and weather events. It leads to ice melt and causes sea levels to rise. Sea level rise is increasing,

[1] See IPCC, 'Summary for Policymakers' in Core Writing Team, H Lee, and J Romero (eds), *Climate Change 2023: Synthesis Report. Contribution of Working Groups I, II and III to the Sixth Assessment Report of the Intergovernmental Panel on Climate Change* (2023).

[2] ibid 10–11.

[3] WMO, 'Past Eight Years Confirmed to Be the Eight Warmest on Record' (11 January 2023) <https://wmo.int/news/media-centre/past-eight-years-confirmed-be-eight-warmest-record> accessed 29 August 2023.

[4] WMO, 'Global Temperatures Set to Reach New Records in Next Five Years' (15 May 2023) <https://wmo.int/news/media-centre/global-temperatures-set-reach-new-records-next-five-years> accessed 29 August 2023.

[5] Lijing Cheng and others, 'Another Year of Record Heat for the Oceans' (2023) 40 Advances in Atmospheric Sciences 963.

[6] National Oceanic and Atmospheric Administration (NOAA) National Centers for Environmental Information, 'Monthly Global Climate Report for July 2023' (2023) <www.ncei.noaa.gov/access/monitoring/monthly-report/global/202307/supplemental/page-1> accessed 29 August 2023.

[7] IPCC, 'Summary for Policymakers' in HO Pörtner and others (eds), *IPCC Special Report on the Ocean and Cryosphere in a Changing Climate* (Cambridge University Press 2019) 8, 20.

with estimates of between 0.28 to 1.01 metres increase relative to 1995–2014 levels by the end of the century, depending on GHG emission scenarios.[8] Some of these changes cannot be undone, with sea levels expected to remain elevated for thousands of years. Yet the extent of impacts depends on how deep and sustained mitigation efforts are, and, as a result, how much warming occurs. For example, one study suggests that 2°C warming would yield sea level rise reaching several metres over 50 to 150 years.[9]

Climate impacts are not linear and accelerate and intensify as warming increases. Less warming will lead to fewer or less extreme impacts, whereas more warming means reaching tipping points and increasing likelihoods of species extinction and irreversible loss of biodiversity across ecosystems.[10] Both the scale and rate of change therefore pose serious threats to ecosystems and people. Multiple and concurrent changes are expected with each increment of warming, producing more frequent compound heatwaves and droughts; intensification and greater variability of the global water cycle, monsoon precipitation, and very wet and dry weather climate events; and increases in fire weather and aridity, global mean sea level rise, ocean acidification, and deoxygenation.[11]

Furthermore, while the effects of climate change are often discussed in global terms, they are not felt uniformly. As the case studies of Bangladesh, the Pacific Islands, and the Sahel will illustrate, certain countries and regions face significant risks from climate impacts.[12] Risks are the potential for adverse consequences to human or ecological systems; they are the threat of harm to people and nature.[13] Some risks come by virtue of geography. Low-lying coastal areas, for example, face sea level rise, storm surges, and shifts in rainfall patterns that are expected to cause a variety of adverse effects, including submergence, flooding, salinization, and erosion.[14] These impacts threaten the habitability and territorial integrity of small island states, such as those in the Pacific Islands. They are also predicted for countries in the Western Sahel and Bangladesh. Rising temperatures are further contributing to desertification, droughts, and land degradation. Many in the Sahel are living in arid areas, or drylands. Droughts and desertification are expected to intensify

[8] IPCC, 'Summary for Policymakers' in *Climate Change 2023: Synthesis Report* (n 1) 18.

[9] J Hansen and others, 'Ice Melt, Sea Level Rise and Superstorms: Evidence from Paleoclimate Data, Climate Modeling, and Modern Observations That 2°C Global Warming Could Be Dangerous' (2016) 16 Atmospheric Chemistry and Physics 3761.

[10] IPCC, 'Summary for Policymakers' in *Climate Change 2023: Synthesis Report* (n 1) 13.

[11] The IPCC is virtually or highly certain of most of these changes; ibid 12–13.

[12] Chapters 7–9.

[13] See IPCC, 'Annex II: Glossary' in HO Pörtner and others (eds), *Climate Change 2022: Impacts, Adaptation, and Vulnerability. Contribution of Working Group II to the Sixth Assessment Report of the Intergovernmental Panel on Climate Change* (Cambridge University Press 2022) 2921.

[14] However, risks vary due to differences in exposure, topography, reef slope, and elevation of groundwater levels. Michelle Mycoo and others, 'Small Islands' in *Climate Change 2022: Impacts, Adaptation and Vulnerability* (n 13) 2053–55.

and expand impacts in drylands, which cover 41 per cent of the world's landmass and are home to one third of the global population.[15] Urban areas also face unique challenges. Heatwaves are often intensified in cities, which can disrupt infrastructure, aggravate air pollution, and increase health risks. Rapidly expanding urban settlements can increase exposure of people in these areas, and global trends in urbanization create specific impacts and different regional risk profiles. Most of the world's population—over 4.2 billion people—live in urban areas, with expanding urban growth taking place in less developed regions. By 2050, an additional 2.5 billion people are predicted to be living in urban areas, with up to 90 per cent of this increase expected in regions in Asia and Africa.[16]

B. Risk and vulnerability to climate change

The risks from climate change are more than a matter of geography. They also arise from the interaction of impacts with societal conditions, which can expose certain places and people within them to greater harm. Risk is therefore not experienced uniformly, and it varies across places, people, and time. One of the means to understand and measure differential risk is through its connection to vulnerability. Vulnerability is frequently cited in analysis of climate change. It is a concept that is described by the IPCC to refer to the 'predisposition to be adversely affected' that encompasses a variety of elements, including susceptibility to harm and lack of capacity to cope or adapt.[17] The two concepts are connected: as vulnerability increases so too do the risks people face and the likelihood that these risks will result in harm. Within this book, vulnerability is understood as a multi-layered concept that is relational rather than defined only by the internal characteristics of a person or place.[18] This understanding of vulnerability draws on its use in international climate treaties, human rights law, and scientific assessments. These sources help establish three different and interacting layers of vulnerability that can increase exposure to risks for people.

[15] Alisher Mirzabaev and others, 'Desertification' in PR Shukla and others (eds), *Climate Change and Land: An IPCC Special Report on Climate Change, Desertification, Land Degradation, Sustainable Land Management, Food Security, and Greenhouse Gas Fluxes in Terrestrial Ecosystems* (Cambridge University Press 2019) 16–18; Stefanie M Herrmann and Charles F Hutchinson, 'The Scientific Basis: Links between Land Degradation, Drought and Desertification' in Pierre Marc Johnson, Karel Mayrand, and Marc Paquin (eds), *Governing Global Desertification: Linking Environmental Degradation, Poverty and Participation* (Routledge 2006) 14.

[16] D Dodman and others, 'Cities, Settlements and Key Infrastructure' in *Climate Change 2022: Impacts, Adaptation and Vulnerability* (n 13) 909.

[17] IPCC, 'Annex II: Glossary' (n 13) 2927.

[18] For a critique of a notion of vulnerability that relies on internal characteristics, see Carol Farbotko and John Campbell, 'Who Defines Atoll "Uninhabitability"?' (2022) 138 Environmental Science & Policy 182, 184; see also Alexandra Timmer and others, 'The Potential and Pitfalls of the Vulnerability Concept for Human Rights' (2021) 39 Netherlands Quarterly of Human Rights 196 (recognizing vulnerability as socially induced and contextual).

The first layer is geographic. As the previous section describes, specific places such as low-lying coastal areas or drylands are exposed to greater risks from the impacts of climate change. Likewise, urban conditions, particularly in informal settlements or slums, can increase vulnerability. Geographic features and urban conditions can combine to further heighten vulnerability, as is the case for cities in low-lying areas. More than a billion people in low-lying urban areas are expected to confront flooding risks by 2050, from sea level rise, more frequent storm surges, and extreme precipitation.[19] Climate treaties recognize that vulnerability can arise from geography. The United Nations Framework Convention on Climate Change (UNFCCC) acknowledges that certain countries are particularly vulnerable to the adverse effects of climate change, including small island countries, countries with low-lying coastal areas, arid or semi-arid areas, or areas liable to floods, drought, and desertification.[20]

The second layer of vulnerability considers the economic conditions and constraints on resources within a country. This layer is state-based and tied to the distinction made in climate treaties between developed and developing countries. Climate treaties consistently highlight the difficulties faced by developing countries, underscore their needs, and call for developing countries that are vulnerable or 'particularly vulnerable' to be prioritized in climate finance, assistance, and implementation of commitments.[21] The UNFCCC connects this layer with the first, highlighting the needs of developing countries with risks due to geography, as well as those with areas of high urban atmospheric pollution, liable to forest decay, or with fragile ecosystems.[22] The Paris Agreement refers to small island developing states as an example of developing countries that are particularly vulnerable and that require consideration of their special circumstances.[23] Least developed countries (LDCs) are also singled out for special consideration.[24] The emphasis in these treaties is warranted: between 3.3 and 3.6 billion people with 'considerable development constraints' live in low to lower middle-income countries that are highly vulnerable to climate change.[25] This vulnerability stems in part from contextual conditions that are connected to historical and systemic issues largely created by external forces and that give rise to and perpetuate developmental differences between countries. It is no coincidence that those countries

[19] Dodman and others (n 16) 909.

[20] UNFCCC (adopted 9 May 1992, entered into force 21 March 1994) 1771 UNTS 107, art 4(8).

[21] See UNFCCC, preamble, arts 3(2), 4(4), (10); Kyoto Protocol to the UNFCCC (adopted 11 December 1997, entered into force 16 February 2005) 2303 UNTS 162, art 12(8); Paris Agreement (adopted 12 December 2015, entered into force 4 November 2016) 3156 UNTS 79, preamble, arts 6(6), 7(2), 9(5), 11(1).

[22] UNFCCC, art 4(8).

[23] Paris Agreement, arts 4(6), 9(4), 11(1), 13(3).

[24] See UNFCCC, art 4(9).

[25] IPCC, 'Summary for Policymakers' in *Climate Change 2023: Synthesis Report* (n 1) 5, 31; J Birkmann and others, 'Poverty, Livelihoods and Sustainable Development' in *Climate Change 2022: Impacts, Adaptation and Vulnerability* (n 13) 1174, 1251.

that contributed little to the causes of climate change, but that carry a disproportionate burden of its adverse effects, are also those reckoning with colonial pasts, discrimination and racism, and resource exploitation and expropriation. As the IPCC recognizes, colonialism and other patterns of inequity exacerbate vulnerability to climate change.[26] Systemic and structural issues can affect access to and the quality of public services. Developing countries in Africa and South Asia, for example, already experience significant challenges due to climate-related diseases and shortcomings in adequate healthcare, compounding impacts to people from malnutrition, vector-borne diseases, and diarrhoea linked to poor water and food security.[27]

The third layer of vulnerability is social. It focuses on people and their relationships and social conditions. It builds on the other two: the impacts of climate change combine with societal conditions and pre-existing inequities to make certain areas and people or groups within them particularly vulnerable. Those already at the margins will bear the brunt of harm from climate change as a result, and they may also have the fewest resources to prepare for, respond to, or recover from its impacts. This can include children; the elderly; persons with disabilities; migrants; the poor; Indigenous peoples; and persons marginalized due to ethnicity, race, or gender. The disproportionate impact of climate change on certain people and the differential risks caused by these interacting layers are recognized across the human rights and climate regimes. For example, the UN Human Rights Council (UNHRC) and the Office of the UN High Commissioner for Human Rights (OHCHR) recognize that the negative impacts of climate change will disproportionately affect people who are in vulnerable situations due to their 'geography, poverty, gender, age, indigenous or minority status, national or social origin, birth or other status and disability'.[28] Vulnerability can also be tied to 'multiple and intersecting forms of discrimination, inequality and structural and societal dynamics' that lead to unequal access to resources and enjoyment of rights.[29] Human rights treaty bodies repeatedly state that particular attention must be given to vulnerable people or marginalized groups, including in participatory processes

[26] See IPCC, 'Summary for Policymakers' in *Climate Change 2023: Synthesis Report* (n 1) 31.

[27] World Bank, 'World Development Report 2010: Development and Climate Change' 95.

[28] UNHRC, 'Human Rights and Climate Change' (19 June 2017) UN Doc A/HRC/35/L.32, preamble; UNHRC, 'Human Rights and Climate Change' (23 June 2014) UN Doc A/HRC/26/L.33, preamble; OHCHR, 'Analytical Study on the Impacts of Climate Change and the Human Right of Everyone to the Enjoyment of the Highest Attainable Standard of Physical and Mental Health, Report of OHCHR' (Study on Climate Change and the Right to Health) (6 May 2016) UN Doc A/HRC/32/23, paras 17, 23; see also Decision 1/CP.16, 'The Cancun Agreements: Outcome of the work of the Ad Hoc Working Group on Long-term Cooperative Action under the Convention' (15 March 2011) FCCC/CP/2010/7/Add.1, preamble.

[29] OHCHR, 'Principles and Practical Guidance on the Protection of the Human Rights of Migrants in Vulnerable Situations, Report of the United Nations High Commissioner for Human Rights' (3 January 2018) UN Doc A/HRC/37/34, para 13.

and providing essential goods and services.[30] Likewise, the Paris Agreement preamble and article on adaptation both incorporate vulnerable persons; the former due to their situation—while at the same time identifying the rights of particular groups—and the latter specifying that vulnerable groups should be considered in adaptation action.[31]

Unsurprisingly, the distribution of risk arising from the interaction of these layers is not equal. People on the margins often live in environmentally fragile or polluted areas, facing conditions and social and economic inequities that leave them more vulnerable and at greater risk of harm. Lower income communities, for example, are at higher risk of adverse health conditions and constrained access to health services, and children within these communities may be at even greater risk. Climate change increases the incidence and reach of diseases like malaria and diarrhoea that disproportionately affect the poor and that are already the main causes of illness and death in children. Malnutrition in children is expected to worsen as climate impacts increase food insecurity, particularly in developing countries.[32] Women may face greater risks depending on their access to resources and entitlements. In many developing countries, women do not possess land rights despite accounting for over 40 per cent of the agricultural workforce.[33] Overlapping experiences of poverty, lack of control over land and resources, and exclusion from decision-making can compound risks and worsen challenges for women, as can intersecting vulnerabilities for girls, older women, and women with disabilities. Differential risks like these show that vulnerability is influenced by a confluence of historical and social factors, which can amplify environmental impacts and shift over time. It also accords with the recognition that disasters are not natural, and that governance and social factors play a significant role in the risks posed by environmental hazards and climate change.[34]

Climate change is therefore recognized as a risk multiplier that interacts with and exacerbates vulnerability.[35] Although there are people facing risks from the

[30] As discussed in Chapters 4 and 5. See eg Human Rights Committee (HRC), 'General Comment No 36: Article 6 (Right to Life)' (3 September 2019) UN Doc CCPR/C/GC/36, para 23; Committee on Economic, Social and Cultural Rights (CESCR), 'General Comment No 14: The Right to the Highest Attainable Standard of Health (Art 12)' (11 August 2000) UN Doc E/C.12/2000/4, para 43; CESCR, 'General Comment No 15: The Right to Water (Arts 11 and 12 of the Covenant)' (20 January 2003) UN Doc E/C.12/2002/11, para 37.

[31] Paris Agreement, preamble, art 7(5).

[32] See OHCHR, 'Study on Climate Change and the Right to Health' (n 28) paras 5, 24, 26; OHCHR, 'Analytical Study on the Relationship between Climate Change and the Full and Effective Enjoyment of the Rights of the Child' (4 May 2017) A/HRC/35/13, paras 4, 10.

[33] Christian Nellemann, Ritu Verma, and Lawrence Hislop (eds), 'Women at the Frontline of Climate Change: Gender Risks and Hopes' (UNEP 2011).

[34] See eg Matthew Scott, *Climate Change, Disasters, and the Refugee Convention* (Cambridge University Press 2020) ch 2; see also United Nations Office for Disaster Risk Reduction (UNDRR), 'Sendai Framework for Disaster Risk Reduction 2015–2030' (2015) para 6.

[35] See eg Rachel Bezner Kerr and others, 'Food, Fibre and Other Ecosystem Products' in *Climate Change 2022: Impacts, Adaptation and Vulnerability* (n 13) 811; G Cissé and others, 'Health, Wellbeing and the Changing Structure of Communities' in ibid 1128; Birkmann and others (n 25) 1176, 1216.

impacts of climate change in all countries, understanding vulnerability through an interaction of layers—geographic, geopolitical, and social—can help establish priorities for adaptation. It enables a narrower focus in this book, on certain developing countries and those individuals who are most at risk within them. Such a focus is, in part, due to an analysis premised on international law and its influence on state action and interaction. This book explores state obligations. It argues that all states are obliged to adapt, but some must be supported in their efforts. It also argues that vulnerable people should be prioritized in adaptive action. Yet it is important not to essentialize individuals or to consider their communities as spaces rife with deficit and deprivation.[36] Indeed, vulnerability does not inherently equate with victimhood, and human experiences are not captured by simple dichotomies. People can hold multiple and seemingly contradictory positions; they can participate in decision-making while also facing situations of disadvantage that inform their greater vulnerability.[37] This book strives for an approach that both recognizes vulnerability and conceives of individuals as active participants entitled to choice and support, with their own capacities, insights, and resilience. As the case studies in subsequent chapters explore, the individuals and communities considered most vulnerable are also some of those at the forefront of advocating for their future, and they must drive the decision-making that will shape how their lives and homes will adapt to the impacts of climate change.

III. The impacts of climate change and human mobility

In addition to their general relationship to vulnerability, the interactions between climate impacts, contextual conditions, and individual circumstances influence human mobility. As the IPCC recognizes, 'the impacts of climatic drivers on migration are highly context-specific and interact with social, political, geopolitical, and economic drivers.'[38] Climate change is thus rarely the exclusive cause of movement, although its role is more significant under certain conditions. Sudden-onset events like tropical cyclones, for example, can more directly contribute to displacement—or movement that is predominately forced—by damaging infrastructure and homes. In contrast, slower-onset events tend to be indirect drivers of human mobility. Salinization and saltwater intrusion from rising seas create risks for drinking water, agriculture, and food security in deltas and estuaries,

[36] See generally Eve Tuck, 'Suspending Damage: A Letter to Communities' (2009) 79 Harvard Educational Review 409.

[37] See Silja Klepp and Libertad Chavez-Rodriguez, 'Governing Climate Change: The Power of Adaptation Discourses, Policies, and Practices' in Silja Klepp and Libertad Chavez-Rodriguez (eds), *A Critical Approach to Climate Change Adaptation: Discourses, Policies, and Practices* (Routledge 2018) 20–21 (discussing gender and intersectionality).

[38] HO Pörtner and others, 'Technical Summary' in *Climate Change 2022: Impacts, Adaptation and Vulnerability* (n 13) 52.

while drought and desertification cause reductions in crop and livestock productivity that can lead to losses of rural income.[39] Such impacts increase the likelihood that people will move. While their gradual occurrence is often thought to induce more voluntary forms of migration, as these interactions show, most movement falls on a continuum between forced and voluntary, with varying degrees of compulsion and choice involved.[40] Climate events can also overlap and accumulate to influence vulnerability and whether and how mobility is undertaken. For example, sea level rise is gradually eroding land in coastal areas, but it is its combination with storm surges, flooding, and other impacts that may ultimately make an area uninhabitable.

How human mobility unfolds is further influenced by the measures in place to reduce risks from and respond to disasters. Most displacement following sudden-onset disasters occurs in developing countries, suggesting a connection between institutional capacity, exposure to risks, poverty, and displacement.[41] Most internal migration—or movement within countries—is to urban centres, and nearly 95 per cent of urban growth in the next two decades will also take place in developing countries.[42] As the Bangladesh case study explores in Chapter 7, conditions in urban areas affect the risks these migrants face, risks that can be mitigated by adaptation planning and implementation.

Likewise, most mobility associated with environmental and climate change occurs internally.[43] Absent concerted climate action, the same is expected to continue, although international mobility will be increasingly considered as climate impacts leave more areas in a country uninhabitable, and as internal mobility pushes those who can move abroad out of the country.[44] Most climate-related

[39] ibid 52; IPCC, *Global Warming of 1.5°C. An IPCC Special Report on the impacts of global warming of 1.5°C above pre-industrial levels and related global greenhouse gas emission pathways, in the context of strengthening the global response to the threat of climate change, sustainable development, and efforts to eradicate poverty* (V Masson-Delmotte and others (eds), Cambridge University Press 2018) 231–33; Mirzabaev and others (n 15) 16.

[40] See OHCHR, 'The Slow Onset Effects of Climate Change and Human Rights Protection For Cross-Border Migrants' (22 March 2018) UN Doc A/HRC/37/CRP.4, paras 12–14 (study undertaken by the author on behalf of OHCHR); Graeme Hugo, 'Environmental Concerns and International Migration' (1996) 30 The International Migration Review 105, 107.

[41] See Internal Displacement Monitoring Centre (IDMC), 'Global Estimates 2015: People Displaced by Disasters' (2015) 9 <https://api.internal-displacement.org/sites/default/files/publications/documents/20150713-global-estimates-2015-en-v1.pdf> accessed 20 November 2023; Margaretha Wewerinke-Singh and Tess Van Geelen, 'Protection of Climate Displaced Persons under International Law: A Case Study from Mataso Island, Vanuatu' (2018) 19 Melbourne Journal of International Law 666, 691.

[42] W Neil Adger and others, 'Human Security' in Christopher B Field and others (eds), *Climate Change 2014: Impacts, Adaptation, and Vulnerability. Part A: Global and Sectoral Aspects. Contribution of Working Group II to the Fifth Assessment Report of the Intergovernmental Panel on Climate Change* (Cambridge University Press 2014) 767; IDMC and Norwegian Refugee Council, 'Global Report on Internal Displacement' (2019) 67.

[43] Dodman and others (n 16) 930; Cissé and others (n 35) 1045, 180–82; B O'Neill and others, 'Key Risks Across Sectors and Regions' in *Climate Change 2022: Impacts, Adaptation and Vulnerability* (n 13) 2414, 2428.

[44] This 'domino' effect might push those with means to migrate internationally. Jane McAdam, *Climate Change, Forced Migration, and International Law* (Oxford University Press 2012) 171–72;

international movement occurs between countries with neighbouring borders.[45] However, people are crossing and will continue to cross oceans and undertake long, dangerous journeys in search of asylum or opportunity elsewhere. It is a focus on and fear of growing numbers of people displaced or seeking international refuge that occupies much of the media coverage and discussion about the implications of climate change on population movements.[46] Yet discussions centred on mass movement and large numbers of 'climate refugees' reinforce security narratives and justify border controls, obscuring the complexities and different ways in which climate change contributes to human mobility.[47] Indeed, the potential for vast displacement is one of the reasons that climate change is being discussed in the UN Security Council.[48]

Displacement on a large scale is a long-standing concern. The IPCC's First Assessment warned that 'the gravest effects of climate change may be on human migration as millions are displaced'.[49] Early predictions put the number at risk of displacement in the hundreds of millions to over a billion.[50] Subsequent scholarship criticized these numbers, questioning their methodological bases and the ways that they are used.[51] But the focus on displacement persists. Climate change

see also UNHRC 'Protection of and Assistance to Internally Displaced Persons', Report of the Special Rapporteur on the human rights of internally displaced persons, Chaloka Beyani (9 August 2011) UN Doc A/66/285, para 19.

[45] Pörtner and others (n 38) 52; Cissé and others (n 35) 1045, 1080–82.

[46] See eg 'There Could Be 1.2 Billion Climate Refugees by 2050. Here's What You Need to Know' <www.zurich.com/en/media/magazine/2022/there-could-be-1-2-billion-climate-refugees-by-2050-here-s-what-you-need-to-know> accessed 20 November 2023; 'The Climate Refugee Crisis Is Landing on Europe's Shores—and We Are Far from Ready' (*POLITICO*, 20 February 2023) <www.politico.eu/article/climate-refugee-crisis-europe-policy/> accessed 20 November 2023; Hanna Hett, 'As the Planet Warms, Canada Faces an Influx of Climate Refugees' *Wired* (September 2022) <www.wired.com/story/as-the-planet-warms-canada-faces-an-influx-of-climate-refugees/> accessed 20 November 2023; see also Maria Sakellari, 'Media Coverage of Climate Change Induced Migration: Implications for Meaningful Media Discourse' (2022) 18 Global Media and Communication 67.

[47] See Lauren Nishimura, '"Climate Change Migrants": Impediments to a Protection Framework and the Need to Incorporate Migration into Climate Change Adaptation Strategies' (2015) 27 International Journal of Refugee Law 107; Ingrid Boas and others, 'Climate Migration Myths' (2019) 9 Nature Climate Change 901.

[48] Introduction Chapter, n 1 and accompanying text.

[49] IPCC, 'Climate Change: The 1990 and 1992 IPCC Assessments, IPCC First Assessment Report Overview and Policymaker Summaries and 1992 IPCC Supplement' (IPCC 1992) 103.

[50] See eg Norman Meyers, 'Environmental Refugees' (1997) 19 Population and Environment 167 (150 to 200 million could be displaced); Nicholas Stern, *The Economics of Climate Change: The Stern Review* (Cambridge University Press 2007) (replicating Meyer's estimate); Daniel Bodansky, Jutta Brunnée, and Lavanya Rajamani, *International Climate Change Law* (Oxford University Press 2017) 313 ('guestimates' about scale of displacement vary from 25 million to 2 billion).

[51] See eg François Gemenne, 'Why the Numbers Don't Add up: A Review of Estimates and Predictions of People Displaced by Environmental Changes' (2011) 21 Global Environmental Change. Part A, Human and Policy Dimensions S41; Ilan Kelman, 'Imaginary Numbers of Climate Change Migrants?' (2019) 8 Social Sciences 1; David Durand-Delacre and others, 'Climate Migration Is About People, Not Numbers' in Steffen Böhm and Sian Sullivan (eds), *Negotiating Climate Change in Crisis* (Open Book Publishers 2021).

is now recognized as a driver of displacement in all regions of the world.[52] Special rapporteurs with related human rights mandates connect climate change to the displacement of millions, although they acknowledge that actual numbers of people displaced across borders are not known.[53] A UN Children's Fund (UNICEF) report links weather-related events between 2016 and 2021 to the internal displacement of 43 million children, highlighting that this is equivalent to approximately 20,000 child displacements per day.[54] Other analyses predict that displacement will continue to increase without urgent and sustained action to address climate change. Reports released by the World Bank, for example, estimate that between 44 to 216 million people across six regions could move within their countries by 2050. Yet while the upper figure can be cited as a possible outcome, these reports find that the number of people forced to move varies greatly depending on global temperature rise, demographic factors, and development scenarios. Up to 80 per cent of projected displacement can be avoided through action to reduce GHG emissions, plan for impacts, and foster resilient and inclusive development.[55]

Displacement is therefore not inevitable, nor does it represent the full picture of the myriad ways people move. There is now a more nuanced understanding of the relationship between climate change and human mobility than when it arose as a concern over three decades ago. Successive IPCC reports have shifted analysis from emphasis on vast displacement to an assessment of climate impacts on different types of human mobility: displacement, relocation, immobility, and adaptive migration. These reports build on studies documenting links between climate events and a range of mobility experiences and outcomes.[56] They establish that climate impacts are influencing temporary, seasonal, and permanent migration. In some circumstances, impacts are upending patterns of movement and decreasing migration flows, in others they are increasing recourse to migration as a strategy to diversify labour, access new resources, or provide remittances. Climate change can also undermine resources and limit mobility, although immobility can reflect an inability to move, be a deliberate choice made despite risk, or involve elements of choice and constraint.

[52] IPCC, 'Summary for Policymakers' in *Climate Change 2022: Impacts, Adaptation and Vulnerability* (n 13) 11.

[53] See UNHRC, 'Report of the Special Rapporteur on the Human Rights of Migrants, Felipe González Morales' (19 July 2022) UN Doc A/77/189, paras 14–15; UNHRC, 'Providing Legal Options to Protect the Human Rights of Persons Displaced across International Borders Due to Climate Change: Report of the Special Rapporteur on the Promotion and Protection of Human Rights in the Context of Climate Change, Ian Fry' (2023) UN Doc A/HRC/53/34, para 1. Special Rapporteur Fry argues that the lack of a total assessment of numbers is a 'significant oversight' that does not correlate with climate change as a 'very clear' driver of displacement. UNHRC, 'Providing Legal Options' para 12.

[54] UNICEF, 'Children Displaced in a Changing Climate: Preparing for a Future Already Underway' (2023).

[55] See Kanta Kumari Rigaud and others, 'Groundswell: Preparing for Internal Climate Migration' (World Bank 2018); Viviane Clement and others, 'Groundswell Part 2: Acting on Internal Climate Migration' (World Bank 2021).

[56] See Cissé and others (n 35) 1079–82.

The kind of adaptation advocated for in this book, which addresses the causes of vulnerability and foreseeable risks to people, can help avoid both displacement and forced immobility by providing a means to safely move elsewhere. It can likewise enable people to choose to remain where feasible, and with continued access to essential resources. Decisions to stay, or to move, are influenced by perceived risks, preferences, networks, and social conditions. Specific individuals or groups may make choices informed by their connections to the land, livelihoods, or cultural ties.[57] Some Indigenous communities are opting to stay in their homes and traditional lands rather than relocate elsewhere, with full awareness of the changes to ecosystems that make it increasingly difficult to remain. Policies and plans that account for the deep connections to home and the costs of moving are a priority, to better enable people to adapt and stay in place for as long as possible. While these efforts are paramount, planned relocation is a further strategy already being pursued. Planned relocation involves the organized movement of individuals and communities away from areas that are highly exposed to climate hazards. The Pacific Islands case study in Chapter 8 explores issues of immobility and the relationship between choosing to stay and planning to relocate. It describes planned relocation as an option of last resort, but one that requires adequate preparation and processes led by communities to succeed.

In addition, there is growing emphasis on migration as a potentially effective form of adaptation.[58] Moving to adapt or avoid harm is not a new strategy. People have moved to avoid harm from disaster or environmental change throughout human history. Yet evidence of the benefits of migration as adaptation is mixed, in part because the onus of migration is currently on the migrant and their networks. When individuals or their communities are expected to marshal resources alone to 'simply move out of harm's way', outcomes vary widely depending on socioeconomic conditions.[59] Often, the success of migration is driven by an individual's ability to integrate into the labour market and the level of remittances they can provide. An approach to adaptive migration that replicates these individual responsibilities and markers of success can come at the expense of investment in building capacity.[60] It also ignores the link to the causes of climate change and the obligations to address its impacts and support those most vulnerable. A proactive approach to adaptive mobility could, however, promote more voluntary movement and agency for

[57] See Caroline Zickgraf, 'Immobility', in Robert McLeman and François Gemenne (eds), *Routledge Handbook of Environmental Displacement and Migration* (Routledge 2018); Carol Farbotko and others, 'Relocation Planning Must Address Voluntary Immobility' (2020) 10 Nature Climate Change 702.

[58] See eg UN General Assembly (UNGA) Res 73/195 (19 December 2018) 'Global Compact for Safe, Orderly and Regular Migration' UN Doc A/RES/73/195, Annex; The Nansen Initiative, 'Agenda for the Protection of Cross-Border Displaced Persons in the Context of Disasters and Climate Change', vol 1 (2015).

[59] See Cissé and others (n 35) 1116.

[60] See Giovanni Bettini and Giovanna Gioli, 'Waltz with Development: Insights on the Developmentalization of Climate-Induced Migration' (2016) 5 Migration and Development 171.

migrants, which improves the likelihood of positive outcomes. Such an approach should include anticipatory state action to ensure continued access to adequate resources and support before, during, and after movement. This stands in contrast with more reactive, and potentially maladaptive, migration that puts the burden on the individual and focuses on survival.[61] Indeed, a proactive, anticipatory approach to mobility facilitated by state action and support offers an alternative to precarious, irregular migration. The chapters that follow offer a legal basis for such an approach, which seeks to compel and shape adaptive action that addresses foreseeable risks. In some circumstances, this will require states to enable human mobility as part of a broader process of adapting to climate change. This is a new way to consider state action and legal protection for people on the move, and it represents a departure from current approaches to climate change and human mobility.

IV. Current approaches, limitations, and gaps in international legal protection

Most of the international law and policy approaches aimed at addressing climate change and human mobility begin with mobility as their conceptual starting point. They therefore consider frameworks that address people on the move and whether they can provide legal protection, including access to basic resources, entitlements, and rights. Legal approaches explore protection through existing international legal regimes or by proposing new frameworks that would create a category of climate-related migrant. Policy approaches also call for increased protection and humanitarian action, but they do not create any legal rights or obligations. Each of these current approaches contributes to how we conceive of and consider climate change and human mobility, the role international law can play, and offers insights into when protection is accessible. Nonetheless, these approaches include key features that can limit their scope and create gaps in legal protection. This section focuses on approaches based in international refugee law, laws on statelessness, and human rights law. It discusses their application to climate change, offering protection for a subset of people on the move but falling short in many cases.

A. Refugee law

As a descriptor, the term 'climate refugee' is often used to refer to people fleeing the effects of climate change.[62] However, its use is controversial, particularly with

[61] See eg Alexander Betts, *Survival Migration: Failed Governance and the Crisis of Displacement* (Cornell University Press 2013); Kira Vinke and others, 'Migration as Adaptation?' (2020) 8 Migration Studies 626.

[62] See eg n 46.

academics and legal analysts. Critics point to pushback by affected communities and argue that it frames migrants as victims, while proponents emphasize that the term captures both the plight of and responsibility owed to those moving. Yet, both concede that the term does not align well with current legal conceptions, which require a refugee to meet the definition set out in relevant legal instruments.[63] Under international law, the Refugee Convention defines a refugee as any person who 'owing to a well-founded fear of being persecuted for reasons of race, religion, nationality, membership of a particular social group or political opinion' is outside their country (of nationality or habitual residence) and 'is unable or, owing to such fear, is unwilling' to return.[64]

Several aspects of this definition limit its scope. First, a refugee must be outside their country. The definition therefore does not apply to the significant internal mobility that is linked to climate change. Second, refugee status is based on meeting the criteria set out in the definition, which requires establishing a well-founded fear of being persecuted and a connection to one of the five enumerated grounds. Persecution, or 'being persecuted', is not defined in the Refugee Convention, but it has been interpreted to require serious harm. Serious harm can be evinced by showing an individual faces sustained, systemic, or severe denial of their human rights, although such a risk need not be exclusive and can be shared by groups or communities.[65] The risk of serious harm will increase as more people and entire communities are exposed to the adverse effects of climate change.[66] However, even a liberal understanding of being persecuted—with a more expansive view of what

[63] Compare François Gemenne, 'The Refugees of the Anthropocene' in Benoît Mayer and François Crépeau (eds), *Research Handbook on Climate Change, Migration and the Law* (Edward Elgar 2017); Avidan Kent and Simon Behrman, *Facilitating the Resettlement and Rights of Climate Refugees: An Argument for Developing Existing Principles and Practices* (Routledge 2018) ch 2 (Why 'climate refugees') with McAdam, *Climate Change, Forced Migration* (n 44) 33, 40, 203; Robyn Eckersley, 'The Common but Differentiated Responsibilities of States to Assist and Receive "Climate Refugees"' (2015) 14 European Journal of Political Theory 482 (Kiribati's president considered the refugee label undignified because people do not wish to move or be treated like refugees).

[64] Convention Relating to the Status of Refugees (adopted 28 July 1951, entered into force 22 April 1954) 189 UNTS 137, art 1(A)(2); Protocol Relating to the Status of Refugees (adopted 31 January 1967, entered into force 4 October 1967) 606 UNTS 267, art 1(2) (together the 'Refugee Convention').

[65] See UN High Commissioner for Refugees (UNHCR), 'Legal Considerations Regarding Claims for International Protection Made in the Context of the Adverse Effects of Climate Change and Disasters' (2020) paras 7–9; James C Hathaway and Michelle Foster, *The Law of Refugee Status* (Cambridge University Press 2014) 182–211, 288–319 (being persecuted involves serious harm plus a failure of state protection, with human rights providing benchmarks for such harm); McAdam, *Climate Change, Forced Migration* (n 44) 43 (persecution involves human rights violations that are 'particularly serious'); *AF (Kiribati)* [2013] NZIPT 800413 (New Zealand Immigration and Protection Tribunal (NZPIT)) para 53 (adopting Hathaway's definition of persecution as 'the sustained or systemic violation of core human rights, demonstrative of a failure of state protection').

[66] For further on timing of the risks of serious harm and its relationship to persecution in refugee status determinations, see Adrienne Anderson and others, 'Imminence in Refugee and Human Rights Law: A Misplaced Notion for International Protection' (2019) 68 International and Comparative Law Quarterly 111 (arguing against the need for imminent risks); Scott (n 34) ch 5 (on the temporal scope of being persecuted).

amounts to a requisite denial of human rights—must be accompanied by some form of discrimination.

There are circumstances where discrimination can arise in the context of climate change. For example, a person who is denied access to basic resources, services, or assistance in a discriminatory manner in the aftermath of a climate event, and who then crosses an international border, could qualify as a refugee. Here, however, refugee protection is not triggered by the harm caused by climate impacts, but rather the discriminatory acts that followed. Refugee movement may also arise when conflict occurs alongside or is exacerbated by the impacts of climate change. The interaction of adverse effects like drought with conflict can contribute to famine and food insecurity. Certain ethnic or religious communities may experience disproportionate harm from this interaction that could constitute persecution, thereby offering an avenue for refugee protection should individuals move across state borders.[67] There is also potential to expand the reach of refugee status beyond these examples, through an interpretation of the definition that allows for discrimination to contribute to an individual's exposure to serious harm, rather than being the direct cause of the harm. Discrimination can expose some individuals to greater risk of harm from climate impacts, creating this predicament by virtue of their existing marginalization. As a result, impacts that are not themselves discriminatory become so in effect. Such an interpretation would shift the focus from the risks connected to an identifiable persecutor to understanding *why* a person is vulnerable to serious harm and whether it is linked to discrimination they experienced.[68] However in this case, as with all refugee status determinations, the discrimination a person faces must be tied to one of the grounds enumerated in the Refugee Convention: their race, religion, nationality, membership of a particular social group, or political opinion. Without a connection to one of these, they will not be considered a refugee under international refugee law.

Regional instruments could eliminate the need for such a connection through expanded refugee definitions. A refugee as defined in the Organization of African Unity (OAU) Convention in Africa and the Cartagena Declaration in the Americas includes individuals compelled to leave their country of origin due to events seriously disturbing public order (OAU Convention and Cartagena Declaration) or who flee because of generalized violence or massive violation of human rights

[67] See UNHCR, 'Legal Considerations Regarding Claims' (n 65) paras 11–12; UNHCR, 'Legal Considerations on Refugee Protection for People Fleeing Conflict and Famine Affected Countries' (2017).

[68] See Lauren Nishimura, 'Climate Change and International Refugee Law: A Predicament Approach' (*RefLaw*, 18 October 2018) <https://michiganreflaw.wordpress.com/2018/10/19/climate-change-and-international-refugee-law/#_ftn11> accessed 20 November 2023; see also OHCHR, 'Slow Onset Study' (n 40) para 70. Matthew Scott argues that discrimination must be a contributory cause of exposure to serious denials of human rights, which can occur in the context of climate change. He proposes a refugee definition that, while reformulated, 'remains narrow'. See Scott (n 34) ch 7.

(Cartagena Declaration).[69] These broader definitions could apply to displaced persons facing serious harm from the impacts of climate change. Indeed, the UN High Commissioner for Refugees (UNHCR) argues for an evolutionary approach to interpreting these instruments—much as this book does in Chapter 3 for adaptation obligations—to understand how events seriously disturbing public order can include the adverse effects of climate change. Doing so could enable people displaced in the context of climate change to be considered refugees under regional criteria.[70] However, these regional frameworks remain focused on people who are forced or compelled to move and would not apply to more voluntary or pre-emptive movement.

There are also further limitations created by where and when refugee status is determined. Generally, status determinations are made on an individual basis, which can lead to inconsistences and subject asylum seekers to jurisdictional constraints. Some states are not party to the Refugee Convention, for example, and domestic law and practice may not reflect international or regional refugee criteria.[71] Whether an individual is considered a refugee could therefore vary wildly, depending on where they seek protection. There is also no guarantee that the people, courts, and bodies making such decisions will interpret the refugee criteria generously, although how they do will shape the understanding of refugee law and its application to climate change going forward. Furthermore, and crucially, protection based on refugee status tends to be granted only *after* people have moved and crossed borders, and only in the face of serious harm. This creates a temporal limitation, with protection not contemplated before or during a person's movement. The burden is also on the individual to establish their claim, with claimants thus far having little success convincing judicial bodies that climate change provides grounds to qualify as a refugee.[72]

[69] OAU Convention Governing the Specific Aspects of Refugee Problems in Africa (adopted 10 December 1969, entered into force 10 September 1969) 1001 UNTS 45, art 1(2); Cartagena Declaration on Refugees (adopted by the Colloquium on the International Protection of Refugees in Central America, Mexico, and Panama, 22 November 1984) Annual Report of the Inter-American Commission on Human Rights (1984–85) OAS Doc OEA/Ser.L/V/II.66/doc.10, rev 1, 190–93 para III(3).

[70] UNHCR, 'Legal Considerations Regarding Claims' (n 65) paras 13–18.

[71] There are 146 parties to the 1951 Refugee Convention as of November 2023. A number of states—many in the Middle East and in South and Southeast Asia—are not signatories. See UN Treaty Collection, 'Status of the 1951 Convention Relating to the Status of Refugees' <https://treaties.un.org/pages/ViewDetailsII.aspx?src=TREATY&mtdsg_no=V-2&chapter=5&Temp=mtdsg2&clang=_en> accessed 20 November 2023; David James Cantor, 'Law, Policy and Practice Concerning the Humanitarian Protection of Aliens on a Temporary Basis in the Context of Disasters' (The Nansen Initiative 2015) Background Paper 18 (noting the 'prevailing view among States . . . that disasters do not as such engage the expanded Cartagena refugee definition', but that it remains an open question if man-made disasters could trigger protection).

[72] See eg *AF (Kiribati)* (n 65) para 74 (no evidence that environmental conditions amount to serious harm for the purposes of the Refugee Convention); upheld by *Teitiota v The Chief Executive of the Ministry of Business, Innovation and Employment* [2013] NZHC 3125 (High Court of New Zealand (NZHC)); *Teitiota v The Chief Executive of the Ministry of Business, Innovation and Employment* [2015] NZSC 107 (Supreme Court of New Zealand (NZSC)).

B. Laws on statelessness

The loss of habitable land and potential submergence of large portions of some states present unprecedented challenges, including the possibility that people will become stateless. This risk is particularly acute for those living in low-lying small island states.[73] Legal scholarship that contemplates the loss of territory due to climate change has focused, in part, on issues of statelessness, as well as questions of sovereignty and statehood.[74] Yet claims for legal protection as a stateless person are undermined by how people are likely to move, both before territory disappears entirely or a state ceases to exist.

Under international law, a stateless person is an individual who is 'not considered as a national by any State under the operation of its law'.[75] This definition focuses on the lack of legal recognition of nationality, rather than the physical loss of a state. Most people who move in the context of climate change will not fit this legal—or *de jure*—definition of a stateless person. It is unlikely that the laws of a government will be invalidated, for example, even as land is lost. A shift in government for island states is not contemplated or wanted, nor does a state necessarily disappear with the loss of land. International law presumes the continuity of states, even when criteria of statehood are not met.[76] Furthermore, even if a state no longer exists as a matter of international law, the legal status of its population remains unclear. Whether people have a right to nationality elsewhere is an open question: the status of a stand-alone right to nationality—as set out in the Universal Declaration of Human Rights—is subject to debate, as it is not included in the International Covenant on Civil and Political Rights (ICCPR), and there is no complementary obligation to confer nationality.[77]

International law on statelessness, however, aims to give effect to the right to nationality. But the two international conventions on statelessness—on protection of stateless persons and prevention of statelessness—are not widely ratified.[78] In addition to sparse participation, an approach to climate change and human mobility

[73] See eg Etienne Piguet, 'Climatic Statelessness: Risk Assessment and Policy Options' (2019) 45 Population and Development Review 865.

[74] See eg Rosemary Rayfuse, 'International Law and Disappearing States: Utilising Maritime Entitlements to Overcome the Statehood Dilemma' [2010] UNSW Law Research Paper No 2010-52; Maxine Burkett, 'The Nation Ex-Situ: On Climate Change, Deterritorialized Nationhood and the Post-Climate Era' (2011) 2 Climate Law 345; Mark Nevitt, 'Climate Change and The Specter of Statelessness' (2023) 35 The Georgetown Environmental Law Review 331.

[75] Convention Relating to the Status of Stateless Persons (adopted 28 September 1954, entered into force 6 June 1960) 360 UNTS 117, art 1(1).

[76] See James Crawford, *The Creation of States in International Law* (2nd edn, Clarendon Press 2006) 667–99.

[77] See Universal Declaration of Human Rights (adopted 10 December 1948) UNGA Res 217 A(III) art 15; see further McAdam, *Climate Change, Forced Migration* (n 44) 138–43.

[78] See Convention on the Reduction of Statelessness (adopted 30 August 1961, entered into force 13 December 1975) 989 UNTS 175. To date, there are 89 parties to the 1954 Convention (n 75) and 68 parties to the 1961 Convention.

based on these conventions shares similar limitations to the refugee regime. The protections they offer for stateless persons are considered modest. There is no right of entry into a new state, for example, which means that a person must find themselves in a state that offers protection to stateless people. Legal status would also be accessible only *after* a person is denied a nationality. For citizens of small island states, this means that international legal protections may be triggered well after they move abroad. Like refugee status, which is meant to provide a sort of surrogate protection for people who are not willing or able to get such protection from their own state, legal protection for stateless persons is intended to fill a governmental gap. However, as discussed in the Pacific Islands case study, most states want to continue to provide protection to their people, and many residents of these states wish to continue to be recognized as nationals of their states should they move.

Thus, international law governing statelessness offers little recourse, although it may retain political importance in highlighting the need for international protection and the risk of *de facto* statelessness.[79] Efforts to prevent this risk from being realized are important, but they do not provide a legal basis to address vulnerabilities before movement or provide for the kind of planning and anticipatory mobility that might be necessary to ensure people safe, dignified lives elsewhere.[80]

C. Human rights law

Human rights law is relevant to human mobility in several ways. First, it provides standards for the minimum conditions and resources that must be ensured for all people under a state's jurisdiction, including migrants. International human rights instruments also enshrine specific rights that are important for people on the move. For movement within a country, the rights to liberty of movement and freedom to choose one's residence provide an internal right of mobility, which can protect against forced internal displacement.[81] These rights are accompanied by state duties to respect rights. If people want to move from vulnerable locations, for

[79] As argued by Benoît Mayer and Christel Cournil, 'Climate Change, Migration and Human Rights: Towards Group-Specific Protection?' in Ottavio Quirico and Mouloud Boumghar (eds), *Climate Change and Human Rights: An International and Comparative Law Perspective* (Routledge 2015) 181–82.

[80] See also Susin Park, 'Climate Change and the Risk of Statelessness: The Situation of Low-Lying Island States' (UNHCR 2011) (outlining options to prevent statelessness); James C Hathaway, 'Afterword' in Benoît Mayer and François Crépeau (eds), *Research Handbook on Climate Change, Migration and the Law* (Edward Elgar 2017) 438 (suggesting a protocol to the statelessness regime).

[81] ICCPR (adopted 16 December 1966, entered into force 23 March 1976) 999 UNTS 171, art 12(1). See also Convention on the Elimination of All Forms of Discrimination Against Women (adopted 18 December 1979, entered into force 3 September 1981) 1249 UNTS 13 (CEDAW) art 15(4); International Convention on the Elimination of All Forms of Racial Discrimination (adopted 7 March 1966, entered into force 4 January 1969) 660 UNTS 195 (ICERD) art 5(d)(i); International Convention on the Protection of Rights of All Migrant Workers and Members of Their Families (adopted 18 December 1990, entered into force 1 July 2003) 2220 UNTS 3 (ICRMW) art 39(1).

example, states cannot inhibit their movement or choice of residence. For international mobility, everyone has a right to leave any country, including their own. This provides a legal means to exit a country. Migrants are further entitled to return to their country, subject to restrictions that are not arbitrary.[82]

There are, however, limitations to rights related to mobility. Both the right to liberty of movement and to choice of residence are reserved for those who are lawfully present within a state.[83] This mainly presents an obstacle for foreigners not lawfully residing in a country, which is the case for irregular migrants who cross borders while moving away from the impacts of climate change.[84] Most mobility rights can also be limited if 'necessary to protect national security, public order, public health or morals or the rights and freedoms of others'.[85] This could justify evacuation when it is necessary to protect people's lives, health, or safety. In addition, there is no a general right to enter another country, which has enabled the proliferation of barriers to international movement and efforts to regulate or restrict people from leaving their own country.[86] These obstacles impede free movement and have long been accompanied by calls for a better way to manage international mobility.[87] Nonetheless, the right to move, safely and with a degree of freedom and choice, is both an underlying premise that motivates this book's analysis and a right that underpins the kind of anticipatory, proactive adaptive mobility it advocates.

Second, human rights law creates positive duties for states that are critical to addressing the impacts of climate change. As subsequent chapters argue, these duties can be integrated into adaptation obligations to clarify their content and shape adaptive action. Doing so underscores the role that human rights can play in proactive, anticipatory climate action, to make their enjoyment a priority in planning processes. Human rights law also includes principles that are important

[82] ICCPR, art 12(2), 12(4). See also Convention on the Rights of the Child (adopted 20 November 1989, entered into force 2 September 1990) 1577 UNTS 3 (CRC) art 10(3); ICERD, art 5(d)(ii); ICRMW art 8; HRC, 'General Comment No 27: Article 12 (Freedom of Movement)' (2 November 1999) UN Doc CCPR/C/21/Rev.1/Add.9, paras 7, 21 (on arbitrariness, there are few, if any, circumstances where it is reasonable to limit the right).

[83] ICCPR, art 12(1).

[84] There are exceptions to this limitation. The HRC, for example, recognizes that a non-citizen may retain their mobility rights based on considerations of non-discrimination, prohibition of inhuman treatment, and respect for family life. HRC, 'General Comment No 15: The Position of Aliens Under the Covenant' (11 April 1986) UN Doc HRI/GEN/1/Rev.1 18 (1994) para 5.

[85] See ICCPR, art 12(3). Restrictions must be proportionate, appropriate to achieve their protective function, and the least intrusive means. HRC, 'General Comment No 27' (n 82) para 14.

[86] These include policies and practices that criminalize entry or stay, erect fences, close borders, and have resulted in the use of violence, pushbacks and interceptions, and administrative detention. See OHCHR, 'Situation of Migrants in Transit' (2016) A/HRC/31/35.

[87] See eg Jaya Ramji-Nogales, 'Migration Emergencies' (2017) 68 Hastings Law Journal 609; Betts (n 61); Susan F Martin, Sanjula Weerasinghe, and Abbie Taylor (eds), *Humanitarian Crises and Migration: Causes, Consequences and Responses* (Routledge 2014); see also Colin Harvey and Robert P Barnidge Jr, 'Human Rights, Free Movement, and the Right to Leave in International Law' (2007) 19 International Journal of Refugee Law 1.

to ensuring that migrants do not face discrimination in accessing resources and enjoying rights, including the principles of equality and non-discrimination.[88]

Third, and for climate-related mobility specifically, human rights law is invoked as a means of protection for migrants. For cross-border mobility, the principle of *non-refoulement* is often raised, as it protects against the forced return of refugees and migrants to life-threatening situations; serious violations of human rights; or cruel, inhuman or degrading treatment. It is a fundamental principle of the Refugee Convention that also developed under human rights law.[89] In *Teitiota v New Zealand*, the Human Rights Committee (HRC) confirmed that *non-refoulement* could be triggered by climate change, raising the possibility of future claims for legal protection based on human rights law.[90] To date, however, there is little indication by the HRC or judicial bodies that climate impacts currently pose the risks necessary to meet the threshold of potential harm required for protection against *refoulement*. Significant restrictions of resources and potable water are not enough, raising questions about how dire circumstances must be before relief is granted.[91] Furthermore, even if individuals could establish that they face serious harm, approaches based on *non-refoulement* suffer from some of the same limitations already highlighted for the statelessness and refugee regimes. Migrants must, for example, already be outside their countries, making claims for legal protection possible only *after* they cross international borders. Individuals must also show that the risk they face is personal in nature and not only derived from the general conditions in their country of origin.[92] Given this, and given the need to meet a high threshold of harm, protection is unlikely to extend to those who move in advance of the worst impacts. As such, the principle would not apply to the kind of anticipatory rights-protective adaptive mobility advanced in this book. States' willingness to adapt and efforts to reduce vulnerabilities also weigh against claims for relief.[93]

[88] Discussed in Chapter 4.

[89] See 1951 Refugee Convention, art 33. Under human rights law, the principle is derived from the ICCPR, the Convention Against Torture, and customary international law. See ICCPR, art 7; Convention against Torture and Other Cruel, Inhuman or Degrading Treatment or Punishment (adopted 10 December 1984, entered into force 26 June 1987) 1465 UNTS 85, art 3; Cathryn Costello and Michelle Foster, 'Non-Refoulement as Custom and Jus Cogens- Putting the Prohibition to the Test' in *Netherlands Yearbook of International Law 2015* (TMC Asser Press 2016) 282–83, 285 ('overwhelming majority of scholarly opinion' views *non-refoulement* as a customary norm); but see James C Hathaway, 'Leveraging Asylum' (2010) 45 Texas International Law Journal 503 (arguing against its status as customary law).

[90] HRC decisions are influential but not legally binding. *Teitiota v New Zealand*, 'Views Adopted by the Committee under Article 5(4) of the Optional Protocol, Concerning Communication No 2728/2016' (HRC, 23 September 2020) UN Doc CCPR/C/127/D/2728/2016, para 9.11. New Zealand also recognizes this possibility. See *AC (Tuvalu)* [2014] NZIPT 800517-520 (NZIPT) para 70; *Teitiota v The Chief Executive of the Ministry of Business, Innovation and Employment* (NZSC) (n 72) [13].

[91] See eg *Teitiota v The Chief Executive of the Ministry of Business, Innovation and Employment* (NZHC) (n 72) [55]; *Teitiota v New Zealand* (HRC) (n 90) para 9.12.

[92] See HRC, 'General Comment No 36' (n 30) para 30; *Teitiota v New Zealand* (HRC) (n 90) para 9.3.

[93] For example, the HRC noted that the Republic of Kiribati had time to 'relocate its population' if necessary and 'was taking adaptive measures' in rejecting a claim that New Zealand erred in removal. *Teitiota v New Zealand* (HRC) (n 90) para 9.12.

For internal mobility, the Guiding Principles on Internal Displacement (Guiding Principles) are put forth as a model to develop legal protection in the context of climate change.[94] The Guiding Principles are not legally binding, but they draw on international human rights and humanitarian law. Their description of internally displaced persons (IDPs) includes those forced to flee natural or human-made disasters, which could cover climate impacts.[95] The Guiding Principles led to the development of binding regional law instruments, including the 2006 Great Lakes Protocol on the Protection and Assistance to Internally Displaced Persons and the 2009 African Union Convention for the Protection and Assistance of Internally Displaced Persons in Africa (Kampala Convention).[96] The Kampala Convention explicitly includes climate change, obligating state parties to 'take measures to protect and assist persons who have been internally displaced due to natural or human made disasters, including climate change'.[97] A number of the countries in the Sahel have ratified the Kampala Convention, creating binding obligations on these states.[98] However, an IDP under the Guiding Principles and these regional instruments is a person who is 'forced or obliged to flee', limiting their general applicability. This is central to these instruments, with an IDP characterized by the coercive nature of their flight.[99] The Guiding Principles do encompass movement 'as a result of *or in order to avoid*' disasters, the latter a form of 'anticipatory displacement'. Accordingly, displacements before or after the occurrence of a sudden-onset event, like a tropical storm or flood, are considered easy cases where IDP protections apply.[100] But such displacement, anticipatory or otherwise, is still considered forced.

The Guiding Principles also aim to prevent the conditions that lead to displacement, which could be accomplished through measures that allow people to stay

[94] See eg Elizabeth Ferris, 'The Relevance of the Guiding Principles on Internal Displacement for the Climate-Migration Nexus' in Benoît Mayer and François Crépeau (eds), *Research Handbook on Climate Change, Migration and the Law* (Edward Elgar 2017); Susan F Martin, 'Toward an Extension of Complementary Protection?' in Benoît Mayer and François Crépeau (eds), *Research Handbook on Climate Change, Migration and the Law* (Edward Elgar 2017) 452; Jane McAdam, 'Climate Change Displacement and International Law: Complementary Protection Standards' (UNHCR 2011) 49.

[95] See United Nations Commission on Human Rights (UNCHR), 'Guiding Principles on Internal Displacement' (11 February 1998) UN Doc E/CN.4/1998/53/Add.2, paras 2–3; Walter Kälin, 'Guiding Principles on Internal Displacement—Annotations' (American Society of International Law 2008) 3–5.

[96] See Protocol on the Protection and Assistance to Internally Displaced Persons of the International Conference on the Great Lakes Region (adopted 30 November 2006, entered into force June 2008); African Union Convention for the Protection and Assistance of Internally Displaced Persons in Africa (adopted 23 October 2009, entered into force 6 December 2012) (Kampala Convention).

[97] Kampala Convention, art 5(4).

[98] This includes Burkina Faso, Chad, Mauritania, Niger, Nigeria, Gambia, and Guinea-Bissau. See Chapter 9.

[99] See Walter Kälin and Jörg Künzli, *The Law of International Human Rights Protection* (2nd edn, Oxford University Press 2019) 539; Brookings Institution and University of Bern, 'Protecting Internally Displaced Persons: A Manual for Law and Policymakers' (2008) 12.

[100] See UNCHR, 'Guiding Principles' (n 95) para 2 (emphasis added); Ferris (n 94) 116–17, 123; Bodansky, Brunnée, and Rajamani (n 50) 319.

in place or that enable adaptive mobility.[101] But guidance on how the Guiding Principles can concretely prevent risks from disasters is lacking, and actions that enable adaptive mobility are not yet incorporated into their interpretation. They are, nonetheless, instructive and support arguments made in this book, including the need to minimize the adverse effects of movement, guarantees of information to and seeking consent from affected persons, and their participation in the planning and management of relocation.[102] In general, however, the Guiding Principles are not relied upon to shape and inform anticipatory movement. They are not legally binding, not aimed at more voluntary movement, and thus not equipped to fully address the range of movement that will arise in the context of climate change.[103]

D. Other approaches

Outside the climate change regime, human mobility is gaining attention as part of discussions under the United Nations Convention to Combat Desertification (UNCCD). The preamble of the UNCCD includes explicit recognition that 'desertification and drought affect sustainable development through their interrelationships with important social problems' including 'those arising from migration [and] displacement of persons'.[104] More recently, parties to the UNCCD adopted decisions on the role that measures taken under the UNCCD can play in addressing degradation-related drivers of migration.[105] However, mobility within the UNCCCD is framed as a negative consequence of desertification and drought, with efforts aimed at reducing forced migration.[106]

There are also other approaches to climate change and human mobility not based in existing law. These include policy approaches and proposals for a new

[101] UNCHR, 'Guiding Principles' (n 95) principle 5.

[102] ibid principle 7.

[103] While the distinction between voluntary and involuntary movement is not always clear (see n 40), there is consensus that much movement prompted by slower-onset events does not engage the Guiding Principles. See Sumudu Atapattu, 'Climate Change, Human Rights, and Forced Migration: Implications for International Law' (2009) 27 Wisconsin International Law Journal 607, 618; Bodansky, Brunnée, and Rajamani (n 50) 316; Alice Thomas, 'Protecting People Displaced by Weather-Related Disasters and Climate Change: Experience from the Field' (2013) 15 Vermont Journal of Environmental Law 803, 813–14.

[104] United Nations Convention to Combat Desertification in those Countries Experiencing Serious Drought and/or Desertification, Particularly in Africa (adopted 14 October 1994, entered into force 26 December 1996) 1954 UNTS 3 (UNCCD) preamble.

[105] Decision 28/COP.13, 'The positive role that measures taken under the Convention can play to address desertification/land degradation and drought as one of the drivers that causes migration' (UNCCD, 23 October 2017) ICCD/COP(13)/21/Add.1; UNCCD Secretariat, 'The positive role that measures taken under the Convention can play to address desertification/land degradation and drought as one of the drivers that causes migration' (24 June 2019) ICCD/COP(14)/19.

[106] See Decision 7/COP.13, 'The UNCCD 2018–2030 Strategic Framework', Annex (15 September 2017) ICCD/COP(13)/21/Add.1; International Organization for Migration (IOM) and UNCCD, 'Addressing the Land Degradation-Migration Nexus: The Role of the United Nations Convention to Combat Desertification' (IOM 2019) paras 1, 5.

international framework on climate mobility.[107] The former often seek to go beyond climate change, to consider disasters or environmental factors more generally.[108] Importantly, however, they lack the legal hook necessary to compel state action and must rely on other means for implementation.[109] This book does not engage deeply with these policy approaches, as it aims to offer a legal basis to help fill legal gaps. But some will be referenced in subsequent chapters, as they include principles or best practices that offer guidance for states in undertaking measures related to adaptive mobility.

These other approaches replicate the limitations of existing regimes that govern migration. They tend to focus on protection needs after people move or cross borders, rather than before or during movement. They also focus on forced migration or displacement and do not account for the range of ways climate change influences human mobility, including immobility or temporary, seasonal, and more voluntary migration.[110] Proposals for new legal frameworks that create a category of climate-related migrant are further undermined by difficulties establishing a causal connection between an individual's movements and specific impacts of climate change.

Finally, there are scholars that criticize making any connection between climate change and human mobility for governance purposes. Critiques emphasize the indeterminacy of climate change as a causal factor in mobility and argue that attempts to link migration to climate belie the complexities of human movements.[111] Calum TM Nicholson, for example, argues that any discussion of climate mobility presumes that one should be able to determine a causal relationship between the two—independent of other variables—to '*distinguish* an "environmental/climate migrant" from any other migrant'. Why else, he asks, would one 'single out' climate change 'if we are not interested in determining its causal effects on society'

[107] See eg Bonnie Docherty and Tyler Giannini, 'Confronting a Rising Tide: A Proposal for A Convention on Climate Change Refugees' (2009) 33 Harvard Environmental Law Review 349; David Hodgkinson and others, '"The Hour When the Ship Comes In": A Convention for Persons Displaced by Climate Change' (2010) 36 Monash University Law Review 69; Frank Biermann and Ingrid Boas, 'Towards a Global Governance System to Protect Climate Migrants: Taking Stock' in Benoît Mayer and François Crépeau (eds), *Research Handbook on Climate Change, Migration and the Law* (Edward Elgar 2017).

[108] See eg Brookings Institution, Georgetown University, and UNHCR, 'Guidance on Protecting People from Disasters and Environmental Change through Planned Relocation' (2015); The Nansen Initiative (n 58); UNGA, 'Global Compact on Migration' (n 58).

[109] The Global Compact on Migration, for example, faces such criticisms; see eg Jolanda van der Vliet and Frank Biermann, 'Global Governance of Climate Migrants: A Critical Evaluation of the Global Compacts' in Avidan Kent and Simon Behrman (eds), *Climate Refugees: Global, Local and Critical Approaches* (Cambridge University Press 2022).

[110] See Introduction; see also 'The Peninsula Principles on Climate Displacement within States' (Displacement Solutions 2013).

[111] See eg Benoît Mayer, 'Who Are "Climate Refugees"?' in Avidan Kent and Simon Behrman (eds), *Climate Refugees: Beyond the Legal Impasse?* (Routledge 2018); Calum TM Nicholson, '"Climate Mobility" Is Not a Proper Subject of Research and Governance' in Alexander Zahar and Benoît Mayer (eds), *Debating Climate Law* (Cambridge University Press 2021).

and why single out migration as a societal effect 'if we are not interested in distinguishing the people thus impacted'?[112] In some ways, this book begins from a similar premise, underscoring the problems with efforts to establish a distinct category of climate migrant or that require causally connecting climate impacts to the serious harm migrants face or endure *after* they move. Yet it does not arrive at the same conclusions—that climate change is not a valid subject of inquiry for mobility research, or conversely, that migration should not be a concern for climate governance. Instead, it focuses on the latter to overcome the limitations identified both in this chapter and in critiques concerned with questions about climate change as a causal factor in mobility. That movement is multicausal, or that impacts also contribute to immobility, does not mean that climate change is not affecting human mobility decisions and patterns. Whether directly or indirectly, climate change is altering access to resources, goods, and services, which in turn can lead people to move. This book takes a broader approach, to account for the interaction of climate events with contextual factors and to shift action to an earlier time, *before* people are unable to access their rights and resources and *before* they are either rendered immobile or forced to move. In some cases, adaptive measures will enable people to stay. In other circumstances, states must plan for and facilitate safe, rights-protective adaptive mobility.[113] The necessity to do so is, as the book will show, a legal obligation and one very much tied to climate change.

V. Conclusion

Overall, the current approaches to mobility outlined in this chapter are ad hoc, come too late, and fail to act on the knowledge available about the foreseeable risks that climate impacts pose. They place the emphasis on providing protection after people move or are displaced, rather than enabling people to stay or move safely elsewhere. More fundamentally, these approaches individualize what should be addressed as a collective problem. They make legal status and protection contingent on meeting specific criteria or establishing extreme personal risks, which constructs the problem as one to be solved on an individual basis. This puts the burden on the individual to seek legal protection and to navigate bureaucratic systems. These systems can take time and leave people in limbo while awaiting decisions, with some individuals succeeding in gaining legal status while others in similar situations do not.

A better approach begins by understanding mobility as part of a process of adapting to climate change. Understanding movement in this way, with climate

[112] Nicholson (n 111) 220–21; see also Calum TM Nicholson and Benoît Mayer, 'Introduction' in Calum TM Nicholson and Benoît Mayer (eds), *Climate Migration: Critical Perspectives for Law, Policy, and Research* (Hart 2023).

[113] Concrete examples are provided in the case studies, Chapters 7–9.

change kept squarely in focus, is important because it incorporates the broader responsibility involved in causing and addressing climate impacts and the diverse ways climate affects human mobility.[114] Such an approach enables both recognition of the socioeconomic conditions and systems that shape differential impacts and context-specific, community-led adaptation. It also draws on the existing legal framework on climate change, which provides a legal basis to argue for adaptation that includes adaptive mobility. As the next chapter explains, this framework provides critical temporal and collective elements that could compel anticipatory state action on adaptation before people are forced to move. It further requires support for such action, which can help shift the burden away from individuals or their communities towards the state, and crucially, to those countries that have contributed most to—and benefited most from—climate change. This represents a new approach, to shape adaptation in general and adaptive mobility specifically.

[114] See eg Ingrid Boas, '"Climate Mobility" Is a Proper Subject of Research and Governance' in *Debating Climate Law* (n 111); Boas and others (n 47).

2
A New Approach: Adaptation in the Climate Change Regime

I. Introduction

This chapter provides the foundation for a new approach to climate change and human mobility, one grounded in adaptation in the international climate change regime. The chapter begins by analysing the meaning of climate change adaptation, which Section II explains is a process influenced by timing and context. This process can lead to a broad range of adaptation measures that impact human mobility, some of which can prevent displacement and allow people to stay in place for as long as possible and others that are necessary to plan for or facilitate adaptive mobility. The chapter then turns to adaptation in the climate change regime. Section III describes the evolution of adaptation in the regime, from an initial perception of adaptation as a secondary response to its increasing importance as knowledge about the impacts of climate change grows. Section IV discusses the relationship between adaptation and loss and damage, and how this relationship affects work on human mobility in the regime. Section V analyses adaptation obligations. It identifies three types of adaptation obligations found in climate treaties: obligations to act on adaptation, to assist with adaptation, and to cooperate in adaptation efforts. It discusses the nature of these obligations and their general scope and content. Adaptation obligations are the legal anchor for arguments to compel and shape state action on adaptive mobility, and the chapter closes by recognizing their breadth and the need for their further concretization, as explored in subsequent chapters.

II. Adapting to the impacts of climate change

The concept of adaptation did not often feature in consideration of environmental risks prior to the negotiation of the first international treaty on climate change—the United Nations Framework Convention on Climate Change (UNFCCC).[1] Its

[1] See E Lisa F Schipper and Ian Burton, 'Understanding Adaptation: Origins, Concepts, Practice and Policy' in E Lisa F Schipper and Ian Burton (eds), *The Earthscan Reader on Adaptation to Climate Change* (Earthscan 2009) 1.

Climate Change, Human Rights, and Adaptive Mobility. Lauren Sakae Nishimura, Oxford University Press.
 DOI: 10.1093/9780198930068.003.0003

inclusion in the UNFCCC established it as one of two categories of response to climate change and its effects. Mitigation is the other, which refers to efforts to address the greenhouse gases (GHGs) causing climate change by reducing emissions or enhancing sinks of GHGs.[2] Adaptation is not defined in the UNFCCC or subsequent treaties of the climate regime. Nor is there a universally agreed upon definition outside the climate regime, despite its extensive use in policy and academic literature on climate change.

The general understanding and dominant definition of climate change adaptation, put forth by the Intergovernmental Panel on Climate Change (IPCC), evolved over time. Initially, IPCC reports contemplated adaptation briefly in reference to sea level rise, coastal zone management, or resource use, particularly for the forestry and agricultural sectors.[3] It gained increasing importance and scope with each successive report, as reflected in the conceptual expansion of the term. In 2007, for example, the IPCC's Fourth Assessment Report included definitions of adaptation both as 'measures to reduce the vulnerability of natural and human systems' in its synthesis report and as '[a]djustment' in these systems that 'moderates harm or exploits beneficial opportunities' in Working Group II's report on impacts, adaptation, and vulnerability.[4] By 2014, the IPCC adopted a single definition of adaptation as 'the *process* of adjustment to actual or expected climate and its effects'. It differentiated between adaptation in human systems that aim 'to moderate or avoid harm or exploit beneficial opportunities' and adaptation in natural systems, where 'human intervention may facilitate adjustment to expected climate'. The IPCC recognized that this definition differed 'in breadth and focus' from its prior reports, 'reflecting progress in science'.[5] It also acknowledged that the 'framing of adaptation' has shifted from a focus on biophysical risk to include 'wider social and economic drivers of vulnerability and people's ability to respond'.[6] Recognition of this broader context is important, as the impacts of climate change interact with these wider drivers and other political and cultural factors to

[2] IPCC, 'Annex II: Glossary' in HO Pörtner and others (eds), *Climate Change 2022: Impacts, Adaptation, and Vulnerability. Contribution of Working Group II to the Sixth Assessment Report of the Intergovernmental Panel on Climate Change* (Cambridge University Press 2022) 2915.

[3] See IPCC, 'Climate Change: The 1990 and 1992 IPCC Assessments, IPCC First Assessment Report Overview and Policymaker Summaries and 1992 IPCC Supplement' (IPCC 1992).

[4] See Alfons PM Baede, Paul van der Linden, and Aviel Verbruggen (eds), 'Annex II: Glossary', *Climate Change 2007: Synthesis Report* (Cambridge University Press 2007); IPCC, 'Annex I: Glossary', *Climate Change 2007: Impacts, Adaptation and Vulnerability. Contribution of Working Group II to the Fourth Assessment Report of the IPCC* (Cambridge University Press 2007) 869; see also IPCC, 'Annex B: Glossary of Terms' in RT Watson and the Core Writing Team (eds), *Climate Change 2001: Synthesis Report* (Cambridge University Press 2001) 365 (definition of adaptation adopted by Working Group II in 2007).

[5] IPCC, 'Annex II: Glossary', *Climate Change 2014: Synthesis Report* (IPCC 2014) 1758 (emphasis added).

[6] IR Noble and others, 'Adaptation Needs and Options' in Christopher B Field and others (eds), *Climate Change 2014: Impacts, Adaptation, and Vulnerability. Part A: Global and Sectoral Aspects. Contribution of Working Group II to the Fifth Assessment Report of the IPCC* (Cambridge University Press 2014) 836.

give rise to the need to adapt. Such a framing of adaptation helps align its scientific understanding with its use in climate governance. The interaction of these drivers also raises further questions about the distributional effects of adaptation, which involves decisions about who and what to prioritize, the scope of adaptation, and the types of activities within its ambit.[7]

Climate change adaptation is further influenced by its timing. Adaptive interventions can occur at different moments: in anticipation of expected impacts, in response to ongoing impacts, or in their aftermath. They can also incorporate both reactive and anticipatory elements. Adaptation can include anticipatory actions to prepare for and mitigate climate impacts, as well as actions that will be taken to respond after impacts occur. This is true for both sudden-onset events, such as storms and hurricanes, and slow-onset events, where prolonged and more gradual impacts afford longer time scales for anticipatory action. Changing risks and new information about climate impacts further underscore that adaptation should be continual, with few one-off solutions. This accords with an understanding of adaptation as an iterative process that brings together a range of activities. Adaptation programmes can include planning, feasibility and outcome assessments, risk management, decision-making, disaster responses, legislation, and education. They can also include planning for and undertaking human mobility, through safe, legal pathways or planned relocation. Measures to facilitate adaptive mobility can be taken to help prevent forced immobility and displacement, and they can occur after displacement to ensure people are able to resettle elsewhere. Furthermore, given the foresight that climate science increasingly affords—as outlined in Chapter 1—these measures can and should be proactive and deliberate. Delays can undermine the efficacy and ability of adaptation to protect resources and livelihoods, which may in turn breach international commitments and violate human rights.[8]

In addition to timing, context further shapes the implementation of adaptation. There are certain general factors that can enable or constrain adaptation. Adaptation strategies often share a common goal of reducing vulnerabilities and the risks associated with climate change. However, the extent and interaction of knowledge, capacity, finance, and governance—including through legal systems and legislation—affect whether and how adaptation occurs.[9] Climate governance is also polycentric and can involve a variety of actors and interests. But adaptation is an inherently localized process, and its implementation is most acutely felt locally. Unlike mitigation, which aims to reduce the causes of climate change globally,

[7] Lauren Nishimura, 'Adaptation and Anticipatory Action: Integrating Human Rights Duties into the Climate Change Regime' (2022) 12 Climate Law 99.

[8] As found in Australia's delay in undertaking adaptation, discussed in Chapter 4. See *Billy v Australia*, 'Views Adopted by the Committee under Article 5(4) of the Optional Protocol, Concerning Communication No 3624/2019' (HRC 21 July 2022) UN Doc CCPR/C/135/D/3624/2019.

[9] See New and others, 'Decision-Making Options for Managing Risk' in *Climate Change 2022: Impacts, Adaptation and Vulnerability* (n 2) 2580–600.

adaptation must be tailored to specific conditions and challenges. The varying ways that climate events interact with multiple layers of vulnerability, arising from geography, societal conditions, and individual experiences,[10] translate into the need for adaptation efforts that account for contextual differences. Adaptation strategies must therefore manage tensions raised by context and time. They must consider differential risks and disproportionate effects while also contributing to broader adaptation goals and efforts to address climate change. There is therefore no single or universal approach to adaptation. However, to succeed in their aims, adaptation approaches must themselves be flexible and adaptive.[11]

The success of adaptation approaches raises three additional points. First, there are limits to adaptation. Adaptation limits refer to the point at which adaptive actions are unable to achieve their objectives, either because intolerable risks are unavoidable or options do not yet exist to avoid such risks.[12] Second, adaptation can be maladaptive. Maladaptation refers to actions that lead to increased risk of adverse or inequitable outcomes, greater vulnerability, or reduced welfare, now or in the future.[13] Research highlights that maladaptation can result from inaction, decisions that seek short-term gains at the expense of longer-term resilience, poor planning or implementation, or failure to account for sectors or groups that may be impacted but are not the target of adaptation measures.[14] Third, policymakers and researchers are increasingly calling for adaptation to be more transformative, to change fundamental aspects of systems and move beyond incremental and preservationist action and policy.[15] Transformational adaptation involves difficult decisions and trade-offs, a shift from reactive to anticipatory action, longer-term planning, integration across sectors and levels of government, adequate financing, and policy alignment.

Thus, adaptation can be aimed at specific risks or systemic change. It can address expected or ongoing impacts, and, as subsequent chapters establish, it can require proactive, anticipatory action to address foreseeable risks of harm from

[10] Chapter 1, II.B.

[11] See Margaux J Hall and David C Weiss, 'Avoiding Adaptation Apartheid: Climate Change Adaptation and Human Rights Law' (2012) 37 The Yale Journal of International Law 309.

[12] IPCC, 'Annex II: Glossary' (n 2) 2898.

[13] ibid 2915; LDC Expert Group (LEG), 'National Adaptation Plans: Technical Guidelines for the National Adaptation Plan Process' (2012) 13.

[14] See Pörtner and others, 'Technical Summary' in *Climate Change 2022: Impacts, Adaptation and Vulnerability* (n 2) 85; Noble and others (n 6) 849, 857–58; AK Magnan and others, 'Addressing the Risk of Maladaptation to Climate Change' (2016) 7 Wiley Interdisciplinary Reviews: Climate Change 646, 650–52.

[15] See eg Decision 2/CMA.5, 'Global Goal on Adaptation' (13 December 2023) FCCC/PA/CMA/2023/16/Add.1, para 8; Pörtner and others (n 14) 97–99; Robin Kundis Craig, '"Stationarity Is Dead"—Long Live Transformation: Five Principles for Climate Change Adaptation Law' (2010) 34 Harvard Environmental Law Review 10; Benoît Mayer, 'Climate Change Adaptation Law: Is There Such a Thing?' in Alexander Zahar and Benoît Mayer (eds), *Debating Climate Law* (Cambridge University Press 2021) 313; Jonathan Verschuuren, 'Introduction to Climate Change Adaptation' in Jonathan Verschuuren (ed), *Research Handbook on Climate Change Adaptation Law* (Edward Elgar 2022) 3.

climate change. The latter can be accomplished in multiple ways: through changes to infrastructure, new land use or agricultural policies, or the outright provision of resources, all of which can enable people to adapt and stay in place as long as possible. But if these measures become insufficient—and resources and rights are no longer accessible—people may need to move. In these circumstances, states must plan for and facilitate safe, rights-protective adaptive mobility as a part of the process of adaptation, a process that is mediated and shaped by timing and context. Such an understanding of human mobility provides the conceptual framing for an approach that enables anticipatory action, rather than being exclusively reactive. Its legal basis can be grounded in the climate change regime to help broaden responses beyond interventions only after people are forced to move.

III. The evolution of adaptation in the climate change regime

In the negotiations leading up to the UNFCCC, efforts focused on mitigation and reaching an agreement on the reduction of GHG emissions. Adaptation became a secondary response, one expected to occur naturally or to be undertaken if mitigation efforts failed. Concerns arose that adaptation could distract attention from mitigation or that funding for adaptation might be tantamount to accepting responsibility for climate change, which developed countries sought to avoid.[16] In contrast, many developing countries did not want the UNFCCC to impede their ability to develop by limiting their emissions and were concerned about potential adverse impacts.[17] This tension set the stage for an agreement that avoided attribution of responsibility for climate change, provided funding without links to historical emissions, and limited attention to adaptation.

The UNFCCC is a legally binding treaty that contains a mix of hard legal obligations, soft law, guiding principles, and recommendations. The 'ultimate objective' of the UNFCCC captures its focus on mitigation, with its aim to stabilize GHG concentrations to a level that prevents 'dangerous anthropogenic interference with the climate system' and to do so 'within a time-frame sufficient to allow ecosystems *to adapt naturally* to climate change'.[18] Despite this emphasis, adaptation is incorporated in the UNFCCC both indirectly and explicitly. The latter includes the legal obligations on adaptation that are discussed further in Section V: to act on adaptation, assist with adaptation, and cooperate in adaptation. Adaptation is indirectly referenced elsewhere in the text, through the concession that adverse

[16] See E Lisa F Schipper, 'Conceptual History of Adaptation in the UNFCCC Process' (2006) 15 Review of European, Comparative & International Environmental Law 82, 83–84.

[17] See Daniel Bodansky, 'The United Nations Framework Convention on Climate Change: A Commentary' (1993) 18 Yale Journal of International Law 451, 477–79.

[18] United Nations Framework Convention on Climate Change (adopted 9 May 1992, entered into force 21 March 1994) 1771 UNTS 107 (UNFCCC) art 2 (emphasis added).

effects must be addressed and in the operative principles that guide UNFCCC's implementation. These principles, for example, suggest parties take 'precautionary measures to anticipate, prevent or minimize the causes of climate change and mitigate its adverse effects'. They also call for action 'on the basis of equity' and in accordance with 'common but differentiated responsibilities and respective capabilities' (CBDR-RC), instructing developed country parties to 'take the lead in combating climate change and the adverse effects thereof' and for full consideration to be given to the 'specific needs and special circumstances of developing country Parties'.[19] Other provisions direct parties to consider actions necessary to meet the needs and concerns of certain developing countries identified by specific geographical and ecological vulnerabilities (small island countries; countries with arid, semi-arid, or low-lying coastal areas, that are liable to drought and desertification; or with areas prone to disasters) or identified as least developed countries (LDCs).[20] The principles of precaution and differentiation are central to adaptation, and their wording in the UNFCCC shows recognition by the parties of adverse impacts and that mitigation alone is an insufficient response.

The initial breadth of adaptation, however, was potentially narrowed by how the UNFCCC defined climate change. The definitions describe 'climate change' as directly or indirectly attributable to human activity—change that is additional to natural climatic variability. As a result, the meaning of 'adverse effects of climate change' is circumscribed to those associated with human activity.[21] This definition could create barriers to adaptation efforts. If strictly followed, it might require parsing out human-induced impacts from natural ones to show actions taken to adapt address anthropogenically caused climate impacts. Yet this distinction may no longer be as important as it seems; the current robust workstream on adaptation in the climate regime implies as much, and there is no indication of a need to establish a causal connection between *specific* climate events and human activity. Furthermore, scientific findings continue to strengthen the link between human activity and climate change, making clear that human influence is leading to the rapid, overall warming of the atmosphere, ocean, and land.[22]

The UNFCCC is also meant to be a starting point. It is the foundation for an international regime that has developed over the last three decades, which established a procedural architecture and institutions that contemplated additional work as science and policymaking on climate change evolved.[23] The climate

[19] UNFCCC, art 3(1)–(3).

[20] ibid art 4(8), (9).

[21] ibid art 1(1), (2).

[22] See eg IPCC, 'Summary for Policymakers' in V Masson-Delmotte and others (eds), *Climate Change 2021: The Physical Science Basis. Contribution of Working Group I to the Sixth Assessment Report of the Intergovernmental Panel on Climate Change* (Cambridge University Press 2021) (human influence 'is unequivocal' and changes to the climate system unprecedented).

[23] See Lavanya Rajamani, 'The United Nations Framework Convention on Climate Change: A Framework Approach to Climate Change' in Daniel A Farber and Marjan Peeters (eds), *Climate Change Law*, vol I (Edward Elgar 2016) 206; Catherine Redgwell, 'Multilateral Environmental Treaty-Making'

change regime grew out of this dynamic framework, and it has come to include subsequent agreements, outputs of its institutional bodies, and decisions of the Conference of the Parties (COP). In the first two decades following the adoption of the UNFCCC, however, work on adaptation progressed slowly. Efforts to advance mitigation commitments left adaptation issues relatively neglected for years. The next legal instrument—the Kyoto Protocol—focused almost entirely on mitigation and did not create any new legal commitments on adaptation. It does reiterate one of the UNFCCC's adaptation obligations on adaptation programmes, discussed further in Section V, and provides limited guidance on what could be included in these programmes. It further directs some of the proceeds from certified project activities of the Clean Development Mechanism, an emission-reduction credit scheme, to be used to assist particularly vulnerable developing country parties in meeting the costs of adaptation. And, most significantly, the Kyoto Protocol laid the foundation for the Adaptation Fund.[24]

As the understanding of and need to address climate impacts increased, attention on adaptation within the climate regime followed suit. Work on adaptation in the regime initially focused on planning and support for developing countries, with adaptation framed as a challenge for low-income countries.[25] COP decisions reflect this, aimed first at the short-term adaptation needs of developing countries and the establishment of mechanisms for technical and financial assistance from developed countries. Highlights included the adoption of the LDC work programme and their National Adaptation Programmes of Action (NAPAs), planning documents intended to address LDCs' immediate adaptation needs, and the formal introduction of three adaptation funds in 2001—the LDC Fund (LDCF), the Special Climate Change Fund (SCCF), and the Adaptation Fund.[26] Yet it was not until the establishment of the Nairobi Work Programme (NWP) in 2005 that the climate regime considered adaptation independently, rather than as an aspect of planning or support for LDCs.[27]

in Vera Gowlland-Debbas (ed), *Multilateral Treaty-Making* (Springer 2000) 93. Development of adaptation in the climate regime is discussed further in Lauren Nishimura, 'The Paris Agreement's Contribution to the Development of the Law on Adaptation' in Alexander Zahar (ed), *Research Handbook on the Law of the Paris Agreement* (Edward Elgar 2024).

[24] See Kyoto Protocol to the United Nations Framework Convention on Climate Change 1997 (adopted 11 December 1997, entered into force 16 February 2005) 2303 UNTS 162, arts 10(b), (d), 12(8).

[25] See Alexandra Lesnikowski and others, 'What Does the Paris Agreement Mean for Adaptation?' (2017) 17 Climate Policy 825, 826.

[26] Decision 2—24/CP.7, 'The Marrakesh Accords' (21 January 2002) FCCC/CP/2001/13/Add.1 (Decisions 2, 7, 10). The Adaptation Fund was not launched until 2007. Decision 1/CMP.3, 'Adaptation Fund' (14 March 2008) FCCC/KP/CMP/2007/9/Add.1. The Adaptation Fund now serves the Paris Agreement. Decision 13/CMA.1, 'Matters relating to the Adaptation Fund' (2 December 2018) FCCC/PA/CMA/2018/3/Add.2.

[27] See Decision 2/CP.11, 'Five-Year programme of work of the Subsidiary Body for Scientific and Technological Advice on impacts, vulnerability and adaptation to climate change' (30 March

The focus on adaptation widened further after the NWP to consider longer-term goals and priorities. In 2007, the Bali Action Plan sought long-term co-operative action on climate change, including enhanced action on adaptation. It identified adaptation as one of the five main categories of future activity for the climate regime and a core component in efforts to establish a new climate agreement.[28] The Cancun Agreements and the Cancun Adaptation Framework (CAF) followed in 2010. The CAF established the Adaptation Committee and medium-to long-term planning through National Adaptation Plans (NAPs).[29] The Cancun Agreements affirmed the 2007 call for enhanced and cooperative action and went a step further, pronouncing agreement that adaptation must be addressed with the same priority as mitigation and with appropriate institutional support. They also created an additional funding mechanism: the Green Climate Fund (GCF).[30] The GCF became the first fund to be asked to balance allocation of resources between adaptation and mitigation; it also mobilized more initial resources than any other climate-dedicated fund.[31] Following the Cancun Agreements and in the lead-up to the Paris Agreement, negotiations continued to focus on how to provide enhanced action on adaptation and ensure a balance between mitigation and adaptation.[32]

These negotiations ultimately led to the Paris Agreement, adopted in December 2015. Its adoption under the UNFCCC means that certain provisions, including the UNFCCC's operative principles, continue to apply.[33] Like the UNFCCC, the Agreement is a legally binding treaty that contains a mix of hard and soft law, guidance, and recommendations. This mix is reflected in its provisions on adaptation. Adaptation is given a central role in the Paris Agreement. It is first included in Article 2, which articulates the treaty's goals and relationship to the UNFCCC. This article connects the Paris Agreement's 'aims to strengthen the global response to the threat of climate change' with 'enhancing the implementation of the Convention [UNFCCC], including its objective'.[34] It sets out to accomplish this by,

2006) FCCC/CP/2005/5/Add.1; Adaptation Committee, '25 Years of Adaptation under the UNFCCC: Report of the Adaptation Committee' (UNFCCC 2019) 13–14.

[28] See Decision 1/CP.13, 'Bali Action Plan' (14 March 2008) FCCC/CP/2007/6/Add1.

[29] Decision 1/CP.16, 'The Cancun Agreements: Outcome of the work of the Ad Hoc Working Group on Long-term Cooperative Action under the Convention' (15 March 2011) FCCC/CP/2010/7/Add.1, paras 15, 20.

[30] ibid paras 2(b), 11–12, 102.

[31] See Green Climate Fund, 'Report of the Thirteenth Meeting of the Board, 28–30 June 2016' (2016) GCF/B.13/33; Jorge Gastelumendi and Inka Gnittke, 'Climate Finance (Article 9)' in Daniel Klein and others (eds), *The Paris Agreement on Climate Change: Analysis and Commentary* (Oxford University Press 2017) 242.

[32] See Irene Suárez Pérez and Angela Churie Kallhauge, 'Adaptation (Article 7)' in *The Paris Agreement on Climate Change* (n 31) 202.

[33] See Decision 1/CP.21, 'Adoption of the Paris Agreement' (29 January 2016) FCCC/CP/2015/10/Add.1, para 1.

[34] Paris Agreement (adopted 12 December 2015, entered into force 4 November 2016) 3156 UNTS 79, art 2(1); see also UNFCCC, art 2 (describing the ultimate objective for the UNFCCC and 'any related legal instruments', which would include the Paris Agreement).

amongst other things, '[i]ncreasing the ability to adapt to the adverse impacts of climate change'.[35] This represents a significant shift from the UNFCCC's objective and its purpose of preventing the harmful effects of climate change so that ecosystems could 'adapt naturally'. It elevates the importance of adaptation, an emphasis welcomed by developing countries that long sought parity between mitigation and adaptation.[36]

The importance of adaptation is further signalled in an article dedicated to adaptation—Article 7—that contains a mix of hard and soft commitments, as well as provisions that are hortatory or descriptive. While not prescriptive, these provisions serve a crucial function in that they 'provide context, construct a narrative, capture shared understandings, and generate mutual reassurances about the nature of the adaptation problem and ways of addressing it'.[37] Article 7(4), for example, recognizes 'that the current need for adaptation is significant and that greater levels of mitigation can reduce the need for additional adaptation efforts, and that greater adaptation needs can involve greater adaptation costs'. This provision evinces that mitigation and adaptation are now understood to be interconnected, rather than competing priorities. The link between the two arose as a key issue for many vulnerable countries that sought to highlight this connection, along with the need for higher mitigation ambition.[38] Article 7(1) also links the two, creating the first global goal on adaptation (GGA) of 'enhancing adaptive capacity, strengthening resilience and reducing vulnerability to climate change' to help ensure 'an adequate adaptation response in the context of the temperature goal' of the Agreement, which aims to limit temperature rise to 'well below' 2°C and to strive to limit it to 1.5°C above pre-industrial levels.[39]

Other contextual provisions build on the GGA to explain the parties' perception of adaptation and how it should be undertaken and reviewed. Adaptation is recognized as polycentric, 'a global challenge faced by all with local, subnational, national, regional and international dimensions', which 'is a key component of and makes a contribution to the long-term global response to climate change to protect people, livelihoods and ecosystems'.[40] This connection between adaptation and impacts on people and their livelihoods invokes human rights and its protective purposes—to ensure conditions that allow people to live with dignity and access to livelihoods, basic needs, and resources.[41] The emphasis on developing countries' needs and efforts, begun in the UNFCCC, persists in the Paris Agreement

[35] Paris Agreement, art 2(1)(b).

[36] See Daniel Bodansky, Jutta Brunnée, and Lavanya Rajamani, *International Climate Change Law* (Oxford University Press 2017) 238.

[37] See ibid 237–38; see also Lavanya Rajamani, 'The 2015 Paris Agreement: Interplay Between Hard, Soft and Non-Obligations' (2016) 28 Journal of Environmental Law 337, 356.

[38] See Bodansky, Brunnée, and Rajamani (n 36) 238; Pérez and Kallhauge (n 32) 205.

[39] Paris Agreement, arts 2(1)(a), 7(1).

[40] ibid art 7(2).

[41] See Nishimura, 'Adaptation and Anticipatory Action' (n 7) 107.

and its calls to take into account 'the urgent and immediate needs' of particularly vulnerable developing country parties and its declaration that 'adaptation efforts of developing country Parties shall be recognized'.[42] Likewise, the provisions on international cooperation and support, including the adaptation obligations discussed in Section V, primarily target developing countries.[43] Finally, the Paris Agreement establishes a Global Stocktake (GST), a collective responsibility of the parties to review implementation of the Agreement and its long-term goals every five years, beginning in 2023. The GST is a comprehensive assessment that considers work under the UNFCCC, Kyoto Protocol, and Paris Agreement.[44] It must include adaptation, and it must also recognize the adaptation efforts of developing countries; enhance implementation of adaptation; and review the adequacy of adaptation, support for adaptation, and the GGA.[45]

In the aftermath of the Paris Agreement, work shifted towards its implementation. Efforts have focused on the first GST and fleshing out the GGA. In 2023 at COP28, parties to the Paris Agreement adopted a framework to meet the GGA set out in Article 7. The framework guides action on adaptation, which includes 'transformational and incremental adaptation'.[46] This shift is reflected in efforts on adaptation in the climate regime: support for and submission of planning documents remains a priority, but there is also a turn towards implementation of adaptation measures and assistance for developing countries in undertaking these measures.[47] The parties also, notably, created targets for enhanced adaptation that align with human rights entitlements such as access to adequate food and nutrition, resilience of infrastructure and human settlements, reducing adverse effects on livelihoods, and goals on health and mortality.[48]

IV. Adaptation, loss and damage, and human mobility in the climate change regime

In addition to its development of adaptation, the Paris Agreement also advances work on loss and damage within the climate change regime. The Agreement marks the first instance that loss and damage is explicitly incorporated into a legal agreement. Like adaptation, loss and damage is the focus of a stand-alone article

[42] Paris Agreement, art 7(2), (3).
[43] ibid arts 7(6), (13), 9(1).
[44] See 'Decision 19/CMA.1: Matters relating to Article 14 of the Paris Agreement and paragraphs 99–101 of Decision 1/CP.21' (19 March 2019) FCCC/PA/CMA/2018/3/Add.2, para 9.
[45] Paris Agreement, arts 7(14), 14.
[46] Decision 2/CMA.5 (n 15) paras 6–8.
[47] See UNFCCC secretariat, 'Progress Tracker: Work Programme Resulting from the Relevant Requests Contained in Decision 1/CP.21' (2018) <https://unfccc.int/sites/default/files/resource/Progress%20Tracker%20PAWP.pdf> accessed 17 October 2023; Decision 2/CMA.5 (n 15) paras 9–10. For further developments on adaptation, see Nishimura, 'The Paris Agreement's Contribution' (n 23).
[48] Decision 2/CMA.5 (n 15) para 9.

(Article 8). However, unlike adaptation, the article on loss and damage does not create any new legal obligations. Instead, Article 8(1) recognizes 'the importance of averting, minimizing and addressing loss and damage associated with the adverse effects of climate change', while the decision adopting the Paris Agreement declares the parties' agreement that the article does not provide a basis for liability or compensation.[49]

Loss and damage refers broadly to harm from the risks and impacts of climate change, and harm can be economic or non-economic.[50] It is a separate workstream within the climate regime, although it can affect and overlap with adaptation. The connection between the two is evident in the emergence of the work programme on loss and damage, which the CAF—an adaptation framework—established.[51] This work programme evolved into the Warsaw International Mechanism for Loss and Damage (WIM). The decision establishing the WIM recognized the role of adaptation in addressing loss and damage and highlighted that the issues it raises go beyond the limits of adaptation.[52] The Paris Agreement continues to reflect this relationship and clarifies that loss and damage is connected to, but distinct from, adaptation. Article 8(4), for example, which sets out areas of cooperation on loss and damage, includes reference to early warning systems; risk assessment and management; and the resilience of communities, livelihoods, and ecosystems. These topics are also part of the adaptation agenda and link back to the language of the GGA (strengthening resilience); Article 7(2) (to protect people, livelihoods and ecosystems); Article 7(5) (consideration of vulnerable groups, communities and ecosystems); Article 7(7) (cooperation on planning, early warning systems); and Article 7(9) (adaptation planning and implementation, assessment of climate impacts and vulnerability, building resilience of socioeconomic and ecological systems).[53]

The issue of human mobility is also considered in work on both adaptation and loss and damage. The CAF included the first explicit language on human mobility in the climate regime: it invited parties to 'enhance action on adaptation' by undertaking '[m]easures to enhance understanding, coordination and cooperation with regard to climate change induced displacement, migration and planned relocation, where appropriate, at the national, regional and international levels'.[54] However, work on these issues shifted to the WIM's Task Force on Displacement (TFD), created alongside the adoption of the Paris Agreement.[55] Adaptation remains integral to efforts to avert, minimize, and address loss and damage, for the WIM in

[49] Decision 1/CP.21 (n 33) para 51.
[50] IPCC, 'Annex II: Glossary' (n 2) 2914.
[51] The Cancun Agreements (n 29) para 25.
[52] Decision 2/CP.19, 'Warsaw International Mechanism for Loss and Damage Associated with Climate Change Impacts' (31 January 2014) FCCC/CP/2013/10/Add.1, preamble.
[53] Nishimura, 'The Paris Agreement's Contribution' (n 23).
[54] The Cancun Agreements (n 29) para 14(f).
[55] Decision 1/CP.21 (n 33) paras 48–50.

general and the TFD specifically. The WIM mandate after the Paris Agreement includes slow-onset events, the TFD and its work on human mobility, and measures to reduce loss and damage, issues connected to adaptation.[56] The TFD developed recommendations for approaches to 'avert, minimize, and address displacement', and has clarified that its focus goes beyond addressing displacement to analyse approaches that include migration as an adaptation strategy.[57] These strategies can reduce the risk of displacement: planning for human mobility and facilitating migration are means to address climate risks and impacts, thereby preventing or averting displacement.[58]

Furthermore, the TFD includes a representative from the Adaptation Committee, as the TFD is meant to collectively represent perspectives from adaptation, loss and damage, LDCs, development, humanitarian, and civil society actors.[59] The Adaptation Committee likewise expanded its remit to assist developing country parties in incorporating approaches to avert, minimize, and address displacement into their adaptation planning.[60] These planning processes are important, as they allow states to include adaptation in domestic planning. Doing so can help address vulnerability through concrete measures that, for example, improve access to food and drinking water, increase livelihood opportunities, or that contemplate migration as an adaptation strategy. The Santiago Network on Loss and Damage may also play a role in issues of human mobility that overlap with adaptation actions. The Santiago Network is part of the WIM and was established in 2019 to catalyse technical assistance to implement the functions of the WIM.[61] It became operational in 2023, at COP28. Its functions, at the time of writing, are still developing, but a mandate focused on averting and minimizing loss and damage arguably aligns with and includes adaptation efforts.

In addition to its connection to loss and damage, there is a crucial reason why adaptation must continue to help shape action on human mobility in the climate regime, despite an institutional shift to the WIM and TFD: the existence of obligations on adaptation. As noted, there are currently no legal commitments on loss and damage in the climate regime. In addition, while efforts to avert and minimize

[56] See Executive Committee of WIM, 'Five-Year Rolling Workplan of the Executive Committee of the Warsaw International Mechanism for Loss and Damage Associated with Climate Change Impacts' (2019).

[57] See Task Force on Displacement, 'Report of the Task Force on Displacement, Recommendations' (2018) 7, 9.

[58] See eg The Nansen Initiative, 'Agenda for the Protection of Cross-Border Displaced Persons in the Context of Disasters and Climate Change' (2015) Vol 1; Lauren Nishimura, '"Climate Change Migrants": Impediments to a Protection Framework and the Need to Incorporate Migration into Climate Change Adaptation Strategies' (2015) 27 International Journal of Refugee Law 107.

[59] Task Force on Displacement (n 57) 2.

[60] See Adaptation Committee 'Flexible Workplan of the Adaptation Committee for 2019–2021' (2020) Version 3, Update following COP 25/CMA 2.

[61] Decision 17/CP.26, 'Warsaw International Mechanism for Loss and Damage Associated with Climate Change Impacts' (31 October 2021) FCCC/CP/2021/12/Add.2, para 1.

loss and damage can and should include adaptive action, the consideration of anticipatory, pre-emptive action is a better fit for discussions of adaptation in the first instance. Developing concrete measures on human mobility in its various forms will, however, continue to cut across workstreams in the climate regime. Likewise, the funding mechanisms created in relation to both adaptation and loss and damage can serve an important role in implementing mobility-related measures. This includes funds that target adaptation—like the Adaptation Fund, SCCF, LDCF, and GCF—as well as the nascent loss and damage fund, with the relationship between these funding streams still evolving.[62]

Dynamism is a defining feature of the climate change regime, as evidenced by the continued development of adaptation in the regime. Adaptation is itself an evolutive concept, shifting to suit changing conditions and increasing knowledge about the threats posed by climate impacts. Current understandings of adaptation will therefore evolve and shape new understandings in the future. The issue of human mobility is similarly growing in importance and focus within the regime, but there remains a need to incorporate it into adaptation planning, to take a more proactive, anticipatory approach, and to ensure participatory adaptation processes.

V. Adaptation obligations

The discussion of adaptation thus far highlights its general evolution, both conceptually and within the climate regime. The development of adaptation is also reflected in the obligations on adaptation within climate treaties. Reading across the UNFCCC and the Paris Agreement, three types of obligations on adaptation emerge: to act on adaptation, to assist with adaptation, and to cooperate in adaptation efforts. These adaptation obligations are a potentially powerful legal tool that can shape adaptation. Participation in both the UNFCCC and the Paris Agreement is nearly universal, and neither treaty permits reservations.[63] As a result, adaptation obligations are binding upon almost every country. They also provide the foundation for an approach to human mobility that prompts planning and anticipatory action, before people are rendered immobile, forced to move, or

[62] See Decision 2/CP.27, 'Funding arrangements for responding to loss and damage associated with the adverse effects of climate change, including a focus on addressing loss and damage' (20 November 2022) FCCC/CP/2022/10/Add.1. In 2023, the COP and COP serving as the meeting of the Parties to the Paris Agreement (CMA) decided that the scope of the fund responding to loss and damage includes the provision of funding for 'displacement, relocation, migration' and that support by the fund may include 'promoting equitable, safe and dignified human mobility in the form of displacement, relocation and migration in the cases of temporary or permanent loss and damage'. Decision 5/CMA.5, 'Operationalization of the new funding arrangements, including a fund, for responding to loss and damage referred to in paragraphs 2–3 of Decisions 2/CP.27 and 2/CMA.4' (6 December 2023) FCCC/PA/CMA/2023/16/Add.1, Annex I, paras 6, 9.

[63] Introduction Chapter, n 16 and accompanying text.

are further displaced. However, adaptation obligations are broad and subjective, in part, by design.[64] Their breadth allows for flexibility in their interpretation and implementation. This section describes the nature and general content of these obligations. Subsequent chapters build on the analysis, to further clarify their content through other rules and principles of international law and by their application in concrete contexts.

A. Obligations to act

The first type of adaptation obligation consists of obligations to act on adaptation. This includes Article 4(1)(b) of the UNFCCC, which requires:

> All Parties taking into account their common but differentiated responsibilities and their specific national and regional development priorities, objectives and circumstances shall . . . [f]ormulate, implement, publish and regularly update national and, where appropriate, regional programmes containing . . . measures to facilitate adequate adaptation to climate change.

Two key points about this obligation underscore its importance and operation. First, it creates a positive duty to act on adaptation for all parties. Consequently, all parties—and by virtue of the wide participation in the UNFCCC, all countries—must formulate and implement national programmes on adaptation. Second, the obligation is qualitative and includes language that enables discretion for parties to determine what constitutes 'adequate adaptation'. Adequacy is a subjective term that implies that some threshold must be met. It can incorporate elements of cost efficiency, feasibility, and other practical considerations.[65] These elements are important as adaptation often involves context-specific measures that address local impacts and vulnerabilities. Context raises an additional, and related, feature: this obligation—along with the other commitments in Article 4(1)—is couched in terms of CBDR-RC. CBDR-RC is an operative principle that, like other principles contained in Article 3, guides and shapes the commitments within the climate regime. Generally, the CBDR-RC principle introduces differentiation in how obligations are met, to account for states' varying abilities to undertake climate action and responsibilities to provide support for such action. It strengthens the need for adaptation to be driven by the conditions of time and place. As the case studies in Chapters 7 to 9 show, adequate adaptation can translate into different measures depending on the context.

[64] See eg discussion of ambiguity in support for adaptation, Section V.B.

[65] See Marco Grasso, *Justice in Funding Adaptation under the International Climate Change Regime* (Springer Science & Business Media 2009) 24.

However, because this obligation does not mandate a particular adaptation outcome or action, some scholars view it as a procedural commitment, albeit 'highly significant' as it leads to processes tasked with the identification, assessment, and implementation of adaptation measures.[66] Likewise, the use of the word 'facilitate' could be understood to narrow the focus to preparation for adaptation, rather than undertaking measures to adapt. Yet nothing in the language of the obligation limits it to planning alone, as the climate regime's COP decisions and guidance indicate. The COP is the 'supreme body' of the regime tasked with promoting implementation of the UNFCCC.[67] While not formally binding, COP decisions have significant operational weight; they offer instructions to parties, explain treaty provisions, and provide a means to achieve greater consistency and coherence in strategies and actions. The decisions of the COP are particularly useful in clarifying agreement on provisions within treaties that are ambiguous, and they can influence norms and state compliance.[68] Adaptation is a process, one initially envisaged by the COP in stages and later characterized as an 'iterative adaptation cycle'.[69] Preparatory activities and planning are a logical starting point for adaptation efforts; to anticipate impacts and create appropriate responses, a baseline understanding of existing vulnerabilities and potential risks is necessary. Subsequent stages, however, must include implementation of measures. This accords with a general understanding of the obligation and of the UNFCCC as establishing a framework meant to evolve and increase in ambition over time. Implementation of adaptation is also assumed in other commitments, including an obligation for parties to '[t]ake climate change considerations into account, to the extent feasible, in their relevant social, economic and environmental policies and actions', including any 'projects or measures' taken to mitigate or adapt.[70]

Parties are further required to regularly communicate about their implementation of the UNFCCC, including commitments on adaptation.[71] COP guidance on this reporting obligation is instructive and indicates that adaptation under the UNFCCC includes measures to plan for and implement adaptation. For example, COP guidance to assist non-Annex I parties—LDCs and developing countries—in preparing their National Communications (NCs) describes information that can be included to communicate implementation of the UNFCCC. This guidance points to a range of actions that 'facilitate adequate adaptation', from identifying

[66] See eg Farhana Yamin and Joanna Depledge, *The International Climate Change Regime* (Cambridge University Press 2005) 218.

[67] UNFCCC, art 7.

[68] See Bodansky, Brunnée, and Rajamani (n 36) 19–20; see also Jutta Brunnée, 'COPing with Consent: Law-Making Under Multilateral Environmental Agreements' (2002) 15 Leiden Journal of International Law 1.

[69] See Decision 11/CP.1, 'Initial guidance on policies, programme priorities and eligibility criteria to the operating entity or entities of the financial mechanism' (6 June 1995) FCCC/CP/1995/7/Add.1; Decision 2/CMA.5 (n 15) para 10.

[70] UNFCCC, art 4(1)(f).

[71] ibid art 12.

and assessing vulnerability and adaptation needs to evaluating and undertaking measures and strategies.[72] Initial NCs included this range: most incorporated vulnerability assessments and plans for adaptation activities and policies, and a number included physical, regulatory, and other adaptation measures already underway.[73] Similarly, the UNFCCC's secretariat staff has highlighted a variety of activities planned or being undertaken to act on adaptation in developing countries. A 2007 publication that drew on NCs, NAPAs, and regional workshops found that assessment of vulnerabilities and the steps needed to adapt remained a priority for developing countries, while also illustrating the diversity of concrete actions across sectors in plans or being implemented. Examples included both reactionary and anticipatory measures such as strengthening coping strategies and coordination of legal policies to address resources at national and regional levels.[74] Thus, on its face and bolstered by guidance and evidence in initial state communications, the obligation to act on adaptation in the UNFCCC incorporates both planning for and carrying out adaptation programmes.[75]

Furthermore, the Paris Agreement eliminates any doubt about whether adaptation within the climate regime includes the implementation of measures. Article 7(9) requires that '[e]ach Party shall, as appropriate, engage in adaptation planning processes and the implementation of actions, including the development or enhancement of relevant plans, policies and/or contributions'. This creates a legal obligation to both plan for and implement adaptation actions. It is an obligation that applies to all parties individually ('Each Party'). However, the qualifier 'as appropriate' softens the obligation and, like the UNFCCC, introduces an element of state discretion. Article 7(9) provides an illustrative list of processes, policies, and actions that include implementation of adaptation actions, undertakings and/or efforts; national adaptation plans; assessment of impacts and vulnerabilities; and monitoring and evaluating of plans and actions. These actions align with

[72] See Decision 17/CP.8, 'Guidelines for the preparation of national communications from parties not included in Annex I to the Convention' (28 March 2003) FCCC/CP/2002/7/Add.2.

[73] See eg Government of the People's Republic of Bangladesh, 'Initial National Communication under the UNFCCC' (2002) (full range of measures, from assessment to adaptation underway); Government of the Fiji Islands, 'Climate Change-The Fiji Islands Response: Fiji's First National Communication' (2005) (assessment, adaptation options, and policies and programmes underway); Kiribati Government, 'Initial Communication under the UNFCCC' (1999) (assessment and plans for implementing programmes); Republic of Sudan, 'Sudan's First National Communication under the UNFCCC' (2003) Vol I: Main Communications (assessment and what adaptation should include); Government of the Republic of the Marshall Islands, 'Initial Communication under the UNFCCC' (2000) (assessment and adaptation options); Federal Republic of Nigeria, 'Nigeria's First National Communication' (2003) (same); Republic of Vanuatu, 'Vanuatu National Communication to the COP to the UNFCCC' (1999) (same).

[74] See UNFCCC secretariat, 'Climate Change: Impacts, Vulnerabilities and Adaptation in Developing Countries' (UNFCCC 2007) <https://unfccc.int/resource/docs/publications/impacts.pdf> accessed 17 October 2023.

[75] See Yamin and Depledge (n 66) 216; Roda Verheyen, 'The Legal Framework of Adaptation and Adaptive Capacity' in Joel B Smith, Richard JT Klein, and Saleemul Huq (eds), *Climate Change, Adaptive Capacity and Development* (Imperial College Press 2003) 170–71.

developments within the climate regime on adaptation, which continue to push for planning while shifting focus towards longer-term needs and implementation.

Other aspects of the Paris Agreement can shed further light on what obligations to act on adaptation entail. The GGA, for example, contains three aims: to enhance adaptive capacity, strengthen resilience, and reduce vulnerability. These are connected to the Agreement's objective to increase the ability to adapt and to ensure an 'adequate adaptation response' in light of its stated temperature goals.[76] The adaptation planning and implementation undertaken under Article 7(9) should therefore contribute to these aims, which will require understanding how the process of adaptation can proactively build resilience and capacity while reducing vulnerability. These aims further link adaptation to a standalone article on technology transfer in the Paris Agreement—Article 10—that highlights the 'long-term vision' and role of technology in improving resilience.[77] Article 10 is explicit in its recognition of the importance of technology for adaptation and the need to balance support between mitigation and adaptation.[78]

Achieving the aims of the GGA and ensuring adequate adaptation are also tied to other contextual provisions of Article 7, discussed in Section III. Article 7(5), for example, provides guidance for parties when undertaking 'adaptation action', acknowledging that such action should be participatory and fully transparent; take a gender-responsive, country-driven approach; consider vulnerable groups, communities and ecosystems; be based on and guided by best available science as well as traditional, Indigenous, and local knowledge systems; and aim to integrate adaptation into relevant socioeconomic and environmental policies and actions. The latter recalls the language of the UNFCCC and highlights the connection between socioeconomic conditions, climate impacts, and responses. Other contextual provisions confirm that the protection of people and livelihoods over the longer term is a central focus of adaptation, and that the needs and efforts of developing countries must be prioritized.[79] They bolster arguments to integrate relevant human rights law, as the next chapter outlines. The targets adopted for the GGA at the COP in 2023 likewise help narrow the focus of adaptation efforts. Although framed in voluntary terms, with parties urged to act rather than required to take specific measures, these targets evince that adaptation should prioritize basic resources and the protection of ecosystems. Targets include significantly reducing hazards and increasing resilience related to access to potable water, adequate food and nutrition, essential services and infrastructure, and health services. They also establish the relevance of certain human rights, namely the component rights associated with the right to an adequate standard of living, as does the recognition

[76] See Paris Agreement, arts 2(1), 7(1).
[77] ibid art 10(1).
[78] ibid art 10(2), (6). Discussed further in Nishimura, 'The Paris Agreement's Contribution' (n 23).
[79] Paris Agreement, art 7(2)–(6).

that adaptive action should be 'in alignment with other global frameworks' and 'take into account . . . human rights approaches'.[80]

In addition, state parties are expected to articulate their plans and actions on adaptation, individually through an 'adaptation communication' and collectively via the GST.[81] These go further than the UNFCCC's reporting requirements and are meant to help operationalize the GGA. Each party is encouraged, 'as appropriate', to submit and periodically update an Adaptation Communication as a component of or in conjunction with other communications, such as a country's NC or NAP. The scope of the Adaptation Communication is expansive and can include 'priorities, implementation and support needs, plans and action'.[82] They are intended to be ongoing and continuous, through updates and their incorporation into the five-yearly GST.

In the lead-up to the first GST in 2023, the Conference of the Parties serving as the meeting of the Parties to the Paris Agreement (CMA) sought assessment of adaptation from a variety of sources. The UNFCCC secretariat, the Adaptation Committee, the LDC Expert Group (LEG), and the Executive Committee of the WIM (ExCom) prepared synthesis reports on adaptation, drawing on different sources that included NCs, Adaptation Communications, NAPs, and Nationally Determined Contributions (NDCs). These synthesis reports were intended to complement one another to provide insight into different aspects of adaptation. Together, they offered an overview of the state of adaptation efforts, experiences, priorities, and support.[83] They provided insight into the kinds of adaptation planning and measures being undertaken by states, best practices, and lessons learned, and highlighted areas where further action or support is needed. Key findings of these reports show the increasing attention to adaptation, the continued work of developing countries to plan for adaptation, the range of different adaptation actions planned for or underway, and the uneven progress on adaptation across regions.

Critically for arguments in this book, many states were already identifying anticipated risks from climate change and planning to address these risks through proactive or anticipatory adaptation. Some mention is also made about migration in planning and national reports. For example, human mobility is mentioned in many of the NAPs submitted to date. However, for the most part, these submissions

[80] Decision 2/CMA.5 (n 15) paras 9, 13.

[81] While the GST is a party-driven process intended to increase the ambition of action over time, its outputs do not focus on individual countries. See 'Decision 19/CMA.1' (n 44) paras 10, 14.

[82] Paris Agreement, art 7(10); discussed further in Pérez and Kallhauge (n 32) 211–13.

[83] See UNFCCC secretariat, 'Synthesis Report for the Technical Assessment Component of the First Global Stocktake: State of Adaptation Efforts, Experiences and Priorities' (2022); Adaptation Committee, 'Synthesis Report for the Technical Assessment Component of the First Global Stocktake' (2022); LEG, 'Synthesis Report for the Technical Assessment Component of the First Global Stocktake' (2022); WIM ExCom, 'Synthesis Report for the Technical Assessment Component of the First Global Stocktake' (2022). Discussed further in Nishimura, 'The Paris Agreement's Contribution' (n 23).

view mobility as a potential risk or negative consequence, with only a handful mentioning planned relocation or migration as adaptation, and few offering clear policy plans.[84] Yet adaptation planning is an ongoing process, and there is scope and need to consider adaptive mobility and other in situ adaptation measures that enable people to stay for as long as possible while planning for movement when necessary.[85]

Finally, assessing the adequacy of a state's adaptation efforts leads to questions about whether adaptation is measured by the processes put in place or its outcomes, or, put differently: are obligations to act obligations of conduct or result? And, building on this question, is a state exercising due diligence enough to satisfy its obligations? None of these obligations are prescriptive. They do not mandate a specific adaptation measure, outcome, or course of action. As a result, adaptation obligations—both in the UNFCCC and the Paris Agreement—are viewed as primarily procedural, requiring conduct or effort rather than result.[86] Yet, even though these obligations afford parties a great deal of discretion in choosing a course of action, they still demand *some* action to be met. As subsequent chapters argue, climate change poses foreseeable risks and a good faith interpretation of these obligations—including the language introducing discretion like 'as appropriate' or 'adequate adaptation'—translates into different measures depending on the context. It is thus not a question of whether it is appropriate to act but how action should be planned for and undertaken. The integration of human rights duties narrows discretion and adds a results-based element to the positive duties arising under adaptation obligations. Such an understanding leads to the need to facilitate certain forms of adaptation, including mobility, to satisfy legal commitments. As a result, adaptation obligations take on elements of obligations of both conduct and result.[87] The emphasis, however, is on the former and taking steps to allow individuals at risk to realize their rights, rather than reaching a particular target or outcome.

Likewise, the integration of human rights duties can help clarify how a duty of due diligence attaches to obligations to act on adaptation. Within environmental

[84] See Kayly Ober, Sam Huckstep, and Sarah Miller, '"It's Time for Us to Be Included": An Assessment of Refugee and Displaced People's Participation in National Adaptation Planning' (Refugees International 2023) Annex A <www.refugeesinternational.org/reports-briefs/its-time-for-us-to-be-included-an-assessment-of-refugee-and-displaced-peoples-participation-in-national-adaptation-planning/> accessed 17 October 2023.

[85] For example, the secretariat noted that migration as an adaptation strategy is less pronounced in priorities identified by parties in relation to sea level rise. UNFCCC secretariat, 'Synthesis Report' (n 83) para 158.

[86] See Yamin and Depledge (n 66); Margaretha Wewerinke-Singh and Curtis Doebbler, 'The Paris Agreement: Some Critical Reflections on Process and Substance' (2016) 39 University of New South Wales Law Journal 1486, 1503–04; Lavanya Rajamani, 'Guiding Principles and General Obligation (Article 2.2 and Article 3)' in *The Paris Agreement on Climate Change* (n 31) 138; Rajamani, 'A Framework Approach' (n 23) 208–09.

[87] See Chapter 4; Office of the UN High Commissioner for Human Rights (OHCHR), 'Human Rights and Indicators: A Guide to Measurement and Implementation' (2012) HR/PUB/12/5, 13.

law, due diligence is typically a duty of conduct that requires states to take appropriate measures aimed at preventing harm, but it does not require that all harm be avoided.[88] Harm prevention and due diligence often accompany the principle of precaution, a key operative principle of the regime. Precaution calls for action in the face of uncertainty, particularly when there is a risk of serious or irreversible harm. As certainty about the risks posed by climate change increases, fulfilling duties to prevent harm requires less gravity of harm to trigger adaptive action. This is detailed in Chapter 5, as are related procedural obligations: satisfying an obligation to act can include the processes and procedural guarantees required by due diligence standards. These are akin to and clarified by procedural human rights, necessitating risk and impact assessments, provision of information, and participatory processes. However, the integration of human rights law establishes that adaptation obligations also require more, to help ensure climate impacts do not cause significant harm to people or violate their substantive human rights.

B. Obligations to assist

Obligations to assist are the second type of adaptation obligation in the climate change regime. While all states must adapt to climate change, which will require planning for and facilitating adaptive mobility in certain circumstances, some are owed support in their adaptation efforts. There are several relevant provisions that set out obligations to assist in the UNFCCC and Paris Agreement, most of which focus on developed countries providing financial support to developing countries. Article 4(4) of the UNFCCC obliges 'developed country Parties and other developed Parties included in Annex II'[89] to 'assist the developing country Parties that are particularly vulnerable to the adverse effects of climate change in meeting costs of adaptation to those adverse effects'. Article 4(3) requires the same developed country parties to provide the financial resources 'needed by the developing country Parties to meet the agreed full incremental costs' of implementing measures that include those required for adequate adaptation. And Article 4(5) further obliges Annex II country parties to assist developing countries through access to and transfer of technology and knowledge.

The assistance of developed countries is critical to the success of the UNFCCC and the participation of developing country parties: the performance of developing countries' obligations is conditional, dependent on 'the effective implementation

[88] See *Pulp Mills on the River Uruguay (Argentina v Uruguay)* (Judgment) [2010] ICJ Rep 14, 58 [101], 79 [197]; *Certain Activities Carried Out by Nicaragua in the Border Area (Costa Rica v Nicaragua)* (Judgment) [2015] ICJ Rep 665, 706 [104]; Tim Stephens and Duncan French, 'ILA Study Group on Due Diligence in International Law: Second Report' (International Law Association (ILA) 2016) 23.

[89] Annex II is made up of Organisation for Economic Co-operation and Development (OECD) countries, as membership stood when the UNFCCC was adopted in 1992.

by developed country Parties of their commitments . . . related to financial resources and transfer of technology'.[90] Assistance obligations are further bolstered by the principles of equity and CBDR-RC, and are anchored by the UNFCCC's concern for the needs and circumstances of developing countries.[91] Article 4(9) makes the latter an explicit requirement for the LDCs, declaring that 'Parties shall take full account of the specific needs and special situations' of these countries in funding and transfer of technology. As with obligations to act, there is ambiguity built into assistance obligations. Yet, again, discretion does not extend to whether states can meet their commitments. Developed states *must* provide financial assistance for adaptation costs. How this is accomplished and the role of the regime's financial mechanisms continue to evolve and will be discussed in Chapter 6.

The Paris Agreement reaffirms and expands upon the assistance obligations of the UNFCCC, while also clarifying some of the activities towards which support should be directed. Leading up to the Agreement, the provision of support for both planning and implementing adaptation was a key demand of LDCs and the alliance of small island states.[92] This manifested in an obligation set out in Article 7(13), which declares '[c]ontinuous and enhanced international support shall be provided to developing country Parties' for the implementation of various adaptation activities, including planning and implementation of measures (Article 7(9)), those needed to enhance action and strengthen cooperation on adaptation (Article 7(7)), and for adaptation communications (Articles 7(10) and 7(11)). Lavanya Rajamani characterizes the provision of support to developing countries as 'the centerpiece of the adaptation article'.[93] The importance of support is explicitly recognized in Article 7(6), as is consideration of the needs of developing country parties. The continuous nature of such support underscores that assistance is not one-off but should be accessible as adaptation needs and efforts evolve.

Similarly, the call for enhanced support appears to build on prior commitments, including those in Article 4 of the UNFCCC and efforts to increase climate finance that preceded the Paris Agreement. At the Copenhagen COP in 2009, developed countries committed to a goal of collectively mobilizing USD 100 billion a year by 2020 to address the needs of developing countries.[94] The decision adopting

[90] UNFCCC, art 4(7).

[91] ibid arts 3(1)–(3), 4(8)–(9).

[92] See Wewerinke-Singh and Doebbler (n 86) 1504; Achala Abeysinghe, Brianna Craft, and Janna Tenzing, 'The Paris Agreement and the LDCs: Analysing COP21 Outcomes from LDC Positions' (International Institute for Environment and Development (IIED) 2016) Issue Paper; see also Earth Negotiations Bulletin, 'Summary of the Paris Climate Change Conference: 29 November–13 December 2015' (2015) 12 International Institute for Sustainable Development (some parties highlighted the need for assessment of the adequacy of support from developed countries).

[93] Lavanya Rajamani, 'The Principle of Common but Differentiated Responsibilities and Respective Capabilities in the International Climate Change Regime' in Rosemary Lyster and Robert RM Verchick (eds), *Research Handbook on Climate Disaster Law: Barriers and Opportunities* (Edward Elgar 2018) 56.

[94] Decision 2/CP.15, 'Copenhagen Accord' (20 March 2010) FCCC/CP/2009/11/Add.1.

the Paris Agreement extended this goal to 2025 and linked it to Article 9(3) of the Agreement. This provision suggests that 'developed country Parties should continue to take the lead in mobilizing climate finance from a wide variety of sources' and that '[s]uch mobilization of climate finance should represent a progression beyond previous efforts'.[95] Decisions of the CMA following the Agreement's adoption emphasize the need to increase adaptation finance, calling upon developed countries to at least double their collective provision of finance for adaptation to developing country parties from 2019 levels by 2025.[96]

Article 9(3) also helps clarify who holds the duty to provide support to developing countries under Article 7(13), which is not specified. Article 7(13) explains that support must be provided 'in accordance with the provisions of Articles 9, 10 and 11' (the articles on finance, technology transfer, and capacity-building, respectively). Thus, it must be read alongside these other provisions. Article 9(1), for example, clarifies that financial obligations are incumbent on developed country parties and uses the imperative language 'shall' to set out their duty to 'provide financial resources to assist developing country Parties with respect to both mitigation and adaptation in continuation of their existing obligations under the Convention'. There is some debate on whether Article 9(1) merely restates the UNFCCC's financial obligations or can be understood to build upon them. Its language provides an argument that it does not create a new obligation.[97] However, it does highlight that the Agreement continues the commitments made under the UNFCCC, and developed countries remain obligated to assist developing country parties in meeting the costs of adaptation. Subsequent CMA decisions also indicate that developed countries must do more: for example, the implementation by developing countries of the framework for the GGA is recognized as dependent, in part, on developed country parties' implementation of their support commitments.[98] In addition, the Paris Agreement repeatedly recognizes enhanced or 'scaled-up' financial resources, as well as the need to balance support between mitigation and adaptation.[99]

Contention about support also focused on whether to expand the parties obliged to provide financial assistance beyond those developed countries referenced in the UNFCCC. Since the adoption of the UNFCCC, 14 other countries have joined the OECD, the basis for Annex II.[100] These other countries have the means to provide financial support to developing countries and thus could be required to

[95] Decision 1/CP.21 (n 33) para 53; Paris Agreement, art 9(3).

[96] Decision 1/CMA.3, 'Glasgow Climate Pact' (8 March 2022) FCCC/PA/CMA/2021/10/Add.1, para 18; Decision 2/CMA.5 (n 15) para 31.

[97] Reference to existing obligations enabled the Unites States to accept the compulsory language of the provision. Bodansky, Brunnée, and Rajamani (n 36) 240.

[98] Decision 2/CMA.5 (n 15) para 32.

[99] See eg Paris Agreement, arts 7(13), 9(4), 10(6).

[100] See n 89; OECD, 'List of OECD Member Countries' <www.oecd.org/about/document/list-oecd-member-countries.htm> accessed 21 December 2023.

do so based on their 'respective capabilities'. Indeed, the distinction between developed and developing countries is the only clear differentiation for obligations, guidance, and support for adaptation in the Paris Agreement. Yet the UNFCCC could be understood as intending to limit financial contributions to those developed countries with the greatest historical responsibility for climate change.[101] Ultimately, Article 9 appears to strike a compromise. Article 9(1) imposes a clear obligation on developed countries to provide financial resources and Article 9(3) reiterates that they should take the lead in mobilizing climate finance. Article 9(2) then expands the financing base by encouraging other parties 'to provide or continue to provide such support voluntarily'. Together, they help clarify who holds the duty to provide adaptation support and explain why Article 7(13) and similar provisions are phrased in the passive voice (eg 'support shall be provided'),[102] as the issue is taken up in Article 9. Other parties are encouraged to provide support, but the primary responsibility for providing and mobilizing financial resources remains on developed countries.[103]

C. Obligation to cooperate

The obligation to cooperate is the third type of adaptation obligation. Article 4(1)(e) of the UNFCCC creates a general obligation to 'cooperate in preparing for adaptation to the impacts of climate change'.[104] Like obligations to act, all parties are expected to cooperate, with the initial focus, again, on preparation. This focus is reflected in additional obligations to cooperate in research and exchange of information, to further understanding of climate change and response strategies.[105] Assistance is a form of cooperation, with support for adaptation a means to operationalize the obligation. The two often feature together and evince an agreement by parties to the climate regime that adaptation is not an undertaking for each state on its own.[106]

The nexus between cooperation and assistance is also seen in the language of the Paris Agreement. Article 7(6), for example, recognizes the importance of international cooperation and support for adaptation efforts. Article 7(7) encourages

[101] Gastelumendi and Gnittke (n 31) 243–44.

[102] See also Paris Agreement, arts 4(5), 10(6), 13(14).

[103] See Lavanya Rajamani, 'Ambition and Differentiation in the 2015 Paris Agreement: Interpretative Possibilities and Underlying Politics' (2016) 65 International and Comparative Law Quarterly 493, 513.

[104] This provision also commits parties to 'develop and elaborate appropriate and integrated plans for coastal zone management, water resources and agriculture, and for the protection and rehabilitation of areas, particularly in Africa, affected by drought and desertification, as well as floods'. UNFCCC, art 4(1)(e).

[105] See eg UNFCCC, art 4(1)(g), (h). These response strategies refer to mitigation and its impact on other states. See Decision 5/CP.4, 'Implementation of Article 4.8 and 4.9 of the Convention (Decision 3/CP.3 and Articles 2.3 and 3.14 of the Kyoto Protocol)' (25 January 1999) FCCC/CP/1998/16/Add.1.

[106] See Yamin and Depledge (n 66) 222.

parties to 'strengthen their cooperation on enhancing action on adaptation' for five different areas; one of these areas includes 'assisting developing country Parties in identifying effective adaptation practices, adaptation needs, priorities, support provided and received for adaptation actions and efforts, and challenges and gaps, in a manner consistent with encouraging good practices'. These provisions, however, use the language 'should' or 'recognize' rather than 'shall'. Thus, unlike the UNFCCC, they do not create hard legal obligations. Instead, they create good faith expectations and can, despite being couched in discretionary terms, still influence states' interpretation and implementation.[107] Furthermore, given its connection with support, cooperation is manifested through the implementation of assistance obligations and can be further concretized through the integration of human rights, as Chapter 6 explores.

The expansion of adaptation in the Paris Agreement, including action on adaptation connected to cooperation, is further evidence of how the climate regime continues to evolve. This development is meant to continue, with progression expected across areas of climate action. The Agreement's cross-cutting provisions make that expectation evident.[108] Article 3, for example, creates a general obligation that 'all Parties are to undertake and communicate ambitious efforts' as set out in the provisions on adaptation and finance, amongst others. It also declares that the 'efforts of all Parties will represent a progression over time'. The language creates a collective expectation of ongoing progress,[109] one that can only be met with continued, cooperative, and increasingly ambitious efforts.

Overall, the adaptation obligations set out in the climate regime are both lauded and criticized by scholars. Some highlight the 'impressive list of adaptation duties' created by the UNFCCC and argue that, while broad, they provide 'significant guidance on the extent and timing of adaptation'.[110] Others consider adaptation obligations to be modest or 'relatively few'.[111] They criticize these obligations for being vague and undemanding; those requiring financial support, for example,

[107] See Rajamani, 'The 2015 Paris Agreement: Interplay Between Hard, Soft and Non-Obligations' (n 37) 356; see generally Dinah Shelton, 'Law, Non-Law, and the Problem of "Soft Law"' in Dinah Shelton (ed), *Commitment and Compliance: The Role of Non-Binding Norms in the International Legal System* (Oxford University Press 2003) 17.

[108] See Paris Agreement, arts 3, 13.

[109] Although its legal status is subject to debate, characterized as a 'substantive new obligation' (Rajamani, 'Guiding Principles and General Obligation' (n 86) 139), as establishing an expectation rather than an obligation (Bodansky, Brunnée, and Rajamani (n 36) 234), and as 'an element of the idea of a collective obligation', one which is a 'collective obligation in form' but not 'a legal obligation with legally binding force in substance' (Alexander Zahar, 'Collective Obligation and Individual Ambition in the Paris Agreement' (2020) 9 Transnational Environmental Law 165, 171, 179).

[110] See Jonathan Verschuuren, 'Climate Change Adaptation under the United Nations Framework Convention on Climate Change and Related Documents' in Jonathan Verschuuren (ed), *Research Handbook on Climate Change Adaptation Law* (Edward Elgar 2022) 18; Yamin and Depledge (n 66) 216; Verheyen (n 75) 170–71.

[111] See Bodansky, Brunnée, and Rajamani (n 36) 135.

lack specificity regarding the amount of assistance required.[112] This book agrees with those that see their potential. It views the breadth of adaptation obligations as a feature that affords greater flexibility and that is necessary to encompass difference across contexts, a range of activities, and the iterative processes required to adapt.

VI. Conclusion

This chapter sets out the contours of adaptation obligations, to lay the foundation for the critical role they can play in shaping more anticipatory adaptive action, including on human mobility. It describes climate change adaptation as a process, informed by context and conditions and that can shift over time. The differences in context and timing, alongside the often local nature of this process, means that global frameworks governing adaptation must themselves be adaptive. The evolution of adaptation within the climate regime shows an increasing understanding of the complexities involved in adaptation. It reflects growing recognition of the importance of adaptation planning and implementation, issues of equity and differentiation, and the need for sustained and accessible support. Climate treaties repeatedly highlight the needs and actions of developing countries, and they prioritize financing for their adaptation costs. These treaties also create three types of adaptation obligations: to act, to assist, and to cooperate. Adaptation obligations are broad and do not prescribe specific measures. Yet, as the UNFCCC and Paris Agreement make clear, they require action: to plan for and implement adaptation that contributes to protecting people and their livelihoods; to cooperate; and for developed countries to support developing countries in their adaptation efforts. These obligations can be further concretized, which this book explores next through tools of treaty interpretation and the integration of human rights law.

[112] Mayer, 'Climate Change Adaptation Law: Is There Such a Thing?' (n 15) 317; Benoît Mayer, 'Climate Change Adaptation and the Law' (2021) 39 Virginia Environmental Law Journal 141, 155–57.

3
Bringing Human Rights to Bear: Interpreting Adaptation Obligations

I. Introduction

This chapter provides the central methodology of the book: the use of treaty interpretation to clarify adaptation obligations. It builds upon the analysis in Chapters 1 and 2, which established human mobility as a form of adaptation and introduced obligations on adaptation in the climate change regime. Adaptation obligations require countries to plan for and implement measures, to cooperate, and, in some circumstances, to provide support. These obligations are broad and subjective, but they can be made more concrete through the integration of relevant human rights law. Integration is required both because adaptation obligations are a part of a broader system of international law and under the rule on treaty interpretation. Section II explains that such integration is necessary for an international legal system that strives for coherence and consistency. Coherence is central to efforts to avoid the fragmentation of international law, efforts that are grounded in the principle of good faith and presumption against normative conflicts. This section also explains the decision to take an evolutive approach to interpretation, which aligns with the dynamism of the climate regime and intent that climate treaties continue to develop and reflect changes in science and law. Section III then turns to treaty interpretation and explores each element of the interpretive process: the treaties' terms, context, object and purpose, and the integration of other relevant rules. Together, they lead to the integration of human rights duties into the interpretation and implementation of adaptation obligations. Such integration provides the framework for the remainder of the book, which uses the interpretive process to integrate specific human rights (Chapters 4 to 6) and in three case studies (Chapters 7 to 9) to show when states are required to plan for and facilitate adaptive mobility.

II. Coherence in international law

Integrating human rights law into the interpretation of adaptation obligations involves the incorporation of extrinsic rules into the climate change regime. Such integration is based, in part, on a premise that these rules are a part of a system of

Climate Change, Human Rights, and Adaptive Mobility. Lauren Sakae Nishimura, Oxford University Press.
 DOI: 10.1093/9780198930068.003.0004

international law that should strive for coherence. This premise is central to efforts to address potential conflicts in international law arising from an increasingly connected, yet pluralistic legal landscape.

The International Law Commission (ILC) took up the issue of the fragmentation of international law in the early 2000s. It considered ways to avoid conflict between different treaty regimes, specialized and general international law, and prior and subsequent law. This work culminated in a study and report that provide guidance and recommendations, including recourse to the principle of systemic integration.[1] While the reports and conclusions of the ILC are non-binding, they are influential and have played a significant role in the progressive development of international law, including international environmental law.[2] The ILC fragmentation report remains frequently cited, and it signalled an increasing turn toward understanding the interaction of different international legal regimes and their potential for norm overlap, development, and friction.[3] Coherence is at the heart of these inquiries. The ILC report emphasizes coherence: problems of coherence in international law provided the rationale for its treatment of fragmentation; coherence is sought in the techniques it offers and accomplished through legal reasoning and interpretation.[4] Yet coherence is an abstract concept that must be balanced with context, function, and outcome. It serves as an aim rather than an end that must be achieved. Indeed, a diversity of norms within the international legal system is critical to its efficacy, which means that some conflict is inevitable. The ILC recognized the unavoidability of the 'collision of norms and regimes' and sought to navigate these collisions with tools that offered the promise of greater consistency.[5] One of these tools is the Vienna Convention on the Law of Treaties (VCLT). The VCLT both articulates and codifies principles that underpin a system of international law that strives for coherence and consistency.

Good faith is a principle of international law. The principle has both a general application and one specific to interpretation. Good faith operates generally to ensure parties perform their treaty obligations.[6] It is also part of treaty interpretation

[1] ILC, 'Fragmentation of International Law: Difficulties Arising from the Diversification and Expansion of International Law; Report of the Study Group of the International Law Commission, Finalized by Martti Koskenniemi' (13 April 2006) UN Doc A/CN.4/L.682.

[2] The Statute of the ILC states that it 'shall have for its object the promotion of the progressive development of international law and its codification'. Statute of the International Law Commission (adopted 21 November 1947) Resolution 174 (II) art 1(1). See also Luis Arevalo, 'The Work of the International Law Commission in the Field of International Environmental Law' (2005) 32 Boston College Environmental Affairs Law Review 493.

[3] See eg Margaret A Young (ed), *Regime Interaction in International Law: Facing Fragmentation* (Cambridge University Press 2012).

[4] ILC, 'Fragmentation Report' (n 1) paras 15, 20, 414, 419, 491.

[5] ibid para 493.

[6] See Vienna Convention on the Law of Treaties (adopted 23 May 1969, entered into force 27 January 1980) 1155 UNTS 331 (VCLT) art 26; *Gabčíkovo-Nagymaros (Hungary v Slovakia)* (Judgment) [1997] ICJ Rep 7, 78 [142].

and, as such, requires that interpretation consider all elements of the interpretive process, including the integration of other relevant rules of law.[7] Alongside and connected to the principle of good faith is a presumption against conflicts. New norms are created within a system of pre-existing international law, and it is presumed that these norms are meant to remain consistent with and build upon law within that system.[8] Likewise, the presumption means that states undertake new treaty obligations not intending to act inconsistently with their existing commitments. These principles help ensure the integrity and efficacy of the international legal system. They underlie the understanding of treaty obligations and the integration of other rules of law, irrespective of whether such integration is made express.[9] Moreover, as the International Court of Justice (ICJ) has stated, '[i]t is a rule of interpretation' that an act or text made by a state government 'must, in principle, be interpreted as producing and as intended to produce effects in accordance with existing law and not in violation of it'.[10] For adaptation obligations, this means that human rights law is part of the normative framework and broader international legal system within which climate treaties are interpreted and applied. Such a conclusion might seem obvious but, given the increase of international treaties with overlapping concerns and the potential for conflict, this relationship is not always clear.

The impacts of climate change implicate nearly every area of international law, creating multiple points of interaction and overlap. The relationship between different legal regimes and the scope of obligations trigged by climate change are subjects of ongoing inquiry. Multiple requests for advisory opinions before international and regional bodies seek clarity on the obligations of states to protect the climate system, marine environment, and people from the adverse effects of climate change. These requests raise questions about general obligations under international law and specific obligations under treaties that govern the law of the sea, human rights, and climate change.[11] The request from the UN General Assembly

[7] VCLT, art 31(1); Richard K Gardiner, *Treaty Interpretation* (2nd edn, Oxford University Press 2015) 171–72; *The South China Sea Arbitration (The Republic of Philippines v The People's Republic of China)* (Award) (12 July 2016) Permanent Court of Arbitration (PCA) Case No 2013–19 [476], [784], [1135], [1141]–[1143], [1171]–[1172], [1190]–[2000] (tribunal noted the need to interpret the treaty in good faith, and to act in good faith in fulfilling obligations and complying with the tribunal award).

[8] See further Joost Pauwelyn, *Conflict of Norms in Public International Law: How WTO Law Relates to Other Rules of International Law* (Cambridge University Press 2009) 240–44.

[9] Campbell McLachlan, 'The Principle of Systemic Integration and Article 31(3)(C) of the Vienna Convention' (2005) 54 International and Comparative Law Quarterly 279, 311.

[10] *Case concerning right of passage over Indian territory (Portugal v India)* (Preliminary Objections) [1957] ICJ Rep 125, 142.

[11] See UN General Assembly (UNGA), 'Request for an Advisory Opinion of the International Court of Justice on the Obligations of States in Respect of Climate Change' (1 March 2023) UN Doc A/77/L.58; Commission of Small Island States on Climate Change and International Law (COSIS), 'Request for an Advisory Opinion' (2022) <www.itlos.org/fileadmin/itlos/documents/cases/31/Request_for_Advisory_Opinion_COSIS_12.12.22.pdf> accessed 1 February 2024; 'Request for an Advisory Opinion on the Climate Emergency and Human Rights Submitted to the Inter-American Court of Human Rights by the Republic of Colombia and the Republic of Chile' (2023) <www.corteidh.or.cr/docs/opiniones/soc_1_2023_en.pdf> accessed 16 November 2023.

to the ICJ, for example, highlights a wide range of relevant treaties and principles of law, which includes the United Nations Framework Convention on Climate (UNFCCC), the Paris Agreement, human rights treaties, and duties of prevention and due diligence. Implicit in the framing of each request is the notion that obligations can be understood as part of a meaningful coherent system, their contours discerned through legal reasoning and decision-making that can balance competing principles and priorities and make sense of the complex challenges created by climate change. Answers may therefore shed light on the reach and role of international law, as well as its limits in addressing climate impacts. Submissions by experts and scholars weighing in on the requests call for a systemic understanding of international obligations. For example, in their submission on the advisory opinion request on the UN Convention on the Law of the Sea (UNCLOS), several special rapporteurs on human rights recognized that UNCLOS, international human rights law, and international environmental law address 'distinct areas of international law' but 'share common purposes'. They declared that 'the three legal regimes work together systemically' and consequently obligations in one regime should be interpreted and applied 'through an approach which systemically integrates' the others. Such a 'holistic approach', they argue, avoids the fragmentation of international law.[12] Other submissions on the various advisory opinions similarly call for integration and look to other areas of law to inform the interpretation of obligations.[13]

Likewise, this book takes a systemic approach and integrates other relevant rules of law. To account for the broader international law system, integration—and an understanding of adaptation obligations—requires the inclusion of a sense of unity and purpose that views different legal rules and principles as part of a coherent whole.[14] Yet integration is more than the recognition that other international law is applicable alongside climate treaties. Integration is used to bring clarity and content to adaptation obligations, strengthening and fleshing out the positive duties

[12] 'Request for an Advisory Opinion Submitted by the Commission of Small Island States on Climate Change and International Law: Amicus Brief Submitted to the International Tribunal for the Law of the Sea by the UN Special Rapporteurs on Human Rights & Climate Change (Ian Fry), Toxics & Human Rights (Marcos Orellana), and Human Rights & the Environment (David Boyd)' (International Tribunal for the Law of the Sea (ITLOS) 2023) Case No 31 <https://www.itlos.org/fileadmin/itlos/documents/cases/31/written_statements/4/C31-WS-4-1_Amicus_Brief_UN_Special_Rapporteurs.pdf> accessed 29 September 2023.

[13] See International Union for Conservation of Nature (IUCN), 'Written Statement of the International Union for Conservation of Nature and Natural Resources—World Commission on Environmental Law, Ocean Law Specialist Group' (2023) (submission to ITLOS); Gamze Erdem Türkelli and others 'Collective Input to the Inter-American Court of Human Rights Regarding the Requested Advisory Opinion Submitted by the Republic of Chile and the Republic of Colombia Regarding "Climate Emergency and Human Rights"' (2023) (author co-authored submission); see also Margaret A Young and Markus W Gehring, 'The Climate Regime and Other Areas of Law' (2023) 13 Climate Law 151; Harro van Asselt, *The Fragmentation of Global Climate Governance: Consequences and Management of Regime Interactions* (Edward Elgar 2014) (exploring different kinds of regime interactions in climate governance, including the interaction of soft and hard law).

[14] See ILC, 'Fragmentation Report' (n 1) paras 414–19.

they entail. This does not mean that integration of relevant law is necessary or desirable in all circumstances, nor is it the only means to manage potential conflict or fragmentation, but it is foundational to arguments that bring human rights to bear on mobility in the climate regime. More generally, integration assists in ensuring that treaties can evolve to account for changes in the law, as well as scientific knowledge and environmental conditions, to remain relevant and effective. This is important for both adaptation and adaptation obligations, which must necessarily include accommodations for changing conditions and developments in international law.

Finally, the interpretation of adaptation obligations builds on a related, albeit obvious, observation: climate treaties and the regime that supports them are dynamic. Thus, as the regime evolves, so too does an interpretation of its legal commitments. This accords with the open-ended nature of interpretation, with outcomes that can change over time. Treaty interpretation is not static, and the rules that govern the task show an appreciation of the need for law to account for changing conditions and subsequent practice.[15] From this need flows an evolutive approach to the interpretation of adaptation obligations, one grounded in the intention of the parties that obligations continue to develop. An evolutionary interpretation allows for changing knowledge about impacts and social conditions to be considered in the interpretation of treaty concepts and obligations. Such an approach is reflective of the way that scientific uncertainty is handled in international environmental law more generally, with treaties characterized as institutionalizing change rather than stability.[16] It is also consistent with two further principles, one that guides the interpretive process and the other that shapes its outcome. The first, the principle of effectiveness, promotes an interpretation that allows for the continued applicability of a treaty and its obligations. It is inherent in the elements of interpretation derived from the VCLT.[17] The second is the principle of precaution, a cornerstone principle of the climate regime that enables continued action in the face of uncertainty and bolsters arguments for more anticipatory action on adaptation, including measures that enable adaptive mobility.[18]

As the subsequent arguments aim to demonstrate, an evolutive approach to the interpretation and application of adaptation obligations can lead to adaptive measures that more consistently consider the risks to human rights and better addresses the needs of those most vulnerable to climate change. More specifically, as the case studies establish, the content and clarity provided for adaptation obligations

[15] VCLT, art 31(3)(a)–(c).

[16] Catherine Redgwell, 'Multilateral Environmental Treaty-Making' in Vera Gowlland-Debbas (ed), *Multilateral Treaty-Making* (Springer Netherlands 2000) 91–92; see generally Jacqueline Peel, *Science and Risk Regulation in International Law* (Cambridge University Press 2013).

[17] See *Mamatkulov and Askarov v Turkey* (2005) App nos 46827/99 and 46951/99 (European Court of Human Rights (ECtHR), 4 February 2005); Gardiner (n 7) 329.

[18] Analysed further in Chapter 5, IV.

through integration lead to action to plan for and facilitate rights-based adaptive mobility.

III. Treaty interpretation and the integration of human rights law

In addition to its role in maintaining coherence in international law, the integration of relevant international human rights law into the climate change regime—and adaptation obligations specifically—is an obligatory part of treaty interpretation. This section applies the elements of treaty interpretation to provisions and terms contained in the UNFCCC and the Paris Agreement. These elements are set out in the VCLT, which provides the legal rules on how treaties should be interpreted and applied. While not universally ratified, its 'general rule on treaty interpretation' is considered customary international law and is therefore binding on all states.[19] Courts including the ICJ and appellate bodies like that of the World Trade Organization (WTO) also look to customary law to guide interpretation and thus to the components of the rule set out in the VCLT.

The rule on treaty interpretation requires that a treaty be interpreted in accordance with the ordinary meaning of its terms in their context and in light of the treaty's object and purpose. It further mandates that '[t]here shall be taken into account . . . any relevant rules of international law applicable between the parties'.[20] This element gives effect to the principle of systemic integration.[21] While these elements are part of a unified interpretive process, they are analysed separately to explain how each contributes to the process and to differentiate between the implications of the treaty text and the integration of external rules from human rights law. The treaty terms, context, and object and purpose of the treaty are explored first, followed by analysis of the integration of external rules from human rights law. Together, they provide the interpretive basis to bring human rights to bear on adaptation obligations and to integrate duties from human rights into their interpretation.

A. Treaty terms

Chapter 2 identified three types of obligations on adaptation established by the UNFCCC and developed in the Paris Agreement. The first are obligations for each

[19] See VCLT, art 31; *Territorial Dispute (Libya/Chad)* (Judgment) [1994] ICJ Rep 6, 21 [41]; Gardiner (n 7) 162.

[20] VCLT, art 31(1), (2), (3)(c).

[21] Campbell McLachlan KC, *The Principle of Systemic Integration in International Law* (Oxford University Press 2024) 3; cf van Asselt (n 13) 65 ('more prudent' to consider systemic integration as an 'objective rather than a principle' as it is not yet recognized as such under general international law).

party to act on adaptation, including the need to 'facilitate adequate adaptation' and to 'engage in adaptation planning and the implementation of actions'.[22] While the latter is prefaced by qualifying language—'as appropriate'—that softens the obligation and introduces discretion for states, some action is required. Adaptation encompasses a range of activities that respond to the different risks posed by the impacts of climate change. It some circumstances, adaptation can require addressing risks to people and their rights, including by planning for and facilitating adaptive mobility. The second type of adaptation obligations is obligations to assist in adaptation. These obligations are focused primarily on developed country parties providing financial assistance to developing country parties.[23] The third type is an obligation to '[c]ooperate in preparing for adaptation'.[24] Support can operationalize cooperation and, alongside assistance obligations, indicates that the parties intended that adaptation not be an undertaking for each state on its own. The Paris Agreement bolsters this understanding of adaptation with provisions that highlight the importance of support for and cooperation on adaptation, and 'on enhancing action on adaptation'.[25]

Adaptation obligations are broad and subjective. Yet these obligations do not exist in isolation, either within the treaty and its context or in the concrete contexts that they are interpreted and applied. Their breadth invites further considerations and additional elements of the interpretive process to inform their content.

B. Context and object and purpose

The treaty's text—in this case the adaptation obligations—are the focus of interpretation, which requires that terms be considered in their context and in light of the treaty's object and purpose. The context includes the text, preamble, and any annexes of a treaty.[26] For the integration of human rights the preamble of the Paris Agreement and the operative principles of the climate regime are of particular importance.

The consideration of preambular language in treaty interpretation is explicit in the VCLT and emphasized by the ILC, which observed '[t]hat the preamble forms part of the treaty for the purpose of interpretation is too well settled to require comment'.[27] The Paris Agreement's preamble acknowledges that states 'should,

[22] United Nations Framework Convention on Climate Change (adopted 9 May 1992, entered into force 21 March 1994) 1771 UNTS 107 (UNFCCC) art 4(1)(b); Paris Agreement (adopted 12 December 2015, entered into force 4 November 2016) 3156 UNTS 79, art 7(9).

[23] UNFCCC, art 4(3), (4); Paris Agreement, arts 7(13), 9(1), (3).

[24] UNFCCC, art 4(1)(e).

[25] Paris Agreement, art 7(6), (7)(d).

[26] VCLT, art 31(2).

[27] ILC, 'Draft Articles on the Law of Treaties with Commentaries' (1966) II Yearbook of the ILC 187, 221, para 13.

when taking action to address climate change, respect, promote and consider their respective obligations on human rights'. It also highlights the right to health and the rights of certain groups, including 'the rights of indigenous peoples, local communities, migrants, children, persons with disabilities and people in vulnerable situations'. While the preamble does not in itself create legal rights or obligations, it can help provide meaning to the Agreement's terms, and it is a basis for including human rights law in its interpretive context. Similar reliance on preambular language to add meaning to draw on external sources has occurred in other contexts. Often, this reliance coincides with support for an evolutive interpretation, connected to the intent of the parties and a treaty's objectives. The ICJ refers to preambular language to provide guidance on the object and purpose of a treaty. For example, in its judgment on the preliminary objections in *Oil Platforms*, the ICJ used preambular language to help determine a treaty's objective and inform its interpretation of other provisions.[28]

Likewise, the WTO's Appellate Body has looked to the preamble of agreements to interpret the meaning of provisions and to avoid conflict between environmental and trade concerns.[29] In the *United States–Import Prohibition of Certain Shrimp and Shrimp Products* (*US–Shrimp*), the Appellate Body relied on the preamble of the WTO Agreement to assist in interpretation.[30] The preamble acknowledged 'the objective of sustainable development', which the Appellate Body used to read environmental protection and conservation into the interpretation of 'exhaustible natural resources'. It noted that the term is not static but is 'by definition, evolutionary', and it went on to explain that the preambular language 'gives colour, texture, and shading to the rights and obligations' of members under WTO agreements and to the interpretation of these agreements.[31] The preamble of the Paris Agreement plays an analogous role and supports an ongoing, evolutive interpretation of its terms, recognizing 'the need for an effective and progressive response to the urgent threat of climate change'. Its acknowledgement that states should

[28] *Oil Platforms (Iran v United States of America)* (Preliminary Objection, Judgment) [1996] ICJ Rep 803, 813–15 [27]–[31] (looking to the preamble for the objective of the treaty, which influenced understanding of an article that shed 'light on the interpretation of the other Treaty provisions'); see also *Sovereignty over Pulau Ligitan and Pulau Sipadan (Indonesia/Malaysia)* (Judgment) [2002] ICJ Rep 625, 651–53 [49]–[52] (preamble invoked to determine the object and purpose of the Convention at issue).

[29] See *United States—Standards for Reformulated and Conventional Gasoline*, Report of the WTO Appellate Body (29 January 1996) WT/DS2/AB/R, 18, 30; *United States–Import Prohibition of Certain Shrimp and Shrimp Products*, Report of the WTO Appellate Body (12 October 1998) WT/DS58/AB/R [129]–[131], [152]–[153], [155].

[30] Marrakesh Agreement Establishing the World Trade Organization (adopted 15 April 1994, entered into force 1 January 1995) 1867 UNTS 154.

[31] *US–Shrimp* (n 29) [129]–[130], [153], [155]; citing *Legal Consequences for States of the Continued Presence of South Africa in Namibia (South West Africa) notwithstanding Security Council Resolution 276 (1970)* (Advisory Opinion) [1971] ICJ Rep 16; see also Callum Musto and Catherine Redgwell, 'US—Import Prohibition of Certain Shrimp and Shrimp Products (1998)' in Eirik Bjorge and Cameron Miles (eds), *Landmark Cases in Public International Law* (Hart 2017) 505.

'respect, promote and consider' their human rights obligations informs an understanding of how adaptation should be undertaken.

Furthermore, like the Paris Agreement, acknowledgement of human rights in a treaty's preamble has shaped understanding of its provisions in other legal contexts. Reference to human rights in the preamble of the 1951 Refugee Convention enabled an express link between international human rights and refugee law.[32] This link 'supports the adoption of international human rights standards as an appropriate source of guidance in interpreting the Refugee Convention'.[33] Although initially distinct in their implementation, the two areas of law—international refugee law and international human rights law—have increasingly converged. The former now tends to be viewed as part of the broader framework of the latter, with strong connections that argue for an understanding of refugee law that is consistent with human rights standards.[34] These standards are evolving, potentially altering the interpretation of the Refugee Convention as human rights law develops. Efforts are also ongoing to bring together human rights and the climate change regime, with interaction between the two unavoidable and human rights standards increasingly necessary to help shape climate action.[35]

The preambular reference to human rights in the Paris Agreement alone, however, does not enable human rights law to be fully integrated into the interpretation and application of climate obligations. The Agreement's preamble is heralded as a milestone, one of the first international environmental treaties to explicitly include human rights. While it has been suggested that this preambular language signals a move 'towards a true incorporation of human rights',[36] its language undercuts such a reading. The preamble does not, for example, reference obligations to protect or fulfil human rights, opting instead for 'promote and consider'. It calls for human rights when parties are 'taking action to address climate change', rather than in determining whether action is needed. Additionally, it refers to parties' respective human rights obligations and not to such obligations in general. This is likely the result of compromises in the negotiation process and the belief that such language may have been too 'operative' for a preamble.[37] A draft of the Agreement had, for example, included human rights in Article 2 of the operative text on the treaty

[32] Convention Relating to the Status of Refugees (adopted 28 July 1951, entered into force 22 April 1954) 189 UNTS 137.

[33] Michelle Foster, *International Refugee Law and Socio-Economic Rights: Refuge from Deprivation* (Cambridge University Press 2009) 49; see also James C Hathaway, *The Rights of Refugees under International Law* (Cambridge University Press 2005) 53–54.

[34] Foster (n 33) 49–51.

[35] Chapter 4, II.–III.

[36] Maria Pia Carazo, 'Contextual Provisions (Preamble and Article 1)' in Daniel Klein and others (eds), *The Paris Agreement on Climate Change: Analysis and Commentary* (Oxford University Press 2017) 115; but see Alan Boyle, 'Climate Change, the Paris Agreement and Human Rights' (2018) 67 International and Comparative Law Quarterly 759, 775 (concluding that the preamble 'does nothing of the kind', but it does reinforce the importance of human rights).

[37] Carazo (n 36) 115.

objectives.[38] This reference does not appear in the final text. Thus, on its own, the preambular language could limit the integration of human rights obligations, particularly their positive duties. Fortunately, treaty interpretation does not end with the preamble, which makes other elements of the rule critical to the interpretation of adaptation obligations.

The objectives and operative principles of climate treaties are further aspects that support the inclusion of human rights. Like the preamble, both are part of the interpretive context. The principles guide the interpretation and application of the UNFCCC's and the Paris Agreement's provisions.[39] Because they do not mandate specific action or create stand-alone commitments, their operationalization is accomplished by linking them to specific obligations—such as on adaptation—or to other legal rules. The principles are set out in the UNFCCC, and those most pertinent to the book's analysis include equity, precaution, and common but differentiated responsibilities and respective capabilities (CBDR-RC).[40] As discussed in greater detail in Chapters 5 and 6, these principles can guide and shape the nature and timing of adaptation and support for adaptation. For example, the precautionary principle operates to help mitigate adverse effects of climate change,[41] which includes effects on people and risks to their rights. It influences when adaptation measures are taken, suggesting anticipatory and preventive action while serving as a general catalyst to adapt in the face of foreseeable harm. The principle informs the integration of positive human rights duties that address risks from climate change and the adaptive mobility that flows from such integration.

Likewise, the evolving objectives of climate treaties and work within the climate regime on adaptation argue for the inclusion of human rights. The Paris Agreement is a treaty under the UNFCCC, and certain provisions build on and are connected to the UNFCCC. The Agreement expands on the objective of the UNFCCC explicitly, 'enhancing the implementation of the Convention', including through an aim to 'strengthen the global response' to climate change by 'increasing the ability to adapt'.[42] This is reinforced by the article on adaptation, which recognizes the

[38] See 'Draft Paris Outcome, Proposal by the President, Version 1 of 9 December 2015 (15:00)' <https://unfccc.int/resource/docs/2015/cop21/eng/da01.pdf> accessed 2 September 2023.

[39] UNFCCC, art 3 ('Parties shall be guided' by the principles); see Alan Boyle, 'Soft Law in International Law-Making' in Malcolm D Evans (ed), *International Law* (5th edn, Oxford University Press 2018) 132 (use of 'should' in specific principles qualifies their application); Lavanya Rajamani and Emmanuel Guérin, 'Central Concepts in the Paris Agreement and How They Evolved' in *The Paris Agreement on Climate Change* (n 36) 82 (Paris Agreement's adoption under the UNFCCC implicitly engages the principles).

[40] Chapter 2, III.

[41] UNFCCC, art 3(3).

[42] Paris Agreement to the United Nations Framework Convention on Climate Change (adopted 12 December 2015, entered into force 4 November 2016) UN Doc FCCC/CP/2015/L.9/Rev/1 (Paris Agreement) art 2; see also art 9(1) (developed countries must provide financial support in continuation of their obligations under the UNFCCC). The relationship between the UNFCCC and the Agreement is discussed further in Lavanya Rajamani, 'Ambition and Differentiation in the 2015 Paris Agreement: Interpretative Possibilities and Underlying Politics' (2016) 65 International and Comparative Law Quarterly 493, 506–07.

significant need for adaptation and the central role it plays in the 'long-term global response' to 'protect people, livelihoods and ecosystems'.[43] The latter establishes the general relevance of human rights law and aligns adaptation with its protective purposes that allow people access to livelihoods, basic needs, and resources.[44] Taken together, the Agreement's acknowledgment of human rights; its enhanced objective; and its references to the protection of people, livelihoods, and ecosystems provide a starting point for an evolving interpretation of adaptation obligations that includes human rights.[45] More generally, domestic policies and courts link the Paris Agreement with the protection of human rights.[46] The Brazilian Supreme Court takes this link a step further, and in a 2022 case declared that environmental treaties, including the Paris Agreement, are a type of human rights treaty that enjoy 'supranational' status, meaning they must not be contradicted by domestic law.[47]

C. Systemic integration

The integration of other rules of law further bolsters the need to incorporate human rights duties. Systemic integration is expressed in an element of treaty interpretation. No element is privileged over any other and all are an obligatory part of a single process. The integration of other rules is not always needed, but it must be considered and it may be justified if any other element of the interpretive process suggests that recourse to external sources of law is needed.[48] It is particularly useful for interpreting terms or provisions like adaptation obligations that are unclear, ambiguous, or open-textured.[49]

For many years following the adoption of the VCLT, academics and international judicial bodies largely overlooked this element of treaty interpretation. Three decades passed before a case—1998 for the WTO's Appellate Body and 2003 for the

[43] See Paris Agreement, art 7(2)–(4).

[44] Chapter 2, III.B.

[45] Lauren Nishimura, 'Adaptation and Anticipatory Action: Integrating Human Rights Duties into the Climate Change Regime' (2022) 12 Climate Law 99, 107.

[46] See eg *The State of the Netherlands v Urgenda Foundation* [2019] Case no 19/00135 (Supreme Court of the Netherlands), discussed further in Chapter 4, II; Ministry of Economy, Republic of Fiji, 'Planned Relocation Guidelines: A Framework to Undertake Climate Change Related Relocation' (2018), discussed further in Chapter 8, IV; see also César Rodríguez-Garavito, 'Litigating the Climate Emergency: The Global Rise of Human Rights-Based Litigation for Climate Action' in César Rodríguez-Garavito (ed), *Litigating the Climate Emergency: How Human Rights, Courts, and Legal Mobilization Can Bolster Climate Action* (Cambridge University Press 2022) 25–26 (arguing that the Paris Agreement is part of the 'international common ground' that helps determine relevant legal standards for domestic and regional cases).

[47] See 'PSB et al v Brazil (on Climate Fund)' (Supreme Court, 1 July 2022) (*Climate Change Litigation*) <https://climatecasechart.com/non-us-case/psb-et-al-v-federal-union/> accessed 10 December 2024.

[48] See ILC, 'Fragmentation Report' (n 1) paras 425, 428.

[49] ibid para 467; McLachlan (n 9) 312.

ICJ—significantly relied on systemic integration and referenced Article 31(3)(c).[50] This relatively recent reliance on the provision may be 'a symptom of the lack of awareness of the obligatory character' of the article.[51] Or it may be due to a failure to identify the integration of other international rules as proceeding from a specific element of interpretation, rather than from the general interpretation of a treaty within its normative environment and as part of a system of international law. It may also be that this system only recently began to fully consider potential fragmentation as treaties increased in quantity, subjects covered, and parties included, and as specialized areas of law developed. The emergence of international environmental law to address transboundary and global threats like climate change evinces the kind of specialized law that could conflict with other international law or legal regimes. As Judge Trindade explained in the ICJ's *Whaling in the Antarctic* case, '[w]ith the growth in recent decades of international instruments related to conservation, not a single one of them is approached in isolation from the others; not surprisingly, the co-existence of international treaties of the kind has called for a systemic outlook, which has been pursued in recent years.'[52] There has thus been a 'recent flowering of case law referring to article 31(3)(c)'.[53]

Prior to renewed interest by dispute bodies and the ILC, however, the provision faced questions about its utility and criticism that it offered little instruction on its application.[54] It is true that, like other elements of the rule, its language does not provide clear guidance on how it should be applied. But treaty interpretation is not a prescriptive process. It is likened to an art rather than science, a flexible exercise that can yield more than one outcome.[55] Systemic integration provides a bridge, between rule and system and mutual support of rules between international legal regimes.[56] It can therefore translate into norms—here from the human rights and

[50] See *US–Shrimp* (n 29); *Oil Platforms (Iran v United States of America)* (Judgment) [2003] ICJ Rep 161. However, the European Court of Human Rights relied on the provision in 1975 to interpret an obligation in light of other principles of international law. *Golder v The United Kingdom* (Judgment) (1979) 1 EHRR 524 [35].

[51] Jason M Pobjoy, *The Child in International Refugee Law* (Cambridge University Press 2017) 39.

[52] *Whaling in the Antarctic, Australia and New Zealand (intervening) v Japan* (Judgment) [2014] ICJ Rep 226, 357 [25] (Separate Opinion of Judge Cançado Trindade).

[53] Gardiner (n 7) 290; *Certain Questions of Mutual Assistance in Criminal Matters (Djibouti v France)* (Judgment) [2008] ICJ Rep 177, 219 [112]–[113]; *Demir and Baykara v Turkey* (2009) 48 EHRR 54 [67]–[68]; *National Union of Rail, Maritime and Transport Workers v the United Kingdom* (2014) 60 EHRR 10 [76]; *European Communities—Measures Affecting the Approval and Marketing of Biotech Products*, Reports of the WTO Dispute Panel (29 September 2006) WT/DS291, WT/DS292, and WT/DS293 [7.67]–[7.72].

[54] See Hugh Thirlway, 'The Law and Procedure of the International Court of Justice 1960–1989: Part Three' (1992) 82 British Yearbook of International Law 1, 58 (declaring it 'doubtful whether this sub-paragraph will be of any assistance in the task of treaty interpretation'). However, even Thirlway recognizes that a treaty must be 'interpreted against the background of general law'; ibid 60.

[55] See ILC, 'Fragmentation Report' (n 1) para 464; Martti Koskenniemi, *From Apology to Utopia: The Structure of International Legal Argument* (Cambridge University Press 2006) 340–42.

[56] See Panos Merkouris, *Article 31(3)(c) VCLT and the Principle of Systemic Integration: Normative Shadows in Plato's Cave* (Brill 2015) 6; Spyridon Aktypis, Emmanuel Decaux, and Bronwen Leroy, 'Systemic Integration between Climate Change and Human Rights at the United Nations?' in Ottavio

climate change regimes—reinforcing one another. It offers a means to help avoid, or to address, potential conflicts when these regimes interact so that adaptation does not create or exacerbate the kinds of risks to people discussed in Chapter 4 that threaten lives, health, homes, and livelihoods.

The limits of systemic integration are equally important. That external rules are to 'be taken into account', as the provision requires, does not equate with the application of different rules alongside one another, although they may well operate in parallel. Nor should use of the provision result in external rules displacing or superseding the provision being interpreted.[57] Integration is instead part of the interpretive process that can circumscribe, clarify, or flesh out a treaty's content.[58] It is an obligatory element of treaty interpretation, which can be supported by treaty context. The Paris Agreement's preambular reference to human rights, for example, can be viewed as 'an invitation to practice systemic integration in the interpretation of Parties' obligations'.[59] Furthermore, as it is the process that is mandated and not the result, the outcome varies and is shaped both by the treaty and its context and by the context of the interpreter. For adaptation obligations, integrating the positive duties from human rights law can help provide concrete content to the obligations. As the case studies in Chapters 7 to 9 show, this allows human rights law to provide clarity as to what adaptation obligations entail and to help compel and shape proactive, concrete adaptive measures, including measures enabling mobility.

Yet systemic integration is not the panacea for all fragmentation concerns. It does not promise to solve all potential conflicts between norms or laws. Integration is instead a tool that promotes the coherence of rules across international law, not the displacement of one area of law by another. Accordingly, and because it is part of an interpretive process and not the application of an external treaty, outside rules can only help provide meaning to terms and should not override them.[60] The use of outside rules is also circumscribed by the language of the element itself, which requires the rules be 'relevant' and 'applicable in relations between the parties'. This raises two further questions: the first asks at what point in time does

Quirico and Mouloud Boumghar (eds), *Climate Change and Human Rights: An International and Comparative Law Perspective* (Routledge 2015) 232.

[57] Philippe Sands, 'Treaty, Custom and the Cross-Fertilization of International Law' (1998) 1 Yale Human Rights and Development Journal 85, 103.

[58] See eg *Golder v The United Kingdom* (Judgment) (n 50) [35]–[36]; *Oil Platforms* (n 50); *US–Shrimp* (n 29); see also Gardiner (n 7) 320; Alexander Orakhelashvili, 'Restrictive Interpretation of Human Rights Treaties in the Recent Jurisprudence of the European Court of Human Rights' (2003) 14 European Journal of International Law 529, 537.

[59] Annalisa Savaresi, 'Climate Change and Human Rights: Fragmentation, Interplay, and Institutional Linkages' in Sébastien Duyck, Sébastien Jodoin, and Alyssa Johl (eds), *Routledge Handbook of Human Rights and Climate Governance* (Routledge 2018) 36.

[60] Pauwelyn (n 8) 254; see also McLachlan, 'The Principle of Systemic Integration and Article 31(3) (C)' (n 9) 318–19.

interpretation look to external rules and the second asks what rules are relevant and between which parties should these relevant rules apply.

(1) An evolutive approach to interpretation

There is no universal answer as to when an interpreter should look to other rules of international law to aid in interpretation. The question is part of the broader issue of intertemporal law, and whether law and its existence, interpretation, and application should be understood at the time of its creation or contemporary to its analysis.[61] The text of Article 31(3)(c) does not explicitly address the issue, nor does the VCLT's preparatory work provide a definitive answer. Such silence is intentional. The ILC drafted articles that formed the basis of the VCLT's text. As commentary on these draft articles indicated, interpretation is not limited to the time the treaty was negotiated or adopted, as this restriction was contemplated and removed from the final text of the VCLT. Instead, it concluded that temporal issues would primarily be resolved by looking to the intent of a treaty's parties and exercising good faith in interpretation.[62]

This book takes an evolutive approach to interpretation, one that allows for the meaning of treaty terms to change over time. Such an approach is confirmed by looking to the intent of the parties and whether the terms were meant to be fixed or to evolve.[63] The dynamism built into climate treaties with their developing concepts and goals, and the inherent need for adaptation to evolve, evince an intent that the content of adaptation obligations shift over time. Such change is accomplished, in part, by integrating relevant legal developments into the process of interpretation.[64] It is also informed by changes to treaty concepts and to the conditions or context in which a treaty operates. Accounting for these changes helps ensure the continued effectiveness of a treaty.[65] This accords with the ILC's commentary and with the approach international courts and tribunals take when tasked with resolving disputes over treaty provisions. These bodies have looked outside the treaty at issue—including to subsequent developments in international law—to justify an evolutionary interpretation and provide clarification about treaty obligations.

In its *Advisory Opinion on Namibia*, for example, the ICJ explained that it was 'bound to take into account' that concepts of the Covenant of the League of Nations, which it was interpreting, were 'not static, but by definition evolutionary'. As a result, 'an international instrument has to be interpreted and applied within the framework

[61] See Merkouris (n 56) 102–04; Gardiner (n 7) 290–92; *Island of Palmas (Netherlands v United States of America)* (1928) 2 RIAA 829 (PCA) 845.

[62] ILC 'Draft Articles' (n 27) 222, para 16.

[63] Eirik Bjorge, *The Evolutionary Interpretation of Treaties* (Oxford University Press 2014) 143 (temporal element controlled by parties' intent).

[64] Nishimura (n 45) 108; Chapter 4, II.

[65] See n 17 and accompanying text.

of the entire legal system prevailing at the time of interpretation'.[66] Subsequent cases and arbitral awards reiterate this point, including *Gabčíkovo-Nagymaros* (the '[t]reaty is not static, and is open to adapt to emerging norms of international law' for the protection of the environment); *Dispute Regarding Navigational and Related Rights* ('where the parties have used generic terms in a treaty . . . having been aware that the meaning of the terms was likely to evolve over time . . . the parties must be presumed, as a general rule, to have intended those terms to have an evolving meaning'); and *Iron Rhine* (citing *Advisory Opinion on Namibia* and explaining that in addition to conceptual or generic terms, 'new technological developments' are subject to 'an evolutive interpretation, which would ensure an application of the treaty that would be effective in terms of its object and purpose').[67]

Likewise, the WTO's Appellate Body has considered the evolution of terms and looked to external legal sources to aid in its interpretive efforts. In *US–Shrimp*, as Section III.B notes, the Appellate Body found that the key term to the dispute, 'exhaustible natural resources', was 'by definition evolutionary'. It made use of various international environmental law sources to support its conclusion that exhaustible natural resources included living and non-living resources and that sea turtles as such a resource fell into this category.[68] This recognition relied on developments in international environmental law that post-dated the WTO-covered agreements. The Appellate Body also explained that its task in interpreting treaty text included 'seeking additional interpretive guidance, as appropriate, from the general principles of international law'.[69] This built upon its interpretive approach in *US-Gasoline*, where it recognized that one of the WTO agreements—the General Agreement on Tariffs and Trade–could not be read in 'isolation from public international law'.[70]

Adaptation obligations must, similarly, be understood as part of a broader international legal system, one that expands and develops to address new challenges. Their interpretation can therefore look to other rules of law—here human rights law—to clarify content, as these rules evolve and as the conditions triggering these rules change. The concept of adaptation is not meant to be fixed to a specific moment. Much like the provisions at issue in the *US–Shrimp* or *Gabčíkovo-Nagymaros* cases, adaptation is a broad term that is tied to environmental change

[66] *Advisory Opinion on Namibia* (n 31) 30 [53].

[67] *Gabčíkovo-Nagymaros* (n 6) 67 [112]; *Dispute regarding Navigational and Related Rights (Costa Rica v Nicaragua)* (Judgment) [2009] ICJ Rep 213, 243 [66]; *Award in the Arbitration regarding the Iron Rhine ('Ijzeren Rijn') Railway (Belgium v Netherlands)* (2005) 27 RIAA 35 (PCA) [80]; see also *Aegean Sea Continental Shelf (Greece v Turkey)* (Judgment) [1978] ICJ Rep 3, 32 [77].

[68] These sources included statements by the World Commission on Environment and Development, provisions of the Conventions on the Law of the Sea and on Biological Diversity, and the Resolution on Assistance to Developing Countries, adopted with the Convention on the Conservation of Migratory Species of Wild Animals. *US–Shrimp* (n 29) [130]–[134].

[69] ibid [158] (citing VCLT, art 31(3)(c)).

[70] *US–Gasoline* (n 29) 17 (citing VCLT, art 31(1)); see Musto and Redgwell (n 31) 505–06 (analysing systemic integration in the WTO context).

and its meaning is not circumscribed by the treaty at issue. It is 'by definition evolutionary': climate impacts create new risks, and adaptation measures will necessarily shift to accommodate changes to ecosystems and resources. A failure to incorporate these changes into adaptation obligations could interfere with the enjoyment of human rights or violate human rights law, undermining adaptation efforts. Without measures to shift agriculture or irrigation in areas of prolonged drought, for example, violations of rights to adequate food and water, or to the highest attainable standard of health, are likely to occur. As impacts become more severe and desertification and famine set in, if individuals are not able to move, then their continued residence in an uninhabitable area could further violate a right to adequate housing, or worse, a right to life. Thus, an interpretive process that can itself adapt to changing conditions results in different interpretations of what adaptation obligations require over time. Such a process can account for an evolving understanding of human rights law, as human rights obligations develop in the context of environmental harm and climate change.[71] Interpretation could, for example, incorporate a right to a healthy environment, if such a right becomes more widely accepted by parties to the climate treaties.[72] In addition, it could incorporate clarifications by judicial bodies about the scope and content of human rights law, as well as developments across and within other regimes, which themselves might also be affected by the impacts of climate change.

(2) Relevant law applicable between the parties

Treaty interpretation requires the interpreter to take into account '[a]ny relevant rules of international law applicable in the relations between the parties'. This requirement creates two limits to what can be integrated. The first limit is that the external rules must be relevant. Jurisprudence makes little mention of the meaning of 'relevant', and different approaches look to the subject matter at issue and its proximity to the external rules or the use of rules to answer questions not resolved by a treaty.[73] What rules are relevant is, like other elements of treaty interpretation,

[71] See John H Knox, 'Human Rights Principles and Climate Change' in Cinnamon P Carlarne, Kevin R Gray, and Richard Tarasofsky (eds), *The Oxford Handbook of International Climate Change Law* (Oxford University Press 2016) 220–21; see also Alan Boyle, 'Human Rights and the Environment: Where Next?' in Ben Boer (ed), *Environmental Law Dimensions of Human Rights* (Oxford University Press 2015) 220 (noting 'evolutionary character' of human rights treaties).

[72] The UN Human Rights Council has recognized such a right. UN Human Rights Council (UNHRC), 'The Human Right to a Safe, Clean, Healthy and Sustainable Environment' (5 October 2021) UN Doc A/HRC/48/L.23 Rev.1. Such a right also exists in regional instruments, which could affect the interpretation of some states' obligations. See eg Arab Charter on Human Rights (adopted 22 May 2004, entered into force 15 March 2008) art 38; African Charter on Human and Peoples' Rights (adopted 27 June 1981, entered into force 21 October 1986) (1982) 21 ILM 58, art 24.

[73] See *European Communities and Certain Member States—Measures Affecting Trade in Large Civil Aircraft*, Report of the WTO Appellate Body (18 May 2011) WT/DS316/AB/R [864] (other rules of law are relevant if they concern the same subject matter as the provision at issue); *Djibouti v France* (n 53) 216–19 [105]–[114] (general obligations can be relevant rules in interpretation of a subsequent treaty on a different subject matter); Merkouris (n 56) (proposing four kinds of proximity for

subjective, but relevance suggests that they must connect in some way to the provision being interpreted. The general relevance of human rights law is clear. The connection between human rights and climate change is now widely recognized. It is highlighted by human rights bodies, the Human Rights Council, and special rapporteurs, and made explicit through the creation of a Special Rapporteur on the promotion and protection of human rights in the context of climate change. Recognition of the link between the two led to the inclusion of the preambular reference to human rights and is implicit in references to rights-related considerations, participation, and provision of information in the body of the Paris Agreement.[74]

Both treaty and customary human rights law are relevant for integration purposes. Their binding nature and legal basis outside the climate regime underscores their role in treaty interpretation, as do the general agreement and practice of looking to rules beyond general principles of international law.[75] The integration of human rights law has been called upon to develop an understanding of provisions of other areas of law, including shaping who is a refugee and refugee rights.[76] It is also used to argue for the integration of the Convention on the Rights of the Child (CRC) into interpretation of the Refugee Convention, to ensure a 'child-rights framework' for assessing the status of and protection for the refugee child.[77] However, international refugee law targets state treatment of people, while climate treaties govern state behaviour and state-to-state relations. State action will necessarily impact people, but there will be some circumstances where it may be more remote or indirect, particularly for decision-making on mitigation. Consequently, for climate obligations, the relevance of rights depends on the context and the actions being undertaken to address climate change. Additionally, as the case studies show, the potential to reach different results when interpreting adaptation obligations is consistent with the broad nature of adaptation and the activities it can encompass. These differences underscore the inherent flexibility needed in interpretation, to allow for a process that can be tailored to the adaptive goal on a case-by-case basis and remain faithful to the limits of treaty interpretation.

relevance: terminological, subject matter, shared parties, and temporal); Sands (n 57) 102 ('relevance' refers to a subject matter or rules related in some way to the norm being interpreted); McLachlan, 'The Principle of Systemic Integration and Article 31(3)(C)' (n 9) 280.

[74] UNHRC, 'Mandate of the Special Rapporteur on the Promotion and Protection of Human Rights in the Context of Climate Change' (4 October 2021) UN Doc A/HRC/48/L.27.

[75] See ILC, 'Fragmentation Report' (n 1) paras 422, 426; McLachlan, 'The Principle of Systemic Integration and Article 31(3)(C)' (n 9) 290; Gardiner (n 7) 301; *EC–Biotech* (n 53) [7.67] ('no doubt' that treaties and customary law are rules of international law).

[76] James C Hathaway and Michelle Foster, *The Law of Refugee Status* (Cambridge University Press 2014) 8–9; Foster (n 33) 51–53.

[77] Pobjoy (n 51) 34–43.

The second limitation of integration is that relevant rules must be 'applicable in the relations between the parties'. The VCLT does not clarify which parties are necessary, and there are differing opinions as to what this means. Customary law does not present an obstacle as it generally applies to all states.[78] The crux of the disagreement is the degree of consistency between the parties to the treaty at issue and the parties to the treaty containing the relevant rule. Several scenarios have been offered as possible solutions, which range from requiring that the external rule apply to all parties—either parties to the treaty being interpreted or parties to a dispute—to an implicit acceptance or toleration of the rule by the parties to the treaty being interpreted. Scenarios that require all parties are criticized as too narrow or as leading to inconsistent interpretations.[79] If identical participation is required, for example, then international human rights law could not be considered relevant to the interpretation of the Refugee Convention—or to any other treaty—because few, if any, treaties share the same membership.[80] A requirement that all parties be bound by treaty containing the external rule would make it difficult for treaties with wide participation, like the UNFCCC or the Paris Agreement, to look to any other treaty rule that is not already part of customary law. It would allow other less accepted treaty bodies to rely on a broader range of international law sources in interpretation. As the ILC report observed, this would create the 'ironic effect that the more the membership of a multilateral treaty . . . expanded, the more those treaties would be cut off from the rest of international law'. This would lead to 'the isolation of multilateral agreements as "islands" permitting no references *inter se* in their application'.[81] Such an outcome runs counter to the purpose of most multilateral treaties and would undercut their efficacy. It undermines the impact of climate treaties and the use of specific provisions like adaptation obligations. Indeed, their wide acceptance and universal applicability help make these obligations a potentially powerful legal tool.

The analysis of this book, therefore, looks to party acceptance of the external rule. It focuses on rights or duties that are customary or that have been ratified or accepted in some form by most states. It also analyses an interpretive moment that occurs before disputes arise, when states are considering their own obligations. This enables human rights law to play an *ex ante* role in shaping adaptation. While

[78] The possibility of a state that persistently objects to a customary rule could create outliers, although this would be rare. See generally John Tasioulas, 'Custom, Jus Cogens, and Human Rights' in Curtis A Bradley (ed), *Custom's Future: International Law in a Changing World* (Cambridge University Press 2016).

[79] These scenarios and criticisms are discussed in McLachlan, 'The Principle of Systemic Integration and Article 31(3)(C)' (n 9) 314–15; ILC, 'Fragmentation Report' (n 1) paras 471–72; Gardiner (n 7) 312–16.

[80] Foster (n 33) 55.

[81] ILC, 'Fragmentation Report' (n 1) para 471.

it should play this role independently, the practical utility of human rights has been criticized because of a failure to adequately incorporate anticipatory measures or account for pre-emptive movement, as rights are often delineated after violations occur.[82] However, even if the interpretation of adaptation obligations becomes the subject of a dispute, and states are considering their own relevant obligations or those of states to which they owe adaptation-related duties, the rights analysed would continue to apply given their wide acceptance (whether explicitly or implicitly). This is the approach adopted by the ILC's fragmentation report and conclusions. The report noted a preference to allow reference to a treaty when participation was shared by parties to a dispute, while also calling for treaty rules with implicit acceptance to be taken into account. The conclusions highlighted the 'particular relevance' of other rules that offer 'common understanding' of the parties about a treaty term or its object and purpose.[83]

Finally, it is worth noting that recourse to external rules in interpretation is not always directly connected to the requirements of Article 31(3)(c). They may be integrated without reference to the specific provisions of the rule on interpretation, or they may shed light on the meaning of a term in connection to other elements, tied to the object and purpose of the treaty or incorporated by reference in the treaty's context. These other elements help build the case for the incorporation of human rights into climate treaties.[84] They also sidestep some of the requirements of Article 31(3)(c), as a WTO Panel did in *EC-Biotech*. The Panel took a narrow view on Article 31(3)(c) and declared other rules of international law should only be considered when they apply to all parties of a treaty being interpreted or all parties to a dispute, which precluded their consideration of other treaty rules in the case. Nonetheless, the Panel noted that, if informative, it could look to other relevant rules to support an analysis of ordinary meaning under Article 31(1).[85] This choice evinces a sense that regime interaction must be engaged with rather than avoided, albeit through reasoning that faced questions and criticism.[86]

For adaptation obligations, integration can serve multiple functions that help ensure that rights are accounted for during both the interpretation and

[82] See eg Mariya Gromilova, 'Revisiting Planned Relocation as a Climate Change Adaptation Strategy: The Added Value of a Human Rights-Based Approach' (2014) 10 Utrecht Law Review 76; Jane McAdam, 'Climate Change Displacement and International Law: Complementary Protection Standards' (UN High Commissioner for Refugees (UNHCR) 2011) 50.

[83] ILC, 'Fragmentation Report' (n 1) para 472; ILC, 'Fragmentation of International Law: Difficulties Arising from the Diversification and Expansion of International Law; Conclusions of the Work of the Study Group' (18 July 2006) UN Doc A/CN.4/L.702 (conclusion 21). The ILC's approach could be read to endorse international consensus as the proper inquiry, rather than consent or non-consent of parties. Margaret A Young, 'The WTO's Use of Relevant Rules of International Law: An Analysis of the Biotech Case' (2007) 56 The International and Comparative Law Quarterly 907, 916–17.

[84] Section III.B.

[85] *EC–Biotech* (n 53) [7.69]–[7.71], [7.75], [7.92]–[7.94].

[86] See ILC, 'Fragmentation Conclusions' (n 83) para 50; Young (n 83) 918–22.

implementation of obligations, which translates into their consideration at different points in the process of adaptation. In interpretation, relevant human rights law can provide content to adaptation obligations. In doing so, relevant rights help shape adaptive action, establishing the need for certain measures and the limits of permissible activities. For example, action may be required to facilitate adaptive mobility, although such mobility cannot occur through measures that forcibly displace people or that violates their human rights. Integration can also require that people be able to participate and guide the adaptation process, including any movement or planned relocation. And it requires planning for such mobility, which accords with the precaution that underpins climate treaties and bolsters the use of human rights in anticipatory action.

In addition to legal arguments, the integration of human rights is desirable for ethical and practical reasons. It foregrounds the rights of individuals throughout the process of adaptation and adaptive mobility, which provides for their proactive and pre-emptive use. Human rights are universal and apply to a person regardless of whether they are a migrant or their migration status. However, in practice—and as Chapter 1 highlights—the rights of migrants are often contemplated after movement. In contrast, adaptive mobility shaped by positive human rights duties seeks to address foreseeable risks, ideally before risks become harm or violate rights.

IV. Conclusion

The climate change regime is not a self-contained body of law. As this chapter argues, it is part of an international law system that should function as a coherent yet pluralistic whole. Consequently, climate treaties remain linked to other rules and principles of international law, particularly human rights law. The elements of treaty interpretation—a treaty's context, object and purpose, and the systemic integration of relevant law—lead to the same conclusion: that human rights law is relevant, and its integration is required in interpreting adaptation obligations. The chapter also argues for an evolutive approach to interpretation, one that accounts for changes in law, increasing knowledge about risks and climate impacts, and the dynamic nature of the climate change regime. This allows for the continued effectiveness of adaptation obligations and the ability to integrate human rights throughout the process of adaptation. Chapters 4 and 5 will build on this, to establish the relevance of specific rights and the resulting need to integrate positive duties associated with these rights. An interpretation of adaptation obligations that includes such integration goes beyond the mere recognition of the risks to human rights that climate impacts pose. By integrating positive duties from human rights, adaptation can focus on the risks to people, prioritize those most vulnerable, and require action before the worst impacts occur.

4
Shaping Adaptive Mobility: Integrating Substantive Human Rights

I. Introduction

Chapter 3 set out the methodology of the book. It looked at the elements of treaty interpretation to argue that adaption obligations can be clarified through the integration of human rights law. This chapter builds on the general argument for integration and demonstrates how the integration of human rights duties can clarify the first type of adaptation obligation found in climate treaties: obligations to act on adaptation. While climate change affects the enjoyment of many human rights, and the interdependence of those rights is not disputed, this chapter focuses on two substantive rights—the rights to life and to an adequate standard of living. The latter is analysed through its component rights to adequate food, water, housing, and health. The substantive rights analysed in this chapter are the focus, in part, because they are critical to survival and subsistence, and they play a significant role in whether a person is able to stay or move elsewhere. They are also customary or widely accepted treaty law, making them broadly applicable to those countries that are parties to the United Nations Framework Convention on Climate Change (UNFCCC) and the Paris Agreement.

The chapter's analysis of these rights first sets out their sources under international law and describes how their enjoyment is affected by climate change. It then discusses the obligations each right entails. All human rights involve positive and negative duties—to act and to refrain from action. The focus in this chapter is on the positive duties or measures states must take to protect and fulfil human rights. These comprise the content of the rights that are integrated into and shape obligations to adapt. Accordingly, the analysis explains that their integration leads to an interpretation of adaptation obligations that requires countries to act in the face of foreseeable risks from climate impacts. For the rights analysed, such an interpretation translates into the general need to ensure access to essential resources and a life with dignity. More concretely, and as further detailed in the case studies, it will necessitate that countries undertake measures to plan for and facilitate adaptive mobility.

Climate Change, Human Rights, and Adaptive Mobility. Lauren Sakae Nishimura, Oxford University Press.
 DOI: 10.1093/9780198930068.003.0005

II. Integrating the right to life

Human rights instruments include the right to life.[1] The right is also part of customary international law and is generally considered to be *jus cogens* or a peremptory norm.[2] As the Human Rights Committee (HRC) explains, it is 'the supreme right from which no derogation is permitted' even in situations of 'public emergencies which threaten the life of the nation'.[3] The right should not be narrowly interpreted, and it requires states to take positive action to ensure its protection.

Climate change puts at risk all 'material aspects of life and livelihood'.[4] It can constrain access to basic necessities and resources. Such constraint, at its worst, can pose a threat to human life, which can be brought about from rising sea levels, prolonged drought and desertification, extreme heat events, and flooding of low-lying areas.[5] The HRC recognized that climate change constitutes one of 'the most pressing and serious threats to the ability of present and future generations to enjoy the right to life'. It advised states to look to their obligations under international environmental law to inform the content of the right and, conversely, for the right to inform international environmental law obligations—much as this book looks to human rights to inform the content of climate change obligations.[6] Indeed, because of its customary status and broad relevance to climate impacts, the right to life should be integrated into the interpretation of adaptation obligations for all parties to climate treaties.[7]

The HRC also explores the duties required for states to meet their human rights obligations, which provides guidance on the content of the right to life. For example, General Comment 36 includes 'degradation of the environment' in its list of 'general conditions in society' that may threaten the right and give rise to state

[1] See The Universal Declaration of Human Rights, 10 December 1948 (GA Resolution 217 A(III)) art 3; International Covenant on Civil and Political Rights (adopted 16 December 1966, entered into force 23 March 1976) 999 UNTS 171 (ICCPR) art 6; Convention on the Rights of the Child (adopted 20 November 1989, entered into force 2 September 1990) 1577 UNTS 3 (CRC) art 6.

[2] BG Ramcharan, 'The Concept and Dimensions of the Right to Life' in BG Ramcharan (ed), *The Right to Life in International Law* (Martinus Nijhoff 1985); WP Gormley, 'The Right to Life and the Rule of Non-Derogability: Peremptory Norms of Jus Cogens' in *The Right to Life in International Law*; United Nations Commission on Human Rights (UNCHR), 'Report of the Commission on Human Rights on the Protection of Human Rights in Chile' (4 November 1982) UN Doc A/37/564, para 22.

[3] HRC, 'General Comment No 36: Article 6 (Right to Life)' (3 September 2019) UN Doc CCPR/C/GC/36, para 2; see ICCPR, art 4.2 (no derogation from the right).

[4] W Neil Adger and others, 'Human Security' in Christopher B Field and others (eds), *Climate Change 2014: Impacts, Adaptation, and Vulnerability. Part A: Global and Sectoral Aspects. Contribution of Working Group II to the Fifth Assessment Report of the Intergovernmental Panel on Climate Change* (Cambridge University Press 2014) 761.

[5] See Intergovernmental Panel on Climate Change (IPCC), 'Summary for Policymakers' in Core Writing Team, H Lee, and J Romero (eds), *Climate Change 2023: Synthesis Report. Contribution of Working Groups I, II and III to the Sixth Assessment Report of the Intergovernmental Panel on Climate Change* (2023) 6, 15; IPCC, 'Summary for Policymakers' in Field and others (eds) *Climate Change 2014: Impacts, Adaptation, and Vulnerability. Part A: Global and Sectoral Aspects* (n 4) 13.

[6] HRC, 'General Comment No 36' (n 3) para 62.

[7] Chapter 3, III.C.

duties. It sets out means to address these conditions, including through 'measures designed to ensure access without delay by individuals to essential goods and services such as food, water, shelter, healthcare, electricity, and sanitation'. It also highlights the need for measures to be taken to address both direct threats from environmental degradation, including climate change, and the additional threats that degradation creates. Doing so expands the scope of the positive duties associated with the right, in part through its extension of the right to include a 'right to a life with dignity' and recognition of the underlying need for an environment that enables enjoyment of the right.[8] The HRC has justified this extension by pointing to the rules on treaty interpretation in the Vienna Convention on the Law of Treaties, drawing on the preamble of the ICCPR, its recognition of the inherent dignity of all people, and the interconnection of rights, including economic, social, and cultural rights.[9] This creates a further link to subsistence rights and the right to an adequate standard of living, analysed in Section III. Such a relationship accords with the nature of the right to life, its content informed by other human rights, and its protection a prerequisite to enjoy those rights.[10]

Decisions of judicial and human rights bodies can provide further guidance on states' positive obligations and how they can be met. For example, regional case law indicates that the right to life creates substantive and procedural duties for states to take protective measures to address environmental risks, which can include disclosing information, regulating risks, and enforcing environmental laws. These cases also make clear that the right applies not only to specific individuals but also to communities or society as a whole.[11] States are given a margin of appreciation in determining how to meet their duties, but their discretion is increasingly limited as the threat to life becomes more serious or immediate.[12] In *Budayeva v Russia*, for example, the European Court of Human Rights (ECtHR) found that Russia violated the right to life in its handling of a deadly mudslide. The government failed to act despite awareness of the risk of a mudslide before it occurred,

[8] HRC, 'General Comment No 36' (n 3) paras 26, 62; see also *Portillo Cáceres v Paraguay*, 'Views Adopted by the Committee under Article 5 (4) of the Optional Protocol, Concerning Communication No 2751/2016' (HRC, 20 September 2019) UN Doc CCPR/C/126/D/2751/2016, para 7.3; Ginevra Le Moli, 'The Human Rights Committee, Environmental Protection and the Right to Life' (2020) 69 International and Comparative Law Quarterly 735, 745.

[9] See *Billy v Australia*, 'Views Adopted by the Committee under Article 5(4) of the Optional Protocol, Concerning Communication No 3624/2019' (HRC, 21 July 2022) UN Doc CCPR/C/135/D/3624/2019, para 8.4.

[10] HRC, 'General Comment No 36' (n 3) para 2.

[11] See *Budayeva v Russia* App nos 15339/02, 21166/02, 20058/02, 11673/02, and 15343/02 (ECtHR, 20 March 2008) [128]–[138]; *Öneryildiz v Turkey* (2005) 41 EHRR 20 [89]–[90]; *Gorovenky and Bugara v Ukraine* App nos 36146/05 and 42418/05 (ECtHR, 12 January 2012) [32]; *Social and Economic Rights Action Center (SERAC) and the Center for Economic and Social Rights (CESR) v Nigeria* [2001] ACHPR Comm no 155/96 (*SERAC v Nigeria*) [67]; see also Walter Kälin, 'The Human Rights Dimension of Natural or Human-Made Disasters Focus: Disaster Preparedness and Response' (2012) 55 German Yearbook of International Law 119, 135–39.

[12] See Bruce Burson and others, 'The Duty to Move People Out of Harm's Way in the Context of Climate Change and Disasters' (2018) 37 Refugee Survey Quarterly 379, 387–88.

and it did not implement sufficient planning or inform the public about the risk, even after the mudslide had begun. There are parallels to this case and climate change: like the Russian government that did not create the mudslide, many developing countries—including those discussed in the case studies in Chapters 7 to 9—are not responsible for causing climate change. However, these states are responsible for taking measures to protect the right to life. The right creates a positive obligation, and while states retain some discretion in choosing how to protect the right, taking *no* action is insufficient. States must act, and they must do so before environmental risks are realized. They must 'take appropriate steps to safeguard the lives of those within their jurisdiction'.[13] At a minimum, these steps must include measures to create legislative and administrative frameworks that prevent and address threats to life, including those from disasters.[14] Further appropriate action can be shaped by adaptation obligations, and by the context of their interpretation and implementation.

However, it is not always apparent when a risk to the right should trigger state action, and how immediate the risk of harm must be. There is confusion, for example, about whether positive duties might be limited to situations where the threat to life is considered imminent. Standing to bring an individual complaint before the HRC requires an imminent or actual risk to the right. This risk must be more than a theoretical possibility, which the HRC found to be the case where individuals faced 'real predicaments' that they personally experienced from disruptive climate events or slow-onset processes.[15] However, imminence is not the standard required for analysis of the merits. As Jane McAdam and Michelle Foster explain, once the HRC determines a matter is admissible, 'the question of imminence of harm is no longer relevant'.[16] Imminence has also been cited in regional case law, although it is not necessarily imminent harm that determines whether a rights violation occurred. In *Budayeva*, the ECtHR determined that obligations to take positive action were not met where circumstances pointed to 'the imminence of a natural hazard that had been clearly identifiable' and 'where it concerned a recurring calamity affecting a distinct area developed for human habitation or use'.[17] Because Russia was aware that a future mudslide posed increasing risks with

[13] *Cyprus v Turkey* (2001) 35 EHRR 30 [219].

[14] *Budayeva* (n 11) [129], [131], [156].

[15] *Billy v Australia* (n 9) paras 7.9–7.10; see also *Aalbersberg v The Netherlands* Communication No 1440/2005 (HRC, 12 July 2006) UN Doc CCPR/C/87/D/1440/2005. States must be party to the Optional Protocol to the ICCPR (adopted 16 December 1966, entered into force 23 March 1976) 999 UNTS 302 to bring an individual complaint against them.

[16] They also note that the HRC has conflated admissibility and merits in the past, introducing uncertainty about the role of imminence. Michelle Foster and Jane McAdam, 'Analysis of "Imminence" in International Protection Claims: Teitiota v New Zealand and Beyond' (2022) 71 International and Comparative Law Quarterly 975, 978–80; see also Adrienne Anderson and others, 'Imminence in Refugee and Human Rights Law: A Misplaced Notion for International Protection' (2019) 68 International and Comparative Law Quarterly 111.

[17] *Budayeva* (n 11) [137].

potentially devastating consequences, due in part to damage caused by past mudslides that it failed to address, it violated the right to life. As confirmed in other cases, obligations are triggered once a state is on notice of a direct threat to people, irrespective of when that risk would be realized.[18]

A growing body of decisions and guidance underscore the need to act to address risks that are foreseeable. Foreseeable threats to life are those that can be prevented or mitigated and that could lead to an individual's unnatural or premature death or interfere with their enjoyment of a life with dignity.[19] Even long-term risks, when known or foreseeable, create obligations to take appropriate measures that ensure human rights.[20] The HRC highlights the foreseeability of risks in its analysis of substantive claims. Its decisions, while not legally binding, represent an authoritative interpretation of human rights law that influences the reasoning of courts, and they are expected to be followed by states. The HRC has found that the failure to take protective action against foreseeable threats, including those from environmental degradation or harm, violates the right to life and to the enjoyment of a life with dignity.[21] States are therefore obliged to protect against such harm *before* it occurs. A failure to do so can give rise to a violation, which does not require loss of life or an imminent risk. Indeed, recent HRC decisions do not mention imminence in their analysis of the merits, instead focusing on whether the threat to life was real and reasonably foreseeable.[22] That the harm is ongoing and expected to worsen bolsters claims, particularly as impacts become more significant. These decisions signal a shift towards a different standard—one that aligns better with precaution and anticipatory action.

Yet the precise contours of this standard remain uncertain. In *Billy v Australia*, for example, the HRC considered whether Australia's mitigation and adaptation measures, or lack thereof, violated Torres Strait Islanders' human rights. In its consideration of the right to life, the HRC rejected what it characterized as Australia's argument that it is not obligated to prevent foreseeable loss of life from climate change, noting that states' obligations extend to reasonably foreseeable threats and that such threats may include the impacts of climate change. However, it found that Australia had not yet violated the right and that 10 to 15 years, which it accepted as the time frame for when islanders' homes would become uninhabitable, allowed for 'intervening acts' and 'affirmative measures' to be taken to protect lives and, 'where necessary', to relocate people.[23] This echoes its reasoning in *Teitiota*

[18] See *Öneryildiz* (n 11) [98]–[101] (risk of gas explosion was known to authorities for years); *Taşkın v Turkey* App no 46117/99 (ECtHR, 10 November 2004) [107], [111]–[114] (rights obligations apply where environmental pollution might only materialize in decades).

[19] HRC, 'General Comment No 36' (n 3) paras 3–4, 6–7; *Portillo Cáceres v Paraguay* (n 8) para 7.5.

[20] See n 18.

[21] *Portillo Cáceres v Paraguay* (n 8) paras 7.3–7.5.

[22] ibid; *Teitiota v New Zealand*, 'Views Adopted by the Committee under Article 5(4) of the Optional Protocol, Concerning Communication No 2728/2016' (HRC, 23 September 2020) UN Doc CCPR/C/127/D/2728/2016, para 9.4; *Billy v Australia* (n 9) para 8.3.

[23] *Billy v Australia* (n 9) paras 8.3–8.7.

v New Zealand—a decision that found that New Zealand did not err in returning an individual to Kiribati—where the HRC similarly concluded that while climate impacts may expose individuals to a violation of the right absent 'robust national and international efforts', 10 to 15 years provided Kiribati, 'with the assistance of the international community', the time to take measures to 'protect, and where necessary, relocate its population'.[24]

Several pertinent conclusions can be drawn from these decisions. First, the HRC acknowledges the interrelationship of the right to life with other rights, although this acknowledgement is difficult to square with some of its reasoning. The HRC found, for example, no violation of the right to life in circumstances that violated other rights, where people face reduced access to marine resources for food, loss of crops, inundation of villages and ancestral burial grounds, and the prospect of losing their homes. Based on these decisions, an existential threat 10 to 15 years in the future is too remote to violate the right, but equally insufficient are existing risks to communities that rely on natural resources to subsist and where 'alternative means of subsistence and humanitarian aid are unavailable'.[25] Although this does not preclude establishing a foreseeable threat to lives,[26] other rights will be key to the analysis of human rights and climate change, and in shaping adaptation. It is also worth noting that the HRC did not accept Australia's arguments that systemic integration is only possible when the relevant rules from another treaty concern the subject matter of the treaty term at issue. It noted that the relief sought did not require determining compliance with other treaties but rather relied on those treaties to aid in the interpretation of the rights obligations of state parties.[27]

Second, these decisions shed light on the perception of rights-based obligations and human mobility in the context of climate change. They explicitly recognize relocation as a proactive means to address climate impacts, one that can protect the right to life and that will require international assistance. The HRC also declared that climate impacts could trigger *non-refoulement* obligations, which arise from threats to the rights to life or to be free from torture or cruel, inhuman or degrading treatment.[28] The right to be free from torture or cruel, inhuman or degrading treatment is interpreted by the HRC as preventing states from exposing individuals to

[24] *Teitiota v New Zealand* (n 22) paras 9.11–9.12.

[25] *Billy v Australia* (n 9) para 8.12. For a critique of HRC's similar reasoning in *Teitiota* as racist legitimation of a different standard of living for Black, Indigenous, and people of colour, see Dylan Asafo, 'The Racism in Climate Change Law: Critiquing the Law on Climate Change-Related Displacement with Critical Race Theory' (2022) 39 Berkeley Journal of International Law 249.

[26] For example, the Committee on the Rights of the Child concluded that, for the purposes of jurisdiction, the child petitioners face reasonably foreseeable threats to their rights to life, health, and culture. *Sacchi v Argentina*, 'Decision adopted by the CRC under the Optional Protocol to the Convention on the Rights of the Child on a Communications Procedure in Respect of Communication No 104/201' (CRC, 22 September 2021) UN Doc CRC/C/88/D/104/2019, paras 10.2–10.12; see also *The State of the Netherlands v Urgenda Foundation* [2019] Case no 19/00135 (Supreme Court of the Netherlands).

[27] *Billy v Australia* (n 9) paras 6.5, 7.5.

[28] *Teitiota v New Zealand* (n 22) para 9.11.

risk of such treatment upon return to another country through their extradition, expulsion, or *refoulement*.[29] However, the time frame allowed for adaptation and intervening acts (10 to 15 years) cuts against relying on *non-refoulement* as a source of protection. Furthermore, and as Chapter 1 argues, such protection comes too late—only after people move, cross borders, and face significant harm if returned.

Finally, the timeliness of adaptation measures is critical for adequate climate action. The HRC cited the state's delay in taking certain adaptive measures, 'failure to adopt timely adequate adaptation measures', and non-performance of positive obligations to implement adequate adaptation measures as violating human rights.[30] It is also increasingly calling for a 'precautionary approach' to satisfy rights obligations.[31] The precautionary principle is an important operative principle of the climate regime and, when incorporated alongside the integration of human rights, can transform adaptation obligations. As Chapter 5 argues, the principle can help shift the scope and timing of positive duties. It mitigates against the need for certainty, bolstering an approach to adaptive mobility that is not contingent on the immediacy of harm and that aims to prevent rights violations. Instead, such an approach focuses on the foreseeability of risks and their likelihood, which enables arguments for anticipatory action. Foreseeability is more than mere speculation and requires that a state be, or should be made, aware of the risk by exercising due diligence in its risk and impact assessments. Recurring hazards, the ability to mitigate risks, and the vulnerability of communities are factors that weigh in favour of state action.[32] Action must be taken in advance of serious harm and before a violation occurs, in line with the principle and positive human rights duties.

Regional bodies and domestic courts have made similar use of human rights obligations and the precautionary principle when weighing the risks to people and their lives from climate change. In an advisory opinion, the Inter-American Court of Human Rights (IACtHR) explained that to respect and ensure the rights to life and personal integrity, states must take preventative action and mitigate environmental damage in accordance with the precautionary principle. Action is required where 'plausible indications' point to 'severe and irreversible damage to the environment' irrespective of scientific certainty. Like the HRC and regional courts, the IACtHR linked positive obligations associated with the right to life to the accessibility and quality of water, food, and health. It called on states to act with due

[29] HRC, 'General Comment No 20: Article 7 (Prohibition of Torture, or Other Cruel, Inhuman or Degrading Treatment or Punishment)' (10 March 1992) UN Doc HRI/GEN/1/Rev.9 (Vol. I) 200 (2008) para 9; see also Office for the UN High Commissioner for Human Rights (OHCHR), 'Key Messages on Human Rights, Climate Change, and Migration' <www.ohchr.org/Documents/Issues/ClimateChange/Key_Messages_HR_CC_Migration.pdf> accessed 17 May 2017.

[30] *Billy v Australia* (n 9) para 8.14.

[31] HRC, 'General Comment No 36' (n 3) para 62; *Billy v Australia* (n 9) Annexes I, III (concurring and partially dissenting opinions).

[32] Chapter 6; see *Budayeva* (n 11) [137]; Kristian Cedervall Lauta and Jens Elo Rytter, 'A Landslide on a Mudslide? Natural Hazards and the Right to Life under the European Convention on Human Rights' (2016) 7 Journal of Human Rights and the Environment 111, 117; Burson and others (n 12) 385.

diligence to prevent harm, tying this duty to precaution and to procedural rights and obligations.[33] Domestically, the *Urgenda* case marked the first successful suit brought by citizens that invoked human rights obligations to hold their government accountable for not doing enough to address climate change. In its decision, the Supreme Court of the Netherlands noted that although the risks posed by climate change 'may only materialise in the longer term' and 'will only be realized in a few decades', the right to life still offers protection in the present, 'in line with the precautionary principle'.[34]

In summary, when the impacts of climate change pose a foreseeable threat to life or a life with dignity, a duty to act in advance of significant harm is triggered. This duty exists even before human rights are violated, irrespective of whether the state is the cause of the environmental harm.[35] When integrated into adaptation obligations, positive duties require that individuals can access a means to survive that avoids significant environment risks. This is due, in part, to the developing scope and nature of the right, as identified in this section: action must be taken to address the general conditions that give rise to threats of this kind, and such a threat need be only reasonably foreseeable—even over longer timescales—rather than imminent. Doing so translates into action to plan for and implement adaptation that is not tied to a fixed time frame but that itself anticipates the changing nature of climate risks. As a result, there is no need to rely on post-hoc humanitarian interventions or responsive measures alone, and adaptation programmes can instead include proactive, anticipatory planning and action, including action to enable adaptive mobility.

III. Integrating the right to an adequate standard of living

In 2018, the United Nations Committee on Economic, Social and Cultural Rights (CESCR) warned that 'climate change constitutes a massive threat to the enjoyment of economic, social and cultural rights', and in particular to rights associated with an adequate standard of living, which will be affected 'at an increasing pace in the future'.[36] The right to an adequate standard of living is not defined in any international instrument. Instead, it is elaborated through its constituent rights: the rights to food, water, housing, and health. The right to health is a separate right in

[33] Advisory Opinion OC-23/17, Requested by the Republic of Colombia (IACtHR, 15 November 2017) [123]–[125], [142], [175]–[180].

[34] *Urgenda* (n 26) [5.2.2], [5.4.2], [5.6.2].

[35] See John H Knox, 'Bringing Human Rights to Bear on Climate Change' (2019) 9 Climate Law 171 (human rights bodies have clarified that states' obligations create this duty to protect).

[36] CESCR, 'Climate Change and the International Covenant on Economic, Social and Cultural Rights: Statement of the Committee on Economic, Social and Cultural Rights' (8 October 2018) <www.ohchr.org/en/NewsEvents/Pages/DisplayNews.aspx?NewsID=23691&LangID=E> accessed 20 September 2019.

the International Covenant on Economic, Social and Cultural Rights (ICESCR), but it overlaps and is inclusive of the other constituent rights.[37]

Human rights are interdependent, and these constituent rights are critical to life and the determinants of a life with dignity.[38] They are necessary for subsistence and survival, and they are central to the ability and decision to stay or move from a place of environmental change. Like other rights in the ICESCR, these rights are subject to progressive realization, with each state party required to take steps 'to the maximum of its available resources'.[39] States also have core obligations that are of immediate effect and must be maintained, even in circumstances where limitations to rights are otherwise permissible. As the CESCR explains, states are obliged 'to ensure the satisfaction of, at the very least, minimum essential levels of each of the rights'.[40] What this entails is developed in CESCR guidance and derived from a reading of the rights that gives them purpose and efficacy.

The risks from climate change to the enjoyment of rights are well-established. When people are unable to access resources needed to secure 'minimum essential levels' of a right, a state's positive duties are triggered. It is in these situations where the integration of relevant human rights incorporates core obligations. These obligations provide the basic content of the right; the positive duties they require are the least that must be integrated into the interpretation of adaptation obligations, to help guide and shape their implementation. Integration of core obligations results in the need for proactive adaptation measures before rights violations occur and may trigger action sooner than the right to life alone. Within the work of the climate regime, the relevance of these rights to adaptation efforts is clear. Each of the underlying entitlements—to adequate food, water, health, and housing—is captured by one of the seven targets for the global goal on adaptation (GGA). The aim is to meet these targets by 2030 and progressively beyond. They are also meant to be 'in alignment with other global frameworks', and parties to the Paris Agreement are accordingly encouraged to 'take into account . . . human rights approaches' in meeting them.[41] Furthermore, as these rights overlap and are interconnected, it may be the cumulative effect of risks to them that requires action, particularly for measures directed toward adaptive mobility.

[37] International Covenant on Economic, Social and Cultural Rights (adopted 16 December 1966, entered into force 3 January 1976) 993 UNTS 3 (ICESCR) arts 11, 12; see CESCR, 'General Comment No 14: The Right to the Highest Attainable Standard of Health (Art 12)' (11 August 2000) UN Doc E/C.12/2000/4, paras 3–4.

[38] See Section II; HRC, 'General Comment No 36' (n 3) para 26.

[39] ICESCR, art 2(1).

[40] CESCR, 'General Comment No 3: The Nature of States Parties' Obligations (Art 2, Para 1, of the Covenant)' (14 December 1990) UN Doc E/1991/23, para 10; see also Burson and others (n 12) 390–91.

[41] Decision 2/CMA.5, 'Global Goal on Adaptation' (13 December 2023) FCCC/PA/CMA/2023/16/Add.1, paras 9, 13; see Chapter 2, V.A.

A. The right to food

The right to adequate food is a component of an adequate standard of living that is included in several international instruments, although it is most comprehensively set out in the ICESCR.[42] Adequate food is an entitlement, owed to every person. It is necessary for survival, and thus its inclusion is considered to be 'at the core' of the right to an adequate standard of living and the dominant concern for the evolution of any state or society.[43] States must also guarantee 'the fundamental right to everyone to be free from hunger'.[44] It is likely that at least this aspect of the right is customary law, and the right in general 'has been endorsed more often and with greater unanimity and urgency than most other human rights.'[45] Adequate food is also interconnected with other human rights, and the positive duties required to ensure enjoyment of these rights. For example, the right to life is understood to require positive measures to eliminate widespread hunger and malnutrition and to provide essential food.[46] The right to food is further linked to human dignity, making it critical to the enjoyment and fulfilment of other rights.[47]

Climate change is expected to hamper nutrition through the disruption of food systems and sources, loss of livelihoods, and increases in poverty. It will put access to adequate food at risk when, for example, desertification or salinization leads to crop failure or reduces agricultural output. These impacts can be compounded by existing conditions: an estimated half of the world's 854 million hungry people live in already degraded lands, degradation that will be exacerbated by climate change.[48] Many of those at risk are rural farmers and the poor in the Global South, including in areas illustrated in the case studies, where up to 80 per cent of the world's hungry live.[49] The World Bank has estimated that a 2°C increase in temperatures would put between 100 million to 400 million more people at risk of hunger, and more extreme warming could lead to over 3 million additional deaths from malnutrition each year.[50] Furthermore, food insecurity can contribute to

[42] ICESCR, art 11(1); see also CRC, arts 24(2)(c), (e), 27(3) (adequate nutrition as component of rights to health and adequate standard of living); Convention on the Elimination of All Forms of Discrimination Against Women (adopted 18 December 1979, entered into force 3 September 1981) 1249 UNTS 13 (CEDAW) art 12(2) (right to nutrition for pregnant and lactating women).

[43] Asbjørn Eide, 'Adequate Standard of Living' in Daniel Moeckli and others (eds), *International Human Rights Law* (3rd edn, Oxford University Press 2017) 190.

[44] ICESCR, art 11(2).

[45] Philip Alston, 'International Law and the Human Right to Food' in Philip Alston and Katarina Tomaševski (eds), *The Right to Food* (Martinus Nijhoff 1984) 9; OHCHR and Food and Agriculture Organization (FAO), 'The Right to Adequate Food' (2010) Fact Sheet No 34, 9.

[46] HRC, 'General Comment No 36' (n 3) para 26. See also Section III.D (right to health).

[47] See eg *SERAC v Nigeria* (n 11) [65].

[48] UN Human Rights Council (UNHRC), 'Report of the Special Rapporteur on the Right to Food, Jean Ziegler' (10 January 2008) UN Doc A/HRC/7/5, para 51; see also CESCR, '2018 Statement' (n 36) para 4.

[49] Nadia Lambek, 'Respecting and Protecting the Right to Food: When States Must Get Out of the Kitchen' in Nadia Lambek and others (eds), *Rethinking Food Systems: Structural Challenges, New Strategies and the Law* (Springer 2014) 106.

[50] World Bank, 'World Development Report 2010 : Development and Climate Change' 5.

human mobility, which is often precarious when undertaken without adequate resources.[51]

As interpreted by the CESCR, the right to adequate food applies universally and is realized when everyone 'alone or in community with others, has physical and economic access at all times to adequate food or means for its procurement'.[52] Like the HRC, the CESCR's analysis is not binding but is highly persuasive and expected to be followed by state parties. The CESCR's explanation of what the right requires, including its core content, emphasizes availability and accessibility. Food must be available 'in a quantity and quality sufficient to satisfy the needs of individuals, free from adverse substances, and acceptable within a given culture', and it must be accessible 'in ways that are sustainable and that do not interfere with the enjoyment of other rights'. Special attention and priority for the accessibility of food may also be required for disadvantaged groups, '[v]ictims of natural disasters', and those 'living in disaster-prone areas'. Furthermore, the right obliges a state to ensure that everyone under its jurisdiction, including migrants, is provided 'access to the minimum essential food' that is sufficient, safe, nutritionally adequate, and culturally appropriate.[53] The right to be free from hunger likewise creates immediate core obligations, which requires states to mitigate or alleviate hunger, even in times of disaster, which includes climate change.[54]

Thus, when the enjoyment of the right is put at risk by the impacts of climate change, states must act to ensure that people are able to produce, cultivate, or access food—both physically and economically. Safeguarding food security is a 'fundamental priority' recognized by the climate regime, as is 'ending hunger' given the 'particular vulnerabilities of food production systems' to the adverse effects of climate change.[55] The importance of adequate food is accordingly made explicit in the targets for the GGA, one of which seeks to attain 'climate-resilient food and agricultural production', supply, and distribution, and to increase 'equitable access to adequate food and nutrition for all'.[56]

[51] See Chapters 7–9; OHCHR, 'Situation of Migrants in Transit' A/HRC/31/35 (2016) para 15; FAO and International Organization for Migration (IOM), 'Agriculture and Migration in the Context of Climate Change' (2017) <www.fao.org/3/a-i7541e.pdf> accessed 20 September 2019; OHCHR, 'The Slow Onset Effects of Climate Change and Human Rights Protection for Cross-Border Migrants' (22 March 2018) UN Doc A/HRC/37/CRP.4 (Slow Onset Study) para 37.

[52] CESCR, 'General Comment No 12: The Right to Adequate Food (Art 11 of the Covenant)' (12 May 1999) UN Doc E/C.12/1999/5, paras 1, 6.

[53] ibid paras 8, 11, 13–15.

[54] ibid para 6; OHCHR, 'Report of the Office of the United Nations High Commissioner for Human Rights on the Relationship Between Climate Change and Human Rights' (UNHRC 15 January 2009) UN Doc A/HRC/10/61, para 25.

[55] Decision 1/CMA.5, 'Outcome of the First Global Stocktake' (13 December 2023) FCCC/PA/CMA/2023/16/Add.1, preamble.

[56] Decision 2/CMA.5 (n 41) para 9(b).

B. The right to water

The right to water is recognized as implicit in the rights to life, an adequate standard of living, and the highest attainable standard of health. It is also explicitly included in several human rights instruments, although often in connection with another right.[57] The connection drawn between water and other rights in these instruments is intuitive. Like the right to food, water is widely considered a prerequisite to the enjoyment of other human rights. It is fundamental to survival, essential to secure an adequate standard of living, and 'inextricably related' to the rights to adequate food and housing.[58] As the UN General Assembly recognized in a 2010 resolution, the right to safe and clean drinking water and sanitation is a human right 'that is essential for the full enjoyment of life and all human rights'.[59] Yet its status independent of human rights treaties is debatable, with scholars analysing the right under customary international law coming to different conclusions. Some point to the right as increasingly included in national law and constitutions, while others highlight a lack of state practice sufficient to support a claim for customary status.[60] And at least one scholar argues that the core content of the right—access to water for basic needs—is a customary norm.[61]

Climate events affect water quality and availability. Drought can reduce access to water supplies, sea level rise can lead to salinization of fresh water sources, and flooding can impact the quality of water. Consequently, climate change is expected to worsen the accessibility of clean water and basic sanitation, with every degree of warming increasing risks. The widespread lack of drinking water is already considered a crisis, with approximately half of the population—around 4 billion people—subject to severe water scarcity for some part of the year, over 2 billion without access to safe drinking water, and polluted water a significant source of disease and death globally.[62] Varying predictions warn that warming of between 2°C

[57] CEDAW, art 14(2)(h); CRC, art 21(2)(c); Convention on the Rights of Persons with Disabilities (adopted 13 December 2006, entered into force 3 May 2008) 2515 UNTS 3, art 28(2)(a); see also ICESCR, arts 11, 12; CESCR, 'General Comment No 15: The Right to Water (Arts 11 and 12 of the Covenant)' (20 January 2003) UN Doc E/C.12/2002/11, paras 1, 3.

[58] CESCR, 'General Comment No 15' (n 57) para 3; UNHRC, 'Report of the Special Rapporteur on the Right to Food' (n 48) para 18.

[59] UN General Assembly (UNGA), 'The human right to water and sanitation', Resolution 64/292 (28 July 2010) UN Doc A/RES/64/292; see also CESCR, 'General Comment No 15' (n 57) para 1.

[60] See eg Daphina Misiedjan and Scott O McKenzie, 'The Human Right to Water' in James R May and Erin Daly (eds), *Human Rights and the Environment: Legality, Indivisibility, Dignity and Geography*, vol VII (Edward Elgar 2019) 341; Ndjodi Ndeunyema, 'Unmuddying the Waters: Evaluating the Legal Basis of the Human Right to Water Under Treaty Law, Customary International Law, and the General Principles of Law' (2020) 41 Michigan Journal of International Law 455, 476–78.

[61] Sara De Vido, 'The Right to Water as an International Custom: The Implications in Climate Change Adaptation Measures' (2012) 6 Carbon & Climate Law Review 221, 224–25.

[62] See MA Caretta and others, 'Water' in HO Pörtner and others (eds), *Climate Change 2022: Impacts, Adaptation and Vulnerability. Contribution of Working Group II to the Sixth Assessment Report of the Intergovernmental Panel on Climate Change* (Cambridge University Press 2022) 558; United Nations Development Programme (UNDP), 'Human Development Report 2006: Beyond Scarcity—Power, Poverty and the Global Water Crisis'.

to 4°C could lead to a doubling of people without access to safe drinking water and billions facing physical water scarcity.[63] Warming above 1.5°C is likely to test the limits of adaptation due to increasingly limited resources in places that depend on water, including small island states. Water-related risks, as the case studies explore, will also be exacerbated by socioeconomic contexts; this includes a lack of adequate urban or hydrological planning.[64] Hydrological changes, through both slow and rapid onset events, are also affecting human mobility. Floods and droughts can adversely affect agriculture, which can either increase or decrease the risk of human mobility depending on the socioeconomic, political, cultural, and environmental context. Overall, the Internal Displacement Monitoring Centre (IDMC) estimates that floods and droughts contribute to an average 12 million new displacements annually, although the data on slow-onset events is still lacking.[65] Millions more are predicted to be displaced due to water-related risks from climate change, although mobility is increasingly considered an adaptation strategy.[66]

The importance of water and its protection as a right is an example of where the development of environmental and human rights law converge. This is illustrated by the CESCR, which cites the right to water in environmental instruments to provide guidance on the human right.[67] Its general comment is the most robust and thorough interpretation of a stand-alone right to water. It lays out a non-exhaustive list of several core obligations of immediate effect, incumbent on all state parties regardless of capacity. These core obligations are aimed at access, availability, and quality and include ensuring access to a minimum essential amount of water, on a non-discriminatory basis and for marginalized groups in particular; physical and equitable access to safe and regular water services; adoption and implementation of a national water strategy; access to adequate sanitation; and monitoring realization of the right. As with the right to food, special attention should be given to groups or individuals facing difficulties exercising their right, namely women, children, Indigenous peoples, displaced persons and refugees, victims of disaster, and persons living in disaster-prone areas.[68]

In addition to core obligations, the CESCR encourages strategies and programmes to ensure safe and sufficient water, both now and for future generations. These strategies and programmes can include assessing the impacts of, amongst other things, climate change and desertification.[69] They are also consistent with

[63] See Slow Onset Study (n 51) para 39; Caretta and others (n 62) 620.

[64] See Chapters 7–9; Caretta and others (n 62) 557, 649; Pedro Arrojo Agudo, 'Special Thematic Report on Climate Change and the Human Rights to Water and Sanitation by the Special Rapporteur on the Human Rights to Safe Drinking Water and Sanitation: Outlining the Impacts of Climate Change on the Human Rights to Water and Sanitation around the World' (OHCHR 2022).

[65] See IDMC, '2020 Global Report on Internal Displacement' (2020); Caretta and others (n 62) 593–94.

[66] Viviane Clement and others, 'Groundswell Part 2: Acting on Internal Climate Migration' (World Bank 2021); Caretta and others (n 62) 622.

[67] CESCR, 'General Comment No 15' (n 57) para 4.

[68] ibid paras 16, 25, 37.

[69] ibid para 28.

the obligation to plan for adaptation. Water-related adaptation measures account for over half of all adaptation responses in the last decade and feature in most states' reporting to the climate regime, through Nationally Determined Contribution (NDC) pledges and adaptation plans.[70] Water is key to the Paris Agreement's GGA, included in one of its targets to significantly reduce water scarcity and work towards 'a climate-resilient water supply, climate-resilient sanitation and access to safe and affordable potable water for all'.[71] Even prior to the adoption of the Paris Agreement, such adaptation could be tied to rights-based approaches to and support for adaptation measures.[72] In the lead-up to the Agreement, parties to the climate change regime highlighted human rights in general, and the rights to food and water specifically, in their calls for climate action.[73] While the negative aspects of human rights are necessary for climate action—to not take adaptation actions that violate rights—the positive actions states must take to ensure the enjoyment of rights directly affect the access to and quality of water. As the IACtHR advised, states must take positive measures to guarantee the essential minimum of water and food for all, they must provide access for those unable to do so, and they must disseminate information on how to use and protect these resources.[74] The latter connects procedural and substantive rights, as discussed in Chapter 5.

C. The right to housing

The right to adequate housing is a third component of an adequate standard of living, which is implied or included in a number of international instruments.[75] There is little agreement or practice to indicate that the right is customary, aside from those who consider the whole of the Universal Declaration of Human Rights (UDHR) as customary international law.[76]

[70] Caretta and others (n 62) 558, 659.

[71] Decision 2/CMA.5 (n 41) para 9(a).

[72] See De Vido (n 61) 226–27; Catrina de Albuquerque, 'Climate Change and the Human Rights to Water and Sanitation' (2010) Position Paper (Special Rapporteur on the Right to Safe Drinking Water and Sanitation).

[73] See Ad-hoc Working Group on Long-term Cooperative Action, 'Ideas and Proposals on the Elements Contained in Paragraph 1 of the Bali Action Plan: Submissions from Parties (Part I)' (10 December 2008) FCCC/AWGLCA/2008/MISC.5/Add.2, 106–07; Ad-hoc Working Group on Long-term Cooperative Action, 'Additional Views on Which the Chair May Draw in Preparing Text to Facilitate Negotiations among Parties: Submissions from Parties' (30 April 2010) FCCC/AWGLCA/2010/MISC.2; Lavanya Rajamani, 'The Increasing Currency and Relevance of Rights-Based Perspectives in the International Negotiations on Climate Change' (2010) 22 Journal of Environmental Law 391, 400–04.

[74] Colombia Advisory Opinion (n 33) [121] (citing the CESCR on the rights to water and food).

[75] See ICESCR, art 11; see also CRC, art 27; CEDAW, art 14(2); International Convention on the Elimination of All Forms of Racial Discrimination (adopted 7 March 1966, entered into force 4 January 1969) 660 UNTS 195 (ICERD) art 5(e)(iii).

[76] See eg Olivier De Schutter, *International Human Rights Law: Cases, Materials, Commentary* (3rd edn, Cambridge University Press 2019) 59–68; Margaretha Wewerinke-Singh, *State Responsibility, Climate Change and Human Rights under International Law* (Hart 2019) 78; Hurst Hannum, 'The

In general, the adverse effects of climate change threaten land and human settlements, exacerbating precarious housing situations. At least 1.6 billion people are already inadequately housed, nearly all in developing countries. The largest number are in Asia, where over 529 million live in informal settlements. The proportion is also significant in sub-Saharan Africa, where around 60 per cent of urban populations live in informal settlements.[77] Those residing in low-lying and coastal areas face similar risks from sea level rise and storms that threaten to destroy coastal dwellings and inundate urban areas. The risks are particularly acute in high-density areas and marginal lands where many urban poor reside. The Special Rapporteur on adequate housing advised that those most vulnerable must be prioritized in the context of climate change, recognizing gender inequalities and the needs of children.[78] Migrants also face increased risks to accessing adequate housing at multiple points in their journey: not only do they leave behind their homes, without proper adaptation planning or implementation they are also likely to encounter poor housing conditions in transit and upon arrival. Indeed, outcomes for migrants are, in part, affected by policy and planning decisions at regional, national, and local levels that relate to housing and infrastructure.[79]

The CESCR explains that the right to adequate housing applies to everyone, and its enjoyment must not be subject to any form of discrimination. The right is comprised of both freedoms and entitlements that create negative and positive duties for states. These duties require the provision of adequate housing, interpreted to include security of tenure and protection against forced evictions, which are obligations of immediate effect; availability of facilities, materials, services, and infrastructure; and access to affordable housing. States are required to monitor the housing situation within their jurisdiction, which is another obligation of immediate effect. The right necessitates the provision of adequate space, security, privacy, and location, as well as enabling the expression of cultural identity. Its enjoyment goes beyond the provision of shelter to include the right to live in peace, security, and dignity.[80] Additionally, the right not to be subjected to arbitrary or

Status of the Universal Declaration of Human Rights in National and International Law' (2014) 25 Georgia Journal of International & Comparative Law 287, 322–35.

[77] See HO Pörtner and others (eds), 'Technical Summary' in *Climate Change 2022: Impacts, Adaptation and Vulnerability* (n 62) 53; UNHRC, 'Report of the Special Rapporteur on Adequate Housing as a Component of the Right to an Adequate Standard of Living, and on the Right to Non-Discrimination in this Context, Balakrishnan Rajagopal. A Place to Live in Dignity for All: Make Housing Affordable' (15 August 2023) UN Doc A/78/192, para 1.

[78] UNHRC, 'The Right to Adequate Housing: Report of the Special Rapporteur on Adequate Housing as a Component of the Right to an Adequate Standard of Living, and on the Right to Non-Discrimination in this Context, Raquel Rolnik' (9 August 2009) UN Doc A/64/255, paras 21, 31–32.

[79] Pörtner and others (n 77) 52; G Cissé and others, 'Health, Wellbeing and the Changing Structure of Communities' in *Climate Change 2022: Impacts, Adaptation and Vulnerability* (n 62) 1083.

[80] CESCR, 'General Comment No 4: The Right to Adequate Housing (Art. 11 (1) of the Covenant)' (13 December 1991) UN Doc E/1992/23, paras 6–8, 13; CESCR, 'General Comment No 7: The Right to Adequate Housing (Art 11.1): Forced Evictions' (20 May 1997) UN Doc E/1998/22.

unlawful interference with one's privacy, family, and home is considered a component of adequate housing.[81] Regional courts have found that states have a duty to take positive action to secure rights related to housing, including the right to privacy.[82]

Coercive, unlawful evictions are considered one of the most severe violations of the right. They also violate other human rights.[83] Protection against forced evictions includes elements that are enjoyed by both an individual and their community. Such protection has been extended where the right to housing is not explicit but is found through a combination of other rights, as is the case with the African Charter on Human and Peoples' Rights.[84] The CESCR dedicated a separate general comment to forced evictions, recognizing it as a serious violation of human rights while noting that there are some circumstances under which evictions can lawfully occur. It further observed that while unlawful evictions often occur in heavily populated areas, they can also be connected to internal displacement, forced relocation, or mass exodus.[85] These situations raise the separate but related right to be protected against arbitrary displacement, derived from international law sources that include the right to housing.[86] Arbitrary displacement and violation of the right to adequate housing can arise in circumstances where a state fails to respond to the risks created by the impacts of climate change. Some risks will require evacuation or lawful eviction. These responses are not the primary focus of this book, which seeks to help avoid emergency and post-disaster movement through adaptive mobility. Planned relocation will be analysed as a form of adaptive mobility, one generally considered a last resort that must be undertaken in compliance with human rights and in a manner that allows people to live in conditions of safety, health, nutrition, and family unity.[87] If not undertaken in

[81] See ICCPR, art 17; CESCR, 'General Comment No 4' (n 80) para 9.

[82] See eg *López Ostra v Spain* (1995) 20 EHRR 277 [51]; *Taşkın* (n 18) [113]–[119]; *Fadeyeva v Russia* (2007) 45 EHRR 10 [92]–[99].

[83] See ICCPR, art 17(1); CESCR, 'General Comment No 4' (n 80) para 18; CESCR, 'General Comment No 7' (n 80) paras 4, 8; UNCHR, 'Forced Evictions', Resolution 1993/77 in 'Report on the Forty-Ninth Session of the UNCHR' (10 March 1993) UN Doc E/CN.4/1993/122.

[84] *SERAC v Nigeria* (n 11) [59]–[63].

[85] CESCR, 'General Comment No 7' (n 80) para 5.

[86] Under international human rights law, the right to be protected against arbitrary displacement is derived from rights related to freedom of movement, choice of residence, freedom from arbitrary interference with one's home, and adequate housing. See ICESCR, art 11(1); ICCPR, arts 12(1), 17; Walter Kälin and Jörg Künzli, *The Law of International Human Rights Protection* (2nd edn, Oxford University Press 2019) 540–41. It is also included in the Kampala Convention and the Guiding Principles on Internal Displacement. African Union Convention for the Protection and Assistance of Internally Displaced Persons in Africa (adopted 23 October 2009, entered into force 6 December 2012) (Kampala Convention) arts 3(1), 4(1), (4); UNCHR, 'Guiding Principles on Internal Displacement' (11 February 1998) UN Doc E/CN.4/1998/53/Add.2, principle 6.

[87] Chapter 8; Brookings Institution, Georgetown University, and UNHCR, 'Guidance on Protecting People from Disasters and Environmental Change through Planned Relocation' (2015); Georgetown University, IOM and UN High Commissioner for Refugees (UNHCR), 'A Toolbox: Planning Relocations to Protect People from Disasters and Environmental Change' (2017).

compliance with these human rights standards, planned relocation runs the risk of unlawfully evicting or displacing people.[88]

Many aspects of the enjoyment of the right are put at risk by climate change. However, there are several that necessitate positive action by the state, including duties to ensure access to housing, provide security of tenure, and protect against forced evictions. Furthermore, as with the rights to food and water, the CESCR advises that 'disadvantaged groups', including those living in disaster-prone areas or victims of disasters, should be afforded priority in housing considerations. Such priority should apply to '[b]oth housing law and policy' to fully account for 'the special housing needs of these groups'.[89] When integrated into adaptation obligations, these duties help clarify that particularly vulnerable individuals and communities should be prioritized in adaptation measures involving housing. At the same time, the targets of the GGA aim to increase the resilience of human settlements and infrastructure to 'ensure basic and continuous essential services for all'. In some circumstances, this may mean that people will need to relocate, most often within borders.[90] However, international migration and relocation are currently being contemplated in places where remaining in one's home country is becoming increasingly infeasible.[91]

D. The right to health

Originally an aspect of an adequate standard of living in the UDHR, the right to health is enshrined in a separate provision of the ICESCR and included in a number of other international human rights instruments.[92] There is currently no consensus that the stand-alone right is customary law and thus applicable to all states regardless of treaty participation, with the caveat that some consider the UDHR to reflect customary law.[93] Within the ICESCR, it is described as a right of everyone to enjoy 'the highest standard of physical and mental health'.[94] Thus, the right is not one to be healthy per se, but rather to be provided with the conditions that allow a person to lead a healthy life in accordance with their individual potential and preconditions.[95]

[88] For a conceptualization of the notion of 'planned relocations', and its relationship to other concepts of movement and displacement, see Jane McAdam and Elizabeth Ferris, 'Planned Relocations in the Context of Climate Change: Unpacking the Legal and Conceptual Issues' (2015) 4(1) Cambridge Journal of International and Comparative Law 137.

[89] CESCR, 'General Comment No 4' (n 80) para 8(e).

[90] Decision 2/CMA.5 (n 41) para 9(e).

[91] This includes some low-lying small island states, as discussed in Chapter 8.

[92] UDHR (adopted 10 December 1948) UNGA Res 217 A(III) art 25; ICESCR, art 12; see CEDAW, arts 12, 14; CRC, art 24; ICERD, art 5(e)(iv).

[93] See n 76.

[94] ICESCR, art 12(1).

[95] CESCR, 'General Comment No 4' (n 80) paras 4, 8–9.

The determinants of health are inextricably linked to environmental and social conditions, and with the enjoyment of other component rights of an adequate standard of living. Climate change poses a serious threat to health by putting clean air and adequate food, water, and housing at risk. Concern for human health is included in climate treaties. Deleterious effects on health are explicitly named as an adverse effect of climate change under the UNFCCC, and the right to health is highlighted by the preamble of the Paris Agreement.[96] Increases in disease and health issues are linked to changing environmental conditions, which climate impacts will continue to worsen. For example, extreme heat events, wildfires, and storms increase the risks of respiratory issues, asthma attacks, and heart conditions. Warming temperatures allow mosquitos and ticks to thrive and broaden their area of impact, carrying vector-borne diseases further.[97] This could mean that by the end of the century, for example, dengue alone will put the health of billions of people at risk.[98] Foodborne illnesses are on the rise, and frequent flooding and changes in rainfall patterns increase the occurrence of waterborne diseases, including cholera and gastrointestinal infections.[99] The adverse effects of climate change can also constrain access to food, through loss of livelihoods, reduced crop yields, and difficulty accessing clean water. These constraints, in turn, can have a detrimental effect on nutrition and put more people at risk of hunger and malnutrition.[100] Mental health can likewise be adversely affected by climate events, cumulative impacts, and the anticipation of climate change. Anxiety, grief, or depression can accompany threats to or loss of homes, livelihoods, resources, and cultural connections. These mental health challenges will be particularly acute for those already living in vulnerable situations, including refugees and forced migrants.[101]

The effects of climate change will also manifest through impacts on governments, their polices, and their systems of administration. The latter includes public health systems, which climate change can disrupt. Sudden-onset events can overwhelm healthcare capacity, and slow-onset impacts can undermine infrastructure and create access and quality issues over time. Developing countries that are particularly vulnerable to climate change may not have the resources necessary to adequately adapt, with constraints tied to historical and systemic issues that arise from colonialism and resource expropriation.[102] Vulnerable countries in Africa and South Asia already face significant health burdens due to climate-related

[96] United Nations Framework Convention on Climate Change (adopted 9 May 1992, entered into force 21 March 1994) 1771 UNTS 107 (UNFCCC) art 1(1); Paris Agreement (adopted 12 December 2015, entered into force 4 November 2016) 3156 UNTS 79, preamble.

[97] OHCHR, 'Analytical Study on the Impacts of Climate Change and the Human Right of Everyone to the Enjoyment of the Highest Attainable Standard of Physical and Mental Health, Report of OHCHR' (Study on Climate Change and the Right to Health) (6 May 2016) UN Doc A/HRC/32/23, paras 18–19.

[98] Cissé and others (n 79) 1094.

[99] IPCC, 'Summary for Policymakers' in *Climate Change 2022: Impacts, Adaptation and Vulnerability* (n 62) 11.

[100] See n 50 and accompanying text.

[101] Cissé and others (n 79) 1078, 86, 98.

[102] Chapter 1, II.B; IPCC, 'Summary for Policymakers, Climate Change 2023' (n 5) 31.

diseases, which compound impacts to people from malnutrition, vector-borne diseases, and diarrhoea linked to poor water and food security.[103] In addition, adverse health effects can be created by government actions that respond to climate change, including mitigation and adaptation measures. The proximity of people to energy infrastructure, relocation of communities, or shifts in land use can expose people to risks. Depending on how such efforts are implemented, they can generate inequities by protecting some areas and not others, thereby improving the resilience of some people while not accounting for the differential vulnerabilities others experience.

As Chapter 1 highlights, the risks posed by climate change, including those to health, are not distributed equally. Those already marginalized are often more vulnerable and consequently face greater risk of harm from the impacts of climate change. Children and the elderly, for example, are at greater risk of heat-related illnesses or impacts, waterborne diseases, and sensitivity to climate hazards given their physiology or changing immune systems. Women and girls face differential exposure to risks that stem from gendered livelihoods or mobility options. Likewise, Indigenous peoples face disproportionate risks due to their dependence upon and ties to land, the environment, and its resources. These include differences in health risks, spiritual well-being, and social ties that are connected to land and culture.[104] The effects of climate change on resources and health are also significant drivers of human mobility. Migrants, particularly those who undertake precarious journeys, face increased health risks that can result from reduced access to healthcare facilities and services; loss of support networks and assets; exclusion from decision-making processes; and difficulty obtaining food, water, and resources.[105]

Like other economic, social, and cultural rights, the right to health creates negative and positive duties. In general, this means that a state must not interfere with the enjoyment of the right, and it must provide equal access to health services without discrimination and irrespective of migration status. The right is also interpreted to include access to the underlying determinants of health—other component rights of an adequate standard of living—as well as participation of people in health-related decision-making. These component rights, as well as other rights, including the rights to life and freedom of movement, 'address integral components of the right to health'.[106] While progressive realization of the right to health is still a goal for many developing countries, the core obligations of the right are considered non-derogable. These core obligations apply to those affected by climate change and include ensuring access to health facilities and services on a

[103] World Bank, 'World Development Report 2010: Development and Climate Change' 95.
[104] Cissé and others (n 79) 1053–58.
[105] ibid 1052; OHCHR, 'Study on Climate Change and the Right to Health' (n 97) para 28.
[106] CESCR, 'General Comment No 14' (n 37) paras 3, 11, 33–34.

non-discriminatory basis, especially for vulnerable groups; minimum essential food; basic shelter, housing, and sanitation; an adequate supply of potable water; an equitable distribution of health facilities and services; and a national public health strategy.[107] The relevance of these core obligations is affirmed by the targets under the GGA, which include those aimed at potable water, access to adequate food, resilience against climate-related health impacts, and resilient human settlements and health services. The health-specific target prioritises 'the most vulnerable communities' and seeks to reduce morbidity and mortality through adaptive action.[108] Finally, and important in the context of climate change, realization of the right includes steps to prevent, treat, and control disease.[109]

E. Duties, priorities, and concrete adaptive action

As this section has established, the impacts of climate change threaten the enjoyment of the right to an adequate standard of living, and the underlying entitlements the right guarantees—adequate food, water, housing, and health—are relevant to adaptation measures that respond to these impacts and the risks they pose. The integration of the right into the interpretation of adaptation obligations occurs by virtue of this relevance and its applicability to the parties to climate treaties.[110] Some of the core obligations of its constituent rights are considered customary law. These rights have also been accepted in some form by nearly every state, including the states discussed in the case studies. For example, while there are 172 state parties to the ICESCR, there is near universal participation in the CRC and wide participation in CEDAW.[111]

Although the focus is on proactive measures, both the negative and positive duties arising from these rights are relevant for integration purposes. The prohibition against forced eviction, for example, requires states to refrain from moving people without adequate safeguards and conditions. Such a negative duty is crucial for any efforts to relocate people. Likewise, when the risks from climate impacts

[107] ibid paras 43, 47.
[108] Decision 2/CMA.5 (n 41) para 9(a)–(c), (e).
[109] ICESCR, art 12(2)(c).
[110] Chapter 3.
[111] Several Pacific Island states have not ratified ICESCR but are party to CEDAW and CRC. See 'Status of ICESCR' <https://treaties.un.org/Pages/ViewDetails.aspx?src=IND&mtdsg_no=IV-3&chapter=4&clang=_en> accessed 14 March 2024. Some countries—like the United States (US)—signed but did not ratify the treaty, meaning they have an obligation to refrain from acts that defeat the object and purpose of ICESCR. Vienna Convention on the Law of Treaties (adopted 23 May 1969, entered into force 27 January 1980) 1155 UNTS 331, art 18 (an obligation also reflected in customary international law). 'Status of CRC' <https://treaties.un.org/Pages/ViewDetails.aspx?src=IND&mtdsg_no=IV-11&chapter=4&clang=_en> accessed 14 March 2024 (only US not a party); 'Status of CEDAW' <https://treaties.un.org/Pages/ViewDetails.aspx?src=IND&mtdsg_no=IV-8&chapter=4&lang=en> accessed 14 March 2024 (only eight states are not parties, including US). However, the US is a party to ICERD, which includes rights to health and housing.

threaten access to the resources needed to secure 'minimum essential levels' of a right, a state's positive duties are relevant. The integration of these duties translates into adaptation plans and measures that ensure continued access to essential resources. This would require, for example, that states take adaptive action to ensure people can produce or access food when the enjoyment of the right is at risk. Similarly, integration of the right to adequate housing requires states to take positive measures to plan for and provide housing that is affordable, accessible, protects people from hazards, and is culturally appropriate. This may mean enabling people to move. Furthermore, given the cumulative risks that climate change creates, and the overlapping and interconnected nature of human rights, it is likely that multiple duties will be relevant and help shape adaptation plans and programmes. And if a state is unable to meet its core duties to provide basic resources it should, and in some cases it must, request international assistance to do so.[112]

The integration of human rights also helps establish priorities in adaptation. The CESCR and other human rights bodies have repeatedly stated that priority must be given to vulnerable, disadvantaged, or marginalized persons or groups, including in providing essential goods and services. As Chapter 1 explains, this adds another layer to the analysis, one that integrates concern for those most vulnerable into adaptation obligations that already prioritize vulnerable developing countries. Such a focus accords with developments in the climate regime. The Paris Agreement recognizes the rights of people in vulnerable situations and calls for the consideration of vulnerable groups in adaptation.[113] Yet people are more than the vulnerabilities they face, and it is important to treat individuals as 'essential partners in the local, national and international efforts to tackle climate change'.[114] To this end, Chapter 5 explores the procedural rights that must be incorporated into action on adaptation.

Finally, and given that core rights obligations are of immediate effect, a risk to the enjoyment of a right need not be imminent to prompt state action. Like the right to life, the standard by which to assess risks should instead focus on knowledge and reasonable foreseeability: if there is a real risk of serious harm, even one caused by the longer-term consequences of climate change, then positive duties are relevant and triggered. The consideration of human rights in advance of harm can be both necessary and prudent. As the Special Rapporteur on adequate housing

[112] See CESCR, 'General Comment No 12' (n 52) para 17; CESCR, 'General Comment No 4' (n 80) para 10; CESCR, 'General Comment No 15' (n 57) para 34.

[113] Paris Agreement, preamble, art 7(5).

[114] OHCHR, 'Five UN Human Rights Treaty Bodies Issue a Joint Statement on Human Rights and Climate Change.' (19 September 2019) <www.ohchr.org/EN/NewsEvents/Pages/DisplayNews.aspx?NewsID=24998> accessed 20 September 2019; see also Committee on the Elimination of Discrimination against Women, 'General Recommendation No 37 (2018) on the Gender-Related Dimensions of Disaster Risk Reduction in the Context of Climate Change' (18 March 2018) UN Doc CEDAW/C/GC/37, paras 7–8.

has highlighted, in disaster situations human rights are often not adequately accounted for, leading to unintended rights violations:

> Sometimes [violations] result from insufficient resources and capacities to prepare and respond to the consequences of the disasters. More often, they are the result of inappropriate policies, neglect or oversight. These violations could be avoided if both national and international actors took the relevant human rights guarantees into account *from the beginning.*[115]

Integration of relevant duties from human rights into adaptation obligations, and throughout the process of adaptation, can facilitate consideration of rights at the outset in adaptation planning and during the implementation of adaptation measures. In doing so, and by taking more proactive measures, the need for emergency action can be better avoided. This reduces the likelihood of rights violations. It also accords with the principle of precaution,[116] as does the need to accept some degree of uncertainty, due to the prospective nature of risk assessment and the different ways that climate change might unfold. Like other contexts, however, the fact that multiple scenarios are possible does not obviate the need for anticipatory action.[117] Indeed, there can be both an obligation to act to address risks and discretion in choosing how to act, provided such action allows for continued access to resources and rights. These obligations can also be clarified through the interpretive process, rather than as part of an adjudication, allowing for human rights law to play a more pre-emptive, ex-ante role in shaping adaptation.[118]

Concretely, the integration of human rights duties can lead to a range of adaptation measures, some of which can increase resilience and adaptive capacity to allow people to remain in place. For example, when faced with rising sea levels and increased storm surges, state adaptation efforts may initially aim to address impacts by building seawalls, changing irrigation or land use policies, and offering alternative sources of drinking water. As the severity of impacts increases and land becomes less habitable, adaptive mobility measures—actions that facilitate

[115] UNHRC, 'Report of the Special Rapporteur on adequate housing as a component of the right to an adequate standard of living, and on the right to non-discrimination in this context, Raquel Rolnik' (20 December 2010) UN Doc A/HRC/16/42, para 12 (quoting UN Inter-Agency Standing Committee, emphasis added).

[116] Detailed in Chapter 5, IV.

[117] Lauren Nishimura, 'Adaptation and Anticipatory Action: Integrating Human Rights Duties into the Climate Change Regime' (2022) 12 Climate Law 99, 119; see Anderson and others (n 16) 129 ('presence of "multiple" possibilities, each of which involves speculation, does not foreclose the possibility that protection is warranted'); but see *Teitiota v New Zealand* (n 22) para 9.12 (citing potential intervening acts as means to protect); UNHCR, 'Guidelines for International Protection No 12: Claims for Refugee Status Related to Situations of Armed Conflict and Violence under Article 1A(2) of the 1951 Convention and/or 1967 Protocol Relating to the Status of Refugees and the Regional Refugee Definitions' (2016) HCR/GIP/16/12, para 25 (focus of risk assessment on foreseeability of harm alongside other contextual factors).

[118] Chapter 3, III.C.2.

migration or planned relocation—may be necessary to protect people and ensure that adequate food, water, housing, and health remain accessible.

IV. Conclusion

Climate impacts threaten a multitude of human rights. This chapter focuses on rights that are critical to subsistence, survival, and the ability of individuals to stay in or move from a place. Integrating these rights into an interpretation of obligations to act on adaptation leads to measures that not only consider human rights but also proactively ensure enjoyment of rights in the face of foreseeable risks. Integrating their positive duties further informs the kinds of adaptive actions necessary to protect rights, which at a minimum requires the provision of essential goods and resources. In some contexts, as illustrated in the case studies, addressing foreseeable risks means that states must undertake anticipatory and proactive forms of adaptation, including measures that enable adaptive mobility. Thus, integrating relevant rights into the interpretive process can help concretize adaptation obligations. It necessitates action sooner than may be required by human rights duties independently. Alongside the principle of precaution—a focus of the next chapter—such integration can shift action to an earlier point, so that states plan for and put into place the frameworks necessary to protect and ensure human rights over the long term. This can improve outcomes by providing some degree of legal certainty; once risks to rights are identified, then the means to ensure these rights can be explored. Such an approach leads to adaptation measures that better avoid risks to people and their rights, to prioritize those most vulnerable, and to ensure access to essential resources.

5
Participation and Anticipatory Action: Integrating Procedural Human Rights and the Precautionary Principle

I. Introduction

This chapter explores how procedural rights can help shape adaptation and adaptive mobility. Procedural rights include rights to participate in decision-making and to access information. Section II discusses the duties associated with these rights and their development through the interaction of human rights with environmental law. It argues that the integration of procedural rights leads to an interpretation of obligations to act on adaptation that requires countries to provide information and opportunities to participate in the process of adaptation, including decisions about adaptive mobility. This section also links the implementation of adaptation obligations to the principles of non-discrimination and equality. These human rights principles are integral to responses that both address people's vulnerability and ensure their inclusion and participation. Section III considers the role that agency, participation, and perception play in mobility outcomes. It establishes the importance of adaptive mobility measures shaped by participatory processes that provide choice, are led by those affected, and strive for consent. The chapter then turns to the principle of precaution, which is a key operative principle of the climate change regime. Section IV analyses the way that the principle affects adaptation obligations, showing how it can support adaptive action sooner than reliance on human rights alone. This step in the interpretive process is important, because the time when positive obligations must be undertaken under human rights law is not always specified or tends to be clarified after a violation occurs. However, an interpretation that both includes the precautionary principle and integrates human rights can better address the foreseeable risks posed by climate change, enabling human rights benchmarks and safeguards to be incorporated into adaptation planning and anticipatory action.

Climate Change, Human Rights, and Adaptive Mobility. Lauren Sakae Nishimura, Oxford University Press.
 DOI: 10.1093/9780198930068.003.0006

II. Integrating procedural rights

A. Procedural rights

Procedural rights draw on the combined development of international human rights and environmental law principles, establishing the role environmental law plays in broadening and shaping human rights law, and the interconnection between the two. Most human rights instruments include some form of procedural rights, enshrining various rights to participate or to access information.[1] For issues affecting the environment, the Rio Declaration on Environment and Development is more explicit. It includes a principle that provides individuals with three rights: the right to access information on the environment, the opportunity to participate in decision-making processes, and access to a remedy or redress.[2] While not legally binding, this principle has inspired regional legal developments, and participatory rights are cited by judicial bodies to help interpret human rights—including the rights to life, to health, to property, and to respect for private and family life—and to understand their procedural dimensions.[3] The Committee on Economic, Social and Cultural Rights (CESCR) likewise interprets the international human rights to water and health to include access to information and participation in decision-making. These rights create positive procedural duties: providing access to information is central to the right to health, and participation is an integral part of any policy, programme, or strategy

[1] See Universal Declaration of Human Rights (adopted 10 December 1948) UNGA Res 217 A(III) arts 8, 19–21 (rights to information, association, assembly, to receive and impart information, participate in government, and to an effective remedy); International Covenant on Civil and Political Rights (adopted 16 December 1966, entered into force 23 March 1976) 999 UNTS 171 (ICCPR) arts 2(3), 19 (right to an effective remedy, right to receive and impart information); Convention on the Elimination of All Forms of Discrimination Against Women (adopted 18 December 1979, entered into force 3 September 1981) 1249 UNTS 13 (CEDAW) art 7 (right to participate in formulation of government policy); Convention on the Rights of the Child (adopted 20 November 1989, entered into force 2 September 1990) 1577 UNTS 3 (CRC) art 13 (right to information); International Covenant on Economic, Social and Cultural Rights (adopted 16 December 1966, entered into force 3 January 1976) 993 UNTS 3 (ICESCR) art 13 (right to education must enable people to 'participate effectively in a free society').

[2] Declaration of the United Nations Conference on the Human Environment (adopted 14 June 1992) UN Doc A/CONF.151/26, vol 1 (12 August 1992) Annex I (Rio Declaration on Environment and Development) principle 10.

[3] See eg Convention on Access to Information, Public Participation in Decision-Making and Access to Justice in Environmental Matters (adopted 25 June 1998, entered into force 30 October 2001) 2161 UNTS 447 (Aarhus Convention); African Convention on the Conservation of Nature and Natural Resources (Revised) (adopted 7 March 2017, entered into force 4 February 2019); *Tătar v Romania* (Chamber judgment) App no 67021/01 (European Court of Human Rights (ECtHR), 27 January 2009); *Öneryildiz v Turkey*, App no 48939/99 (2005) 41 EHRR 20; *Maya Indigenous Community of the Toledo District v Belize* (Merits) Case 12.053, IACtHR Report No 40/04 (12 October 2014); *Social and Economic Rights Action Center (SERAC) and the Center for Economic and Social Rights (CESR) v Nigeria* [2001] ACHPR Comm no 155/96 (*SERAC v Nigeria*); discussed in Jonas Ebbesson, 'Principle 10' in Jorge E Viñuales (ed), *The Rio Declaration on Environment and Development: A Commentary* (Oxford University Press 2015) 293–94, 298–308.

concerning water.[4] Alongside their derivation from evolving human rights and environmental law, procedural rights also emerged with the increasing emphasis on and enhancement of legitimacy in decision-making. Consequently, they are both embodied in and further the application of already established human rights, to better enable fair, inclusive decision-making processes.[5]

Given their connection, Alan Boyle argues that consideration of human rights in an environmental context must account for environmental rights in other treaties. He cites the interpretive provision for systemic integration in the Vienna Convention on the Law of Treaties to argue that environmental rights, and those that provide for access to information and participation in decision-making specifically, must be considered and may be necessary in interpreting and applying human rights treaties.[6] While this book uses integration in the opposite direction, to shape climate obligations with human rights law, both examples of integration evince the synergistic relationship between the two bodies of international law. Indeed, it is the influence of environmental law on human rights that is read back into adaptation obligations, through the integration of procedural rights and accompanying positive duties. This relationship has been recognized generally in the need to apply procedural human rights to climate action, which results from a 'systemically integrated interpretation' of the rights to access information and participation found in environmental and human rights instruments.[7]

There is thus a cross-fertilization between human rights and environmental issues, one confirmed by regional cases and instruments. Regional courts and human rights bodies have found that access to information and certain procedural guarantees in decision-making are required by states to assess and respond to environmental risks. They have also found that affected persons are entitled to meaningful consultation or to participate in decision-making.[8] As Boyle concludes, decisions like these suggest that:

[4] See CESCR, 'General Comment No 14: The Right to the Highest Attainable Standard of Health (Art 12)' (11 August 2000) UN Doc E/C.12/2000/4, paras 3, 11, 12; CESCR, 'General Comment No 15: The Right to Water (Arts 11 and 12 of the Covenant)' (20 January 2003) UN Doc E/C.12/2002/11, para 48.

[5] See Ebbesson (n 3) 289; John H Knox, 'Human Rights Principles and Climate Change' in Cinnamon P Carlarne, Kevin R Gray, and Richard Tarasofsky (eds), *The Oxford Handbook of International Climate Change Law* (Oxford University Press 2016) 221–22.

[6] Alan Boyle, 'Human Rights and the Environment: Where Next?' in Ben Boer (ed), *Environmental Law Dimensions of Human Rights* (Oxford University Press 2015) 212.

[7] Bridget Lewis, 'Balancing Human Rights in Climate Policies' in Ottavio Quirico and Mouloud Boumghar (eds), *Climate Change and Human Rights: An International and Comparative Law Perspective* (Routledge 2015) 47.

[8] See n 3; *SERAC v Nigeria* (n 3) [52]–[53] (interpreting the African Charter to require 'meaningful opportunities for individuals to be heard and to participate in the development decisions affecting their communities'); *Taşkın v Turkey* App no 46117/99 (ECtHR, 10 November 2004) [118]–[119] (scrutinizing procedural safeguards and the extent individuals' views were taken into account in the decision-making process, and declaring the process must be fair, include appropriate studies that assess risks, and provide information to the public); *Claude-Reyes v Chile* (Merits, Reparations and Costs) IACtHR Series C No 151 (19 September 2006) [81]; *Maya Indigenous Community* (n 3) [154]–[155]

> [T]he most important contribution that existing human rights law has to offer with regard to environmental protection and sustainable development is the empowerment of individuals and groups affected by environmental problems, and for whom the opportunity to participate in decisions is the most useful and direct means of influencing the balance of environmental, social, and economic interests.[9]

Procedural rights are therefore crucial to the implementation of climate action. As the Office of the UN High Commissioner for Human Rights (OHCHR) put it in a 2009 report on climate change and human rights, the 'effective participation of individuals and communities in decision-making processes affecting their lives' is of 'critical importance'. It viewed the human rights framework as one that 'seeks to empower individuals', a framework that offers particular attention to vulnerable or marginalized people.[10] More recently, the Human Rights Committee's (HRC) general comment on the right to life connects measures to address climate change and environmental degradation with, amongst other things, access to information on environmental hazards.[11]

Although broadly worded, climate treaties confirm the importance of procedural rights, with text that emphasizes access to information, public participation, and cooperation. The United Nations Framework Convention on Climate Change (UNFCCC) commitments include an obligation to '[p]romote and cooperate in education, training and public awareness related to climate change and encourage the widest participation in this process'.[12] Greater specificity is provided in a dedicated article on 'education, training and public awareness', which obliges parties to 'promote and facilitate' public access to information on climate change and public participation in addressing climate effects and developing adequate responses.[13] The Conference of the Parties' subsequent work programme on this obligation, however, focused on educating and engaging stakeholders, not on ensuring

(connecting duty to consult with rights to life, family, health, and self-determination); see also *Pulp Mills on the River Uruguay (Argentina v Uruguay)* (Judgment) [2010] ICJ Rep 14 [203]–[206], [215]–[219] (environmental impact assessments may now be a requirement of general international law; parties agreed assessment should include consultation of affected persons); CESCR, 'General Comment No 15' (n 4) para 56.

[9] Boyle (n 6) 216–17.

[10] OHCHR, 'Report of the Office of the United Nations High Commissioner for Human Rights on the Relationship between Climate Change and Human Rights' (2009 Report) (UN Human Rights Council (UNHRC) 15 January 2009) UN Doc A/HRC/10/61, para 81.

[11] HRC, 'General Comment No 36: Article 6 (Right to Life)' (3 September 2019) UN Doc CCPR/C/GC/36, para 62.

[12] United Nations Framework Convention on Climate Change (adopted 9 May 1992, entered into force 21 March 1994) 1771 UNTS 107 (UNFCCC) art 4(1)(i).

[13] ibid art 6.

participation for specific climate actions or policies.[14] The Paris Agreement's preamble repeats the language of the UNFCCC, affirming the importance of 'education, training, public awareness, public participation, public access to information and cooperation at all levels'. The Agreement's provision on capacity-building encourages these procedural duties, and it includes a similar obligation to the UNFCCC, couched in softening language, that parties shall cooperate 'as appropriate, to enhance climate change education, training, public awareness, public participation and public access to information'.[15]

The two treaties are brought together in the decision adopting the Paris Agreement, which calls upon all parties to ensure that the obligations on education, training, and public awareness included in both treaties are considered in their contribution to capacity-building.[16] Like adaptation obligations, these obligations are ambiguous and do not require parties to take particular action. Yet they establish the need for participation, a need that extends to understanding what adaptation entails. As recognized in Article 7 of the Paris Agreement, adaptation should follow a participatory, fully transparent, and gender-responsive approach that considers vulnerable groups, communities, and ecosystems.[17] This is echoed by human rights bodies, which call for adaptation that is participatory, transparent, and that prioritizes people in vulnerable situations.[18]

Indeed, the need to provide information and participation is a thread that continues to bind climate treaties—from the UNFCCC to the Paris Agreement—to human rights law. As Lavanya Rajamani observed, procedural rights fit comfortably in the architecture of the Paris Agreement and will likely be 'progressively mainstreamed and expanded'. They offer an easier entry point than the integration of substantive human rights, at least for the processes necessary to establish levels of ambition for emission reductions and climate action. Following the adoption of the Agreement, this appeared to be the case.[19] The guidance developed on implementing the Paris Agreement, for example, does not mention substantive human rights but does include recommendations that parties provide

[14] See eg Decision 11/CP.8, 'New Delhi work programme on Article 6 of the Convention' (28 March 2003) FCCC/CP/2002/7/Add.1; discussed in Farhana Yamin and Joanna Depledge, *The International Climate Change Regime* (Cambridge University Press 2005) 208–11.

[15] Paris Agreement (adopted 12 December 2015, entered into force 4 November 2016) 3156 UNTS 79, arts 11(1), 12.

[16] Decision 1/CP.21, 'Adoption of the Paris Agreement' (29 January 2016) FCCC/CP/ 2015/10/ Add.1, para 82.

[17] Paris Agreement, art 7(5).

[18] See eg OHCHR, 'Key Messages on Human Rights and Climate Change' <www.ohchr.org/ Documents/Issues/ClimateChange/KeyMessages_on_HR_CC.pdf> accessed 17 May 2017; OHCHR, 'Key Messages on Human Rights, Climate Change, and Migration' <www.ohchr.org/Documents/ Issues/ClimateChange/Key_Messages_HR_CC_Migration.pdf> accessed 17 May 2017; OHCHR, 'Analytical Study on the Relationship between Climate Change and the Full and Effective Enjoyment of the Rights of the Child, Report of OHCHR' (4 May 2017) UN Doc A/HRC/35/13, para 33.

[19] Lavanya Rajamani, 'Integrating Human Rights in the Paris Climate Architecture: Contest, Context, and Consequence' (2019) 9 Climate Law 180, 191–92, 197–98.

information, involve stakeholders, and create opportunities for participation and consultation.[20] Yet the initial emphasis on participatory processes may promote further coherence between the implementation the Paris Agreement and human rights norms.[21] It could also offer a starting point to connect the two. Subsequent decisions of the parties to climate treaties reference general human rights approaches, which signals a continued engagement with rights more broadly.[22]

Furthermore, procedural rights are foundational to the enjoyment of other rights. They require participation throughout the process of adaptation, which in turn can inform and affect the integration of substantive rights. The relationship between the two kinds of rights can be understood as a 'virtuous circle', with implementation of procedural duties enabling the enjoyment of substantive rights.[23] Procedural duties to provide access to information and participation in decision-making, for example, are needed to fully realize the rights to water and health.[24] The enjoyment of cultural rights is also interpreted to require effective participation in decisions affecting members of minority communities.[25] Procedural rights are, as Section III discusses, central to securing the local input, awareness, and approval needed to produce meaningful change. They are thus a necessary component of preventive and proactive climate measures, which help achieve the purpose of adaptation in the regime: to anticipate and minimize the harmful impacts of climate change.

Nonetheless, and despite their importance, there remains some ambiguity surrounding the status of human rights in the climate regime. This lack of clarity makes arguments for their integration via systemic integration even more critical to their incorporation into adaptation. As discussed in Chapter 3, integration of human rights duties into the interpretive process is supported by the text of climate treaties, both through their inclusion in the preamble and through reference

[20] See Decision 17/CMA.1, 'Ways of enhancing the implementation of education, training, public awareness, public participation and public access to information so as to enhance actions under the Paris Agreement' (19 March 2019) FCCC/PA/CMA/2018/3/Add.2; Decision 18/CMA.1, 'Modalities, procedures and guidelines for the transparency framework for action and support referred to in Article 13 of the Paris Agreement' (19 March 2019) FCCC/PA/CMA/2018/3/Add.2.

[21] Sébastien Duyck, 'Delivering on the Paris Promises? Review of the Paris Agreement's Implementing Guidelines from a Human Rights Perspective' [2019] Climate Law 202, 223.

[22] See eg 'Glasgow–Sharm El-Sheikh work programme on the global goal on adaptation referred to in Decision 7/CMA.3 (Draft Decision -/CMA.5)' (2023) FCCC/PA/CMA/2023/L.18; Task Force on Displacement, 'Report of the Task Force on Displacement, Recommendations' (2018); Decision 2/CP.24, 'Local Communities and Indigenous Peoples Platform' (19 March 2019) FCCC/CP/2018/10/Add.1.

[23] UNHRC, 'Report of the Independent Expert on the Issue of Human Rights Obligations Relating to the Enjoyment of a Safe, Clean, Healthy and Sustainable Environment, John H Knox: Preliminary Report' (24 December 2012) UN Doc A/HRC/22/43, para 42; Naysa Ahuja and others, 'Advancing Human Rights through Environmental Rule of Law' in James R May and Erin Daly (eds), *Human Rights and the Environment: Legality, Indivisibility, Dignity and Geography*, vol VII (Edward Elgar 2019) 19.

[24] See n 4.

[25] HRC, 'General Comment No 23: Article 27 (Rights of Minorities)' (8 April 1994) UN Doc CCPR/C/21/Rev.1/Add.5, para 7.

to human rights concerns in the operative text. It is also required by virtue of the relevance of human rights to adaptation. Adaptation raises human rights concerns due to the localized, often direct effect that adaptive action has on people and their access to resources.[26] Adaptation funding, likewise, will require greater control and understanding at local levels in developing countries, thereby implicating human rights norms and principles.[27] Local impacts also make procedural rights central to the initial stages of adaptation, and thus their integration must occur early in the adaptation process. Many of these processes will involve changes to the way land is used, including shifts in agricultural practices, repurposing land, and planned relocation. In each of these cases, procedural rights provide affected persons with rights to access information, participation in decision-making, and an effective legal remedy for any harm suffered. When combined with participatory obligations in climate treaties, adaptation can also entail that states develop and disseminate information, again reflecting the convergence of human rights and environmental law, in this case through an emerging 'right to know'.[28]

Concretely, the integration of procedural rights can translate into specific actions that relate to adaptive mobility. These range from measures that enable participation in decision-making on adaptive mobility and planned relocation, to input on and the shaping of implementation and monitoring. They include access to information, the opportunity to participate in required environmental and human rights impact assessments, and participation in selecting places to move and implementing relocation plans. These measures must also prioritize the participation of vulnerable people and groups in adaptation planning, including in National Adaptation Plan (NAP) and nationally determined contribution (NDC) processes.[29] The need to undertake assessments and provide information is based, in part, on obligations derived from human rights law to ensure the protection of substantive rights.[30] The timing of these procedural measures and assessments also

[26] Chapter 2, II.

[27] Chapter 6, V; Margaux J Hall and David C Weiss, 'Avoiding Adaptation Apartheid: Climate Change Adaptation and Human Rights Law' (2012) 37 The Yale Journal of International Law 309, 321–22.

[28] A right identified in Rebecca Bratspies and Sarah Lamdan, 'The Human Right to Environmental Information' in *Human Rights and the Environment: Legality, Indivisibility, Dignity and Geography* (n 23) 141.

[29] Kayly Ober, Sam Huckstep, and Sarah Miller, '"It's Time for Us to Be Included": An Assessment of Refugee and Displaced People's Participation in National Adaptation Planning' (Refugees International 2023) <www.refugeesinternational.org/reports-briefs/its-time-for-us-to-be-included-an-assessment-of-refugee-and-displaced-peoples-participation-in-national-adaptation-planning/> accessed 17 October 2023 (arguing that planning should include and account for displaced persons and refugees).

[30] See n 8; UNHRC, 'Report of the Independent Expert on the Issue of Human Rights Obligations Relating to the Enjoyment of a Safe, Clean, Healthy and Sustainable Environment, John H Knox: Mapping Report' (30 December 2013) UN Doc A/HRC/25/53, paras 29–35; *Taşkın case* (n 8) [119] (studies and assessments allow states 'to predict and evaluate in advance the effects of those activities which might damage the environment and infringe individuals' rights') (citations omitted); see also *Öneryildiz case* (n 3) [90] (right to information tied to protection of right to life); *Budayeva v Russia* App nos 15339/02, 21166/02, 20058/02, 11673/02, and 15343/02 (ECtHR, 20 March 2008) [132]

matters. They must necessarily occur before adaptation policies are put in place to identify vulnerabilities, risks to rights, the impact of in situ adaptation options, and whether mobility will needed. The knowledge they produce helps flesh out a state's obligation to respond to those climate impacts that pose a threat to life and an adequate standard of living. They also facilitate planning for and implementation of mobility that is proactive rather than reactive, which can better ensure that people are able to move to a suitable, more sustainable location rather than simply wherever is most accessible.

B. Principles of non-discrimination and equality

In addition to the substantive and procedural human rights analysed, two overarching human rights principles are relevant to adaptive mobility: the principles of equality and non-discrimination. That some experience greater risks to their rights—whether based on structural, personal, or intersectional causes—makes these principles integral to adaptation efforts that account for vulnerability.[31] They underlie states' obligations and individuals' exercise of all rights under international human rights law.[32] This includes procedural rights, with several human rights instruments connecting rights to access information and participation with eliminating discrimination, providing equal opportunities to vote, and ensuring access to education on environmental issues.[33] States are 'legally bound' to address vulnerabilities in accordance with the principles.[34] Such an obligation supports and mandates state action to address the risks to those most vulnerable to climate change, who also are often already marginalized.

For adaptation obligations, inclusion of these principles bars state action that discriminates. Such a prohibition does not mean, however, that action cannot be directed at or benefit certain groups or individuals. There is a distributive aspect to equality under human rights law: substantive equality requires not only that a state refrain from treating individuals differently who are in similar situations, but also that it treats differently those whose situations make accessing rights difficult.[35] Consequently, a state must account for differences in circumstances while

(same); *SERAC v Nigeria* (n 3) [53] (right to information and requirement to undertake impact studies linked to right to health and satisfactory environment).

[31] Chapter 1, II.B.

[32] See ICCPR, preamble, arts 2, 26; ICESCR, preamble, arts 2(2), 3.

[33] CEDAW, art 7; International Convention on the Elimination of All Forms of Racial Discrimination (adopted 7 March 1966, entered into force 4 January 1969) 660 UNTS 195 (ICERD) art 5; CRC, arts 24(2)(e), 29(1)(e); see also OHCHR, '2009 Report' (n 10) para 81.

[34] OHCHR, '2009 Report' (n 10) para 42; see also OHCHR, 'Key Messages on Human Rights and Climate Change' (n 18).

[35] HRC, 'General Comment No 18: Non-discrimination' (10 November 1989) UN Doc HRI/GEN/1/Rev.1 26 (1994) para 10.

also ensuring both an individual's dignity and their connection to community. It must do so in a manner that promotes respect for equal dignity, individual differences, and, ultimately, that empowers people and enables their choice and participation.[36] In practice, these principles could mean, for example, that when considering the feasibility or costs of planned relocation, the burden should not be disproportionately borne by those most vulnerable or already marginalized. They also require decision-making processes that first consider who is involved and how their preferences are incorporated, to ensure that people with barriers to participation are informed and included. Likewise, these principles necessitate consideration of the risks migrants encounter, as well as those who may be unable to move due to their vulnerabilities.[37] Integrating these principles could therefore lead to adaptation that affirmatively addresses vulnerability and prioritizes vulnerable individuals and groups in measures to facilitate adaptive mobility. The need to foreground vulnerabilities from the outset is essential, in part because many of those likely to move in the context of climate change have faced 'systemic impoverishment and injustice' while contributing little to its causes.[38]

III. Agency, choice, and mobility rights

Procedural rights require processes that ensure people are informed and can participate in decision-making. These processes can reduce potential conflicts between adaptation plans and human rights. They can also help provide agency for people faced with choices affecting their rights and interests. Participation is a right, and it can lead to greater consensus when affected persons' opinions are considered. However, current international law does not require their consent, which can lead to decisions that are not agreed to by individuals or their communities.

There are normative frameworks and guidance that require consent as a prerequisite to certain state-involved activities, including planned relocation. Free, prior and informed consent (FPIC) is required, for example, by the UN Declaration on the Rights of Indigenous Peoples for decisions affecting the use of Indigenous land and any planned relocation involving Indigenous peoples. Consent is also called for by the Guiding Principles on Internal Displacement, the Nansen Principles, the Peninsula Principles on Climate Displacement, the Sydney Declaration of Principles on the Protection of Persons Displaced in the Context of Sea-Level Rise, and guidance from the International Federation of Red Cross and

[36] Sandra Fredman, 'Redistribution and Recognition: Reconciling Inequalities' (2007) 23 South African Journal on Human Rights 214, 225–27.

[37] See OHCHR, 'Key Messages on Human Rights, Climate Change, and Migration' (n 18).

[38] Julie Koppel Maldonado and others, 'The Impact of Climate Change on Tribal Communities in the US: Displacement, Relocation, and Human Rights' [2013] Climate Change 601, 602.

Red Crescent Societies (IFRC).[39] But none of these instruments is legally binding and, consequently, they do not create any legal commitments for states.

While not legally required, seeking informed consent can lead to better planning processes, particularly for planned relocation. These processes include participation and consultation, which are precursors to consent that fall on a 'participation spectrum' identified by Jane McAdam and Elizabeth Ferris. This spectrum moves from passive participation or information sharing, to soliciting and listening to opinions, to some or full control over decision-making.[40] Yet people do not have uniform preferences and needs, within and across communities. Nor are communities themselves static or homogenous. For these reasons, Erica Bower, Rachel Harrington-Abrams, and Betsy Priem seek a more nuanced understanding of community engagement in climate-related planned relocation. They argue that policymakers, practitioners, and those facilitating relocation should be focused on answering three questions: who constitutes the community, who facilitates planned relocation, and what is meaningful engagement? Answers to these questions will vary across contexts, but interrogating how decisions are made and who makes them can contribute to more inclusive participation throughout the process of relocation.[41] In practice, this can enable explicit consideration of the roles of various actors and how inequities will be addressed, so that a range of affected persons are engaged. In addition to policymakers and relocation facilitators, planning processes could therefore include people who are moving, who choose to remain behind, and who live in or near sites of relocation. Furthermore, and in line with the polycentric nature of adaptation planning and implementation, participatory processes are shaped by subnational actors and intermediaries that bridge external and internal groups in a relocation process.[42] Bringing this back to the book's focus

[39] UN General Assembly (UNGA) Res 61/295 (13 September 2007) 'United Nations Declaration on the Rights of Indigenous Peoples' UN Doc A/Res/61/295, arts 10, 32; see also United Nations Commission on Human Rights (UNCHR), 'Guiding Principles on Internal Displacement' (11 February 1998) UN Doc E/CN.4/1998/53/Add.2, principle 7 (must seek FPIC for those to be displaced); 'Nansen Conference on Climate Change and Displacement in the 21st Century: Chairperson's Summary and Nansen Principles' (2011) principle X (implement policies and responses—including planned relocation–on the basis of *inter alia* participation and consent of those affected); 'The Peninsula Principles on Climate Displacement within States' (Displacement Solutions 2013) principle 10 (aside from exceptional situations relocation must include full, informed consent); ILA, 'Sydney Declaration of Principles on the Protection of Persons Displaced in the Context of Sea-Level Rise' (2018) Resolution 6/2018, principle 6 <https://disasterdisplacement.org/wp-content/uploads/2019/06/ILA-Sydney-Declaration.pdf> accessed 14 September 2019 (unless needed to protect life or safety, planned relocation must be undertaken upon request of community with full, free, informed consent); IFRC, 'Planned Relocation in the Context of Disasters and Climate Change: A Guide for Asia Pacific National Societies' (2021) 26–27 (supporting consent-based planned relocation).

[40] Jane McAdam and Elizabeth Ferris, 'Planned Relocations in the Context of Climate Change: Unpacking the Legal and Conceptual Issues' (2015) 4 Cambridge Journal of International and Comparative Law 137, 146–47.

[41] See Erica Bower, Rachel Harrington-Abrams, and Betsy Priem, 'Complicating "Community" Engagement: Reckoning with an Elusive Concept in Climate-Related Planned Relocation' (2024) 88 Global Environmental Change 102913.

[42] ibid 4–5; see also Chapter 2, II.

on state obligations, adaptation planning should involve national processes that remain attuned to the multiple levels of governance involved, the complexity of communities, and the power dynamics at a local level. A state's procedural legal duties might be met by something less. But without an intentional, people-centred approach, it is more likely that perfunctory participatory processes will result. Such processes can, in turn, exclude critical perspectives and lead to relocations that run afoul of obligations to ensure substantive human rights.

Lessons learned from past relocations also indicate that people's perceptions are equally important to the processes put in place. These relocations show that access to information and participatory processes are central not only to how people move, but also to community buy-in and perceptions over the longer term. Jane McAdam explored past relocations in the Pacific and found that the perceived absence of agency, irrespective of historically accuracy, can be key to the enduring understanding and narratives of relocated communities.[43] Accordingly, at a minimum, those affected should be provided with information about the reasons for moving and way in which relocation will be undertaken. They must also actively participate in the process, and they should not view movement as forced. Meaningful, inclusive participation must also occur at multiple points in the process, from planning and implementation to monitoring after movement. Consent can remain an aim, but it implies a primacy and deference to external control that does not bestow a full sense of empowerment. Indeed, planned relocations are more likely to result in positive outcomes if they involve high levels of engagement or are led by community members.[44] A more promising approach might frame decision-making in terms of a right to self-determination, while recognizing that people's needs and perceptions are not uniform. This would enable processes that allow for collective decision-making on how to proceed in the face of climate risks, which may or may not include migration. More generally, research indicates that the degree of agency under which people make migration decisions and move are strong determinants of whether well-being is improved. The more agency, the greater likelihood of positive outcomes for migrants, their networks, and receiving communities. Likewise, the more choice and range of adaptation options available, the more likely it is that when mobility is undertaken, it occurs with greater agency and is consequently more beneficial.[45]

[43] See Jane McAdam, 'Historical Cross-Border Relocations in the Pacific: Lessons for Planned Relocations in the Context of Climate Change' (2014) 49 The Journal of Pacific History 301; see also Robin Bronen, 'Climate-Induced Community Relocations: Creating an Adaptive Governance Framework Based in Human Rights Doctrine' (2011) 35 NYU Review of Law and Social Change 357; McAdam and Ferris (n 40).

[44] See eg Erica R Bower and others, 'Enabling Pathways for Sustainable Livelihoods in Planned Relocation' (2023) 13 Nature Climate Change 919.

[45] G Cissé and others, 'Health, Wellbeing and the Changing Structure of Communities' in HO Pörtner and others (eds), *Climate Change 2022: Impacts, Adaptation and Vulnerability. Contribution of Working Group II to the Sixth Assessment Report of the Intergovernmental Panel on Climate Change* (Cambridge University Press 2022) 1116.

Choice raises further questions for the human rights framework. It connects with the aims of equality and its redistributive aspect. Sandra Fredman argues that substantive equality should seek to 'break the cycle of disadvantage' for those in certain status-based groups, to facilitate their genuine choice. She also links equality to a person's full participation in society, which includes participating in decisions that affect their lives.[46] Yet choice is context dependent. It does not guarantee an outcome, or that it can help overcome systemic barriers that limit feasible options. In some cases, then, enabling people to exercise choice can lead to a range of different viable strategies to adapt to climate change. But in others, the context may necessitate accommodations to account for vulnerability and marginalization, which are influenced by social, economic, or cultural conditions. Different conditions and preferences can lead to individual choices that are in tension with a state's obligation to ensure substantive rights. Here, questions about the limits of legal arguments are instructive. Can a state intervene if, for example, a person chooses to stay in a place where their life is foreseeably put at risk by the impacts of climate change, provided they are fully informed of those risks? What if staying is the only means to ensure the exercise of other rights, like the rights to self-determination or culture? There may not be clear legal answers, particularly in choices involving mobility. Leaving one's home comes with significant costs and can mean the irreplaceable loss of culture and livelihoods tied to land. Given these prospects, some may choose to stay as a form of 'voluntary immobility'.[47] Such a choice shows the potential collision between individual agency and adaptative mobility plans, and the need to navigate situations that require balancing rights and obligations. The fundamental notion of human dignity should serve as a touchstone, a guide in situations where different choices may arise in the face of similar risks.[48]

In addition, the choice and ability to move implicates mobility rights. As Chapter 1 explains, the right to move safely and with a degree of freedom and choice motivates the analysis in this book, and the anticipatory, proactive adaptive mobility it advocates. Mobility rights include the rights to liberty of movement and freedom to choose one's residence, to leave any country, and to return to one's own country.[49] Yet there are important limits to these rights: there is no general right to enter another country or explicit right to stay, for example, and free movement can be circumscribed, limited to the lawfully present, or constrained if 'necessary to protect national security, public order, public health or morals or the rights and freedoms of others'.[50] These limitations are addressed, in part, by

[46] Fredman (n 36) 226–27; Sandra Fredman, 'Substantive Equality Revisited' (2016) 14 International Journal of Constitutional Law 712, 730.

[47] See Carol Farbotko and others, 'Relocation Planning Must Address Voluntary Immobility' (2020) 10 Nature Climate Change 702.

[48] Chapter 8, IV raises some of these questions and situations in discussing the Pacific Islands.

[49] ICCPR, art 12(1), (2), (4).

[50] See ICCPR, art 12(3); Chapter 1, IV.

grounding adaptation planning, facilitated internal mobility, and legal pathways to migration in adaptation obligations, offering a different legal basis for their implementation than mobility rights. Limitations are also addressed by the adaptive measures themselves. If adaptive mobility is undertaken in the manner set out by the arguments in this book, then it should occur before situations arise that would warrant limitations to mobility rights, and in advance of the kind of public emergency that allows for derogation of rights obligations.[51] Adaptive mobility measures—and their requisite planning for movement before displacement occurs—also provide the benefit of lawful mobility. Along with such mobility comes freedom of movement and choice of residence. Likewise, it can help address or avoid the lack of a general right to enter another state with lawful means to enter and reside. The legal status it can confer protects against expulsion and provides a measure of security that affects access to livelihood opportunities and other fundamental rights.

Furthermore, and although they are not the focus of analysis in this book, mobility rights must be taken into account in the interpretation and implementation of obligations when relevant. Negative and positive duties result. States cannot violate the rights of those who wish to move from vulnerable locations or inhibit their internal choice of residence. Positive measures may also be necessary when individuals cannot access these rights throughout the process of adaptive mobility. Accordingly, if a person is rendered immobile due to foreseeable risks to their lives or access to essential resources—as Chapter 4 explores—then they must be provided with a means to adapt in situ or the opportunity to move to a location where their rights can continue to be ensured. They must also, at a minimum, be able to participate in decision-making and, more advisedly, be able to make an informed choice. This connects mobility rights with other substantive and procedural rights to inform how states facilitate adaptive mobility. For example, the rights to freedom of movement involve an element of choice and thus the active participation of those undertaking adaptive mobility in choosing when and where to move. State obligations to protect against arbitrary displacement create positive duties to prevent foreseeable displacement, and to address vulnerabilities that could lead to such displacement.[52] Likewise, the need for anticipatory action can

[51] Derogation is only permissible under the ICCPR when a public emergency 'threatens the life of the nation'; ICCPR, art 4. See Emanuele Sommario, 'Limitation and Derogation Provisions in International Human Rights Law Treaties and Their Use in Disaster Settings' in Flavia Zorzi Giustiniani and others (eds), *Routledge Handbook of Human Rights and Disasters* (Routledge 2018) 105–06, 111–12; but see Walter Kälin, 'The Human Rights Dimension of Natural or Human-Made Disasters Focus: Disaster Preparedness and Response' (2012) 55 German Yearbook of International Law 119, 128–31 (derogations likely impermissible but limitations to specific rights are possible in disaster settings); Bruce Burson and others, 'The Duty to Move People Out of Harm's Way in the Context of Climate Change and Disasters' (2018) 37 Refugee Survey Quarterly 379 (high threshold for derogation unlikely to be reached by most disasters currently).

[52] Chapter 4, III.C; UNCHR, 'Guiding Principles' (n 39) principles 7, 8.

lead to measures to plan for instances when displacement or evacuation may be unavoidable, particularly for sudden-onset events. Finally, adaptation obligations add a collective element to mobility rights. As discussed in Chapter 6, developed countries are obligated to assist and cooperate in adaptation efforts, including efforts to facilitate adaptive mobility. This distributes the burden of movement, particularly across borders, through the assistance of developed countries. Forms that such assistance may take include financial and technical support, cooperation on planned relocation, and the admittance and stay in another state, as further illustrated in the case studies.

In summary, procedural rights help determine how decisions are made and who must be involved, but they do not require the consent of affected persons. Yet underlying concepts and principles of human dignity and choice demand that individuals drive the decision-making that impacts their resources, livelihoods, and homes, which requires access to accurate information and meaningful processes of participation. For mobility, facilitating a fully informed process that is led by individuals and their communities can improve outcomes for all affected. Such a process must prioritize the participation of those already marginalized and most vulnerable to climate change. And it should also maintain access to mobility rights, which provide freedom of movement, choice of residence, and a right to leave.

IV. Transforming adaptation obligations through the precautionary principle

To this point, the book establishes that the integration of procedural and substantive human rights into the interpretation and implementation of adaptation obligations shapes adaptation measures, their planning and decision-making processes, and what measures may entail. It also provides guidance on when these measures should occur—once the impacts of climate change pose a foreseeable risk of serious harm. The precautionary principle further underscores the importance of the timing of adaptation. It bolsters arguments that adaptive measures are necessary to address foreseeable risks, through proactive, anticipatory action.

A. Precaution as an operative principle

The operative principles of the climate regime are set out in Article 3 of the UNFCCC. They include the requirement that 'Parties should take precautionary measures to anticipate, prevent or minimize the causes of climate change and mitigate its adverse effects'. Furthermore, a 'lack of full scientific certainty' should not postpone action when climate change poses a threat of 'serious or irreversible damage'. Action should be 'cost-effective', 'take into account different socio-economic contexts', 'cover all relevant . . . adaptation', and may be carried out cooperatively.[53]

In general, the precautionary principle encourages preventive action to avoid serious harm or irreversible damage and it precludes lack of certainty as a reason not to act.[54] It builds upon and is interrelated with the principle of prevention, with both principles 'intended to guide or orient initiatives to avoid harm or probable harm to the environment'.[55] The two principles are part of a 'temporal continuum', with prevention targeting more certain, impending risks and precaution aimed at risks with less certainty. The likelihood of harm required for action is reduced through the precautionary principle, particularly when the consequences could be grave. Together, these principles provide a legal framework for protective action to address the risks posed by climate change, even in the absence of conclusive scientific proof of damage.[56] Action is prompted sooner by precaution, before risks are immediate or certain, although prevention is one of the aims of precautionary measures in the climate regime.

The UNFCCC's articulation of the principle is one of the most detailed in an international instrument, but it does not prescribe the measures that must be taken. Instead, it provides guidance, with language that leaves room for interpretation, allowing for its application across time and contexts. The provision does not, for example, specify what constitutes serious or irreversible damage or the degree of certainty needed before acting. Elsewhere, and as Chapter 4 advocates for human rights law, the focus is on foreseeable risks. The International Law Association (ILA), for example, drafted articles on the existing legal principles that relate to climate change, including the precautionary principle. These draft articles declare that the principle necessitates action when there is a 'reasonably foreseeable threat of serious or irreversible damage'.[57] The interaction of the principles of prevention and precaution in due diligence standards also sheds some light: triggering this duty does not require that harm be inevitable, only that there be foreseeable risks, and the greater the certainty that a risk of harm exists, the less severe such harm must be to compel action.[58]

How to determine the cost-effectiveness of climate action is also an open question. Cost may limit the scope of precautionary adaptation measures, but these

[53] UNFCCC, art 3(3).

[54] See eg Declaration of the United Nations Conference on the Human Environment (adopted 16 June 1972) UN Doc A/Conf48/14/Rev 1 (Stockholm Declaration) principle 15.

[55] *Pulp Mills* (n 8) 159–62 [61]–[68] (Separate Opinion of Judge Cançado Trindade).

[56] Antônio Augusto Cançado Trindade, 'Principle 15' in Jorge E Viñuales (ed), *The Rio Declaration on Environment and Development: A Commentary* (Oxford University Press 2015) 422–23; see also ILA, 'Declaration of Legal Principles Relating to Climate Change' (2014) Resolution 2/2014, arts 7A, 7B (obligation of prevention and precautionary principle).

[57] ILA (n 56) art 7B(1).

[58] See Tim Stephens and Duncan French, 'ILA Study Group on Due Diligence in International Law: Second Report' (ILA 2016) 21; Alan Boyle and Catherine Redgwell, *Birnie, Boyle, & Redgwell's International Law and the Environment* (4th edn, Oxford University Press 2021) 170–71.

measures are likely to be less costly than waiting to act until after damage occurs. The Paris Agreement recognizes the importance of timing: mitigation can reduce the need for adaptation and thus its costs, while adaptation can help avert or minimize the catastrophic and costly damage expected to occur in the absence of action.[59] Indeed, some of the latent risks created by climate change are best addressed with early and anticipatory adaptive action. Yet the overall risk climate change creates is unique, posing different challenges than other forms of environmental harm. Climate risks are global but vary locally; unprecedented and rapidly occurring; and accompanied by some degree of uncertainty in timing, extent, and distribution of impacts.[60] There will, then, be some circumstances where a cost-benefit analysis is infeasible or inappropriate. Measures that consider the enjoyment of certain human rights—like the right to life, for example—should not rely on any cost-based rationale. In addition, problems arise in measuring the cost of potential non-economic losses, some of which cannot or should not be easily quantified.

For these complicated and unquantifiable situations, precaution offers what Jonathan Wiener considers to be a general posture for policymaking, with core elements that include the stance that uncertainty does not preclude measures; preference for earlier measures that anticipate and prevent risk; more stringent protection rather than less; and monitoring and assessment of measures over time.[61] Climate impacts pose multiple and changing risks. Such a general, precautionary posture enables a holistic approach, one that weighs benefits and harms to reduce multiple risks at once. Temporal factors can be incorporated into actions that are able to shift over time as knowledge is updated about risks, thereby allowing consideration of long-term impacts or extended timescales. Thus, precautionary actions can themselves become adaptive over time and as more information is acquired.

Furthermore, the indeterminate aspects of the principle require a case-by-case analysis. This allows for the reflection of different values or approaches to risk depending on the context. Uncertainties will necessarily arise from the flexible exercise of the principle, but these simply show that general principles like precaution cannot be flattened in practice. The principle's application enables cultural

[59] See Paris Agreement, arts 7(4), 8.

[60] See Nicolas de Sadeleer, 'The Precautionary Principle and Climate Change' in Daniel A Farber and Marjan Peeters (eds), *Climate Change Law*, vol I (Edward Elgar 2016) 21–22.

[61] Jonathan B Wiener, 'Precaution and Climate Change' in Cinnamon P Carlarne, Kevin R Gray, and Richard Tarasofsky (eds), *The Oxford Handbook of International Climate Change Law* (Oxford University Press 2016) 168–69; Jonathan B Wiener, 'Precautionary Principle' in Emanuela Orlando and Ludwig Krämer (eds), *Principles of Environmental Law*, vol VI (Edward Elgar 2018) 181; but see Cass R Sunstein, *Laws of Fear: Beyond the Precautionary Principle* (Cambridge University Press 2005) 14–15, 50 (principle is incoherent, but it could be useful if focused narrowly on the distribution of risks and effects of climate change).

differences and varying perspectives on risk to factor into decision-making and action. Such flexibility, which can lead to 'interpretational variety', is therefore an asset of the principle.[62] Flexibility, likewise, permits incorporation of different risks across contexts and influences the integration of human rights obligations to address the foreseeable impacts of climate change. These impacts include those that pose risks of serious harm to human health, life, and sources of subsistence. To avoid such risks, precaution favours earlier measures rather than waiting for impacts to occur.

B. Precaution and adaptation obligations

The precautionary principle expands the timing of climate action while prohibiting use of uncertainty to justify not acting. For the interpretation of adaptation obligations, it plays this role by reinforcing the integration of positive duties to protect rights in advance of significant harm. The principle strengthens arguments that duties to protect the right to life require states to act not only when harm is imminent or certain but also when climate impacts present a foreseeable risk of serious harm. Similarly, the principle reinforces the need for adaptive action when those minimum essential resources associated with core obligations from the right to an adequate standard of living are foreseeably threatened or no longer accessible.[63]

Precaution also bolsters a related duty of due diligence and procedural rights. For example, efforts to satisfy obligations to act on adaptation can include processes and procedural guarantees needed to meet due diligence standards. These guarantees include ensuring access to information and opportunities to participate. Integrating procedural rights requires that these guarantees apply to affected persons, while also necessitating assessment of, responses to, and concrete actions that address risks and help prevent harm. Incorporating the precautionary principle alongside such integration better ensures that relevant human rights duties are called upon earlier to incorporate human rights benchmarks and safeguards into planning and anticipatory action, both in the process of decision-making and in the implementation of adaptation. The principle thus serves a transformative role, one that aids in overcoming limits to the use or integration of human rights law alone. This additional interpretive step is important, because it is not always clear when positive obligations must be undertaken under human rights law. An interpretation that supports anticipatory action is consistent with both adaptation

[62] Miriam Haritz, 'Liability with and Liability from the Precautionary Principle in Climate Change Cases' in Michael Faure and Marjan Peeters (eds), *Climate Change Liability* (Edward Elgar 2011) 33.
[63] Chapter 4, II–III.

obligations and their inclusion of planning and preparation and the principle's aim to prevent and reduce harm from the impacts of climate change.[64]

Furthermore, as risks from climate change become increasingly evident, inclusion of the precautionary principle requires more urgent preventive action. Jacqueline Peel observes that climate change presents a straightforward context for the application of the principle, as there is well-developed evidence about the nature of climate risks.[65] The level of foreseeability can also be weighed with the magnitude of potential harm, to better determine actionable risks. Where there is serious risk—and the impacts of climate change pose many—uncertainty is insufficient to excuse inaction, and some form of precautionary measure is required. Together with the integration of rights, the principle could help compel proactive adaptation measures or be used to challenge a state for failing to act in the face of foreseeable risks. It can support demands by people to protect their interests and rights. Threats to people and the enjoyment of their rights is growing in certainty, and precaution dictates that states must take timely action to prevent harm and rights violations. Precautionary action is often in the best interest of affected states as well, particularly in circumstances that climate change presents, where risk assessments may be complicated by multiple factors or catastrophic impact is possible. In some circumstances, decision-makers must choose which risks to prioritize, and a human rights-based approach is one way to help do so.

Precaution also plays a critical role in compelling and shaping mobility. By acting before serious harm occurs—and at a time prior to when human rights law alone may require—adaptation can help avoid or reduce displacement, prevent precarious migration, and facilitate anticipatory, proactive movement elsewhere. Adaptation measures can help avoid the kinds of adverse effects that precautionary action under climate treaties is meant to prevent or mitigate, which includes risks to and violations of human rights. A precautionary approach to mobility is therefore a means to protect rights. It is crucial, for example, that advanced planning and early and ongoing participation occur where planned relocation is contemplated as an adaptive measure, albeit one of last resort. More generally, a precautionary approach can generate protection arguments for migrants. Mariya Gromilova argues that it is necessary to take a precautionary approach following population displacement, even if it is not possible to fully attribute its cause to climate change.[66] Graeme Hugo made a similar argument over a decade ago, explicitly calling for the application of the precautionary principle 'to put in place mechanisms at national

[64] Chapter 2, V; Lauren Nishimura, 'Adaptation and Anticipatory Action: Integrating Human Rights Duties into the Climate Change Regime' (2022) 12 Climate Law 99, 124.

[65] Jacqueline Peel, 'International Environmental Law and Climate Disasters' in Rosemary Lyster and Robert RM Verchick (eds), *Research Handbook on Climate Disaster Law: Barriers and Opportunities* (Edward Elgar 2018) 89.

[66] Mariya Gromilova, 'Finding Opportunities to Combat the Climate Change Migration Crisis: The Potential of the "Adaptation Approach"' (2016) 33 Pace Environmental Law Review 105, 131.

and regional levels which have the capacity to facilitate the process of resettlement of individuals, families, and in some cases entire communities displaced by climate change'.[67]

Thus, when read into an interpretation that integrates human rights law, precaution can further shape and guide adaptation obligations, to allow for various measures that can be tailored to the risks and context. Adaptation is an iterative process, and it provides a means for states to affirmatively act to protect and fulfil human rights. It must also adjust to changing circumstances. Initially, obligations might be satisfied by adaptation in situ, including through efforts to reduce vulnerability and prevent displacement. These and other adaptation measures can allow for a delay in the need for mobility, and more time to assess when conditions in certain areas will put people and the enjoyment of their rights at serious risk. This is consistent with the language in the Paris Agreement, requiring states to implement adaptation 'as appropriate'.[68] States have discretion in what adaptation measures may be appropriate for the given context, although *some* action to satisfy adaptation obligations and protect rights is required. However, human rights law speaks to what may be appropriate, as discretion for obligations of conduct—like obligations to act on adaptation—can be limited if a specific kind of measure is essential to avoid harm.[69] What is necessary and appropriate can also shift over time. Foreseeable risks can therefore require 'a continuum of response from protection in place to community relocation' as climate change alters the landscapes and resources relied upon by people.[70]

Finally, the precautionary principle further strengthens arguments for proactive assistance and cooperation. As Chapter 6 details, the success of adaptation may, in part, hinge on the breadth and timeliness of assistance. Because adaptation is a process, and because adaptation obligations are evolutive, support cannot be a one-off or one-time contribution. Assistance and cooperation should therefore reflect the changing needs of developing countries, to provide ongoing support that allows people to adapt when faced with significant risks to the enjoyment of their human rights.

V. Conclusion

This chapter analyses procedural rights, the positive duties associated with these rights, and their development through interaction with international

[67] Graeme Hugo, 'Lessons from Past Forced Resettlement for Climate Change Migration' in Etienne Piguet, Antoine Pécoud, and Paul de Guchteneire (eds), *Migration and Climate Change* (Cambridge University Press 2011) 267.

[68] Paris Agreement, art 7(9).

[69] Stephens and French (n 58) 7–8.

[70] See Bronen (n 43) 394.

environmental law norms and principles. It constructs an argument that the integration of procedural rights in the interpretation of obligations to act on adaptation requires affected persons be provided with information and opportunities to participate in decision-making processes on adaptation and adaptive mobility. It also connects the implementation of adaptation obligations to the principles of equality and non-discrimination. These principles are key to climate action that accounts for and addresses vulnerability, and that ensures the participation and inclusion of marginalized individuals and groups. While procedural rights and non-discrimination are required, the perception of affected people and their choice in shaping mobility are equally important. The participatory processes in place and agency afforded to individuals will help determine the manner and outcomes of migration. Positive outcomes are more likely when decision-making processes are based on choice and led by those affected. In addition, this chapter explores the role of the precautionary principle in adaptive action. Precaution is a central principle of the climate regime. The principle informs the timing of action and the types of harm that action seeks to address. When included in the interpretive process, the principle reinforces the integration of rights and transforms adaptation obligations. It supports arguments for action in the face of foreseeable risks to people and their rights, despite uncertainty and without the need for immediacy of harm or rights violations. Such an interpretive process leads to proactive adaptation measures, in anticipation of predicted or foreseeable risks to essential resources or continued survival. In some circumstances, as the case studies discuss, these measures must include planning for and facilitating adaptive mobility.

6

Assistance and Cooperation: Strengthening Support for Adaptive Mobility

I. Introduction

Obligations to assist and cooperate in adaptation are often overlooked features of climate treaties. Yet they offer a critical means to enable adaptation and a legal basis for support for adaptive mobility. This chapter focuses on the interpretation and implementation of these adaptation obligations. Section II explores relevant rules from human rights law. It explains how their integration into the interpretation of adaptation obligations helps clarify that adaptation is not an undertaking for each country on its own, and that the support envisaged in the climate regime includes a collective dimension meant to help share the burden of adaptation. Section III then turns to the operative principles of climate treaties, which include principles equity, common but differentiated responsibilities and respective capabilities (CBDR-RC), and special circumstances. These principles provide further guidance on who is owed support and who must provide it. Together with human rights duties, they argue for adaptation obligations that require developed countries to support developing countries in their adaptation efforts, including adaptive mobility measures. Such support is critical to the book's approach to mobility, and Section IV discusses concrete forms of support. At a minimum, support includes financial and technical assistance for adaptive mobility measures. In some contexts, it can and should also include cooperative efforts to facilitate mobility across borders. However, there are barriers to assistance and cooperation that must be overcome. Section V discusses these barriers and raises potential ways forward, as well as open questions that remain. These questions provide fertile ground for future research and areas of advocacy, to fill gaps in financing and support for climate adaptation and adaptive mobility.

II. Integrating relevant rules

Chapters 4 and 5 show that the integration of positive duties from human rights can shape obligations to act on adaptation, leading to adaptive mobility measures that address foreseeable risks to the enjoyment of rights and ensure the participation of those put at risk. Obligations to assist and cooperate in adaptation provide

Climate Change, Human Rights, and Adaptive Mobility. Lauren Sakae Nishimura, Oxford University Press.
 DOI: 10.1093/9780198930068.003.0007

a means to help implement these measures. Assistance obligations include the obligation of developed countries to provide financial support to developing countries.[1] The UNFCCC recognizes the importance of such support, and that the extent of developing countries' effective implementation of their commitments—including on adaptation—is contingent on developed countries meeting their obligations related to financial support and technology transfer.[2] The Paris Agreement continues and builds on these obligations to require '[c]ontinuous and enhanced international support' to developing countries for the implementation of their adaptation obligations, with financial assistance owed from developed countries.[3] Support is needed when, for example, developing countries cannot implement adaptive action to protect life or provide essential resources without assistance. These countries may experience greater difficulty in their adaptation efforts because of weak governance structures, which hamper measures to address the causes of climate vulnerability.[4] In general, support can help improve capacity and provide resources, thereby contributing to efforts to manage climate risks in a comprehensive manner. More specifically, it can help ensure that adaptive mobility is an adaptation option, and that when undertaken it leads to more rights-protective outcomes. Like obligations to act on adaptation, however, adaptation obligations of support lack specificity and detail.[5] They therefore require further development or interpretation to resolve their ambiguity.

International human rights law includes duties of cooperation and assistance that are relevant to an interpretation of adaptation obligations, which strengthen the argument that developed countries are legally required to provide support. International instruments, and the UN Charter and the International Covenant on Economic, Social and Cultural Rights (ICESCR) in particular, contain relevant provisions on cooperation and assistance. The UN Charter is a treaty, with all members bound by its articles.[6] All UN member states are parties to the UNFCCC, and nearly all are also parties to the Paris Agreement. This overlap means that the UN Charter's articles apply to those bound by climate treaties. Cooperation is central to the UN, and the Charter sets out one of its purposes as being '[t]o achieve

[1] See United Nations Framework Convention on Climate Change (adopted 9 May 1992, entered into force 21 March 1994) 1771 UNTS 107 (UNFCCC) art 4(3), (4).

[2] ibid art 4(7).

[3] Paris Agreement (adopted 12 December 2015, entered into force 4 November 2016) 3156 UNTS 79, arts 7(13), 9(1).

[4] Intergovernmental Panel on Climate Change (IPCC), *Managing the Risks of Extreme Events and Disasters to Advance Climate Change Adaptation*, A Special Report of Working Groups I and II of the IPCC (Field CB and others (eds), Cambridge University Press 2012) 451.

[5] Chapter 2, V.B; see also Yulia Yamineva, 'A Legal Perspective on Climate Finance Debates: How Constructive Is the Current Norm Ambiguity?' in Alexander Zahar and Benoît Mayer (eds), *Debating Climate Law* (Cambridge University Press 2021).

[6] See *Certain Expenses of the United Nations (Article 17, Paragraph 2 of the Charter)* (Advisory Opinion) [1962] ICJ Rep 151, 157; Nico J Schrijver, 'The Future of the Charter of the United Nations' in Armin von Bogdandy and Rüdiger Wolfrum (eds), *Max Planck Yearbook of United Nations Law*, vol 10 (2006) 5.

international cooperation in solving international problems of an economic, social, cultural, or humanitarian character, and in promoting and encouraging respect for human rights and for fundamental freedoms for all'.[7] UN members also pledged to cooperate in promoting 'universal respect for, and observance of, human rights and fundamental freedoms'.[8] While general, such a commitment supports the need for states to cooperate in adaptation efforts that seek to avoid or mitigate risks to human rights from climate change.[9]

The ICESCR provisions are more specific. The treaty calls for the progressive realization of rights 'individually and through international cooperation'.[10] This provision applies to all ICESCR rights, including the rights analysed in Chapter 4—to health and to an adequate standard of living. The ICESCR further highlights that its parties recognize 'the essential importance of international co-operation' in realizing the right to an adequate standard of living. For the related right to be free from hunger, states 'shall take, individually and through international co-operation', those measures needed to improve methods of food production, conservation, and distribution and to ensure equitable distribution of food globally according to need.[11] In its general comments interpreting rights, the ICESCR's treaty body—the Committee on Economic, Social and Cultural Rights (CESCR)—further emphasizes that assistance and cooperation play an 'essential role' in the realization of economic, cultural, and social rights. Cooperation is an obligation of all states, but '[i]t is particularly incumbent' on those states—namely developed countries—that possess the ability to assist.[12] This heightened responsibility is highlighted in general comments on the rights to health and to water, with the CESCR reiterating that 'it is particularly incumbent on State parties and other actors in a position to assist to provide international assistance and cooperation ... which enables developing countries to fulfil their core obligations'.[13] The CESCR encourages states to facilitate the realization of the right to water in other countries through the provision of resources, financial and technical assistance, and necessary aid, including to refugees and displaced persons. This duty of assistance falls upon 'economically developed' states, as they 'have a special responsibility and

[7] Charter of the United Nations (adopted 26 June 1945, entered into force 24 October 1945) 1 UNTS XVI, art 1(3).

[8] ibid arts 55, 56.

[9] Lauren Nishimura, 'Adaptation and Anticipatory Action: Integrating Human Rights Duties into the Climate Change Regime' (2022) 12 Climate Law 99, 121.

[10] International Covenant on Economic, Social and Cultural Rights (adopted 16 December 1966, entered into force 3 January 1976) 993 UNTS 3 (ICESCR), art 2(1).

[11] ibid art 11(1)–(2).

[12] CESCR, 'General Comment No 3: The Nature of States Parties' Obligations (Art 2, Para 1, of the Covenant)' (14 December 1990) UN Doc E/1991/23, paras 13–14.

[13] CESCR, 'General Comment No 15: The Right to Water (Arts 11 and 12 of the Covenant)' (20 January 2003) UN Doc E/C.12/2002/11, para 38; CESCR, 'General Comment No 14: The Right to the Highest Attainable Standard of Health (Art 12)' (11 August 2000) UN Doc E/C.12/2000/4, para 45.

interest to assist the poorer developing States in this regard'.[14] More broadly, when discussing international sanctions, the CESCR explains that:

> Just as the international community insists that any targeted State must respect the civil and political rights of its citizens, so too must that State and the international community itself do everything possible to protect at least the core content of the economic, social and cultural rights of the affected peoples of that State.[15]

General comments further clarify that when a state is unable to meet its required core minimum duties to provide certain basic resources, it should—and in some cases must—request international assistance to do so. For example, in interpreting the right to adequate housing, the CESCR suggests that when a state is unable to take 'steps which must be taken immediately' due to resource constraints, 'it is appropriate that a request be made as soon as possible for international cooperation'.[16]

These duties to cooperate and assist apply in the context of climate change. Scholars have interpreted the duty to fulfil rights under the ICESCR as suggesting that wealthy states are obliged to assist poor ones in adaptation.[17] The CESCR, likewise, called on developed countries to support developing countries in adaptation efforts as part of their duties of international cooperation and assistance. It reminded states that they not only owe human rights duties to their own populations but also to those outside their territories, both under the ICESCR and consistent with pledges of cooperation and universal respect for human rights in the UN Charter.[18]

Taken together, the ICESCR's provisions and the CESCR's interpretations lead to an understanding of cooperation and assistance with several key implications. First, developed countries are obligated to assist in adaptation, including in measures that address risks to the enjoyment of fundamental rights. These measures arise from developing countries taking action to satisfy their obligation to act on adaptation. As Chapter 4 establishes, a state must provide the minimum essential resources required for access to adequate food, water, housing, and healthcare,

[14] CESCR, 'General Comment No 15' (n 13) para 34.

[15] CESCR, 'General Comment No 8: The Relationship between Economic Sanctions and Respect for Economic, Social and Cultural Rights' (12 December 1997) UN Doc E/C.12/1997/8, para 7.

[16] CESCR, 'General Comment No 4: The Right to Adequate Housing (Art. 11 (1) of the Covenant)' (13 December 1991) UN Doc E/1992/23, para 10.

[17] Daniel Bodansky, 'Climate Change and Human Rights: Unpacking the Issues' (2010) 38 Georgia Journal of International & Comparative Law 511, 521; Daniel Bodansky, Jutta Brunnée, and Lavanya Rajamani, *International Climate Change Law* (Oxford University Press 2017) 306.

[18] CESCR, 'Climate Change and the International Covenant on Economic, Social and Cultural Rights: Statement of the Committee on Economic, Social and Cultural Rights' (8 October 2018) (2018 Statement) <www.ohchr.org/en/NewsEvents/Pages/DisplayNews.aspx?NewsID=23691&LangID=E> accessed 23 September 2022, paras 5–6.

which can trigger positive action to ensure these resources, including through adaptation and adaptive mobility measures.[19] Moreover, if a state cannot meet its core rights obligations—including by satisfying its obligations to act on adaptation—it should ask for assistance. Other states, particularly developed countries with the means to do so, are obliged to cooperate and assist in meeting those core obligations. These support duties are required by adaptation obligations and clarified by relevant human rights duties. They direct assistance toward the protection of people and help specify one form required of developed states: support to ensure basic necessities for vulnerable populations in developing countries.

Second, relevant human rights duties to cooperate and assist underscore the importance of the right to an adequate standard of living in an analysis that takes a holistic approach, one that accounts for interrelated nature and cumulative impacts to rights. The right to life compels states to assist in certain circumstances, but it is likely to apply only in acute cases, like those that trigger *non-refoulement* obligations.[20] There is also no similar duty to cooperate and assist in the International Covenant on Civil and Political Rights (ICCPR). As interpretations of the right to life evince, however, the conditions required for a dignified life overlap with those protected by the right to an adequate standard of living.[21] The inclusion of the latter makes the integration of related positive duties more direct, as do ICESCR obligations to assist and cooperate. In practical terms, the integration of relevant human rights duties can translate into the incorporation of rights-protective measures in requests for adaptation assistance within the climate regime and a double, or strengthened, duty to provide such assistance.[22] This is also underscored by developments in the climate regime. The framework on the global goal on adaptation (GGA) seeks to catalyse regional and international cooperation to scale up adaptation action and support.[23]

Finally, a duty to assist with international mobility is also supported by other legal sources. For example, the United Nations Convention on the Law of the Sea (UNCLOS) contains a duty to render assistance—which could have extraterritorial impact—although it is limited to assistance at sea.[24] More generally, the International Tribunal for the Law of the Sea (ITLOS) interpreted UNCLOS to include obligations to cooperate in efforts to prevent, reduce, or control marine pollution from the greenhouse gas emissions that cause climate change. ITLOS also advised that state parties have specific obligations to assist developing states, and vulnerable developing states in particular, in their efforts to address marine

[19] See Chapters 7–9.

[20] See Chapter 1, IV.

[21] Chapter 4, II.

[22] Nishimura, 'Adaptation and Anticipatory Action' (n 9) 122.

[23] Decision 2/CMA.5 'Global Goal on Adaptation' (13 December 2023) FCCC/PA/CMA/2023/16/Add.1, para 19.

[24] United Nations Convention on the Law of the Sea (adopted 10 December 1982, entered into force 16 November 1994) 1833 UNTS 396 (UNCLOS) art 98.

pollution and to grant them preferential treatment in funding and technical assistance.[25] These obligations complement and buttress adaptation obligations of support. For international mobility, the principle of 'temporary refuge' offers another means of support, one rooted in customary law. The principle creates time-bound obligations of admission and non-return.[26] A number of states have granted temporary refuge following disasters, although this is often by regularizing the status of migrants already in a new country. By being temporary in nature, this form of protection is unlikely to offer a long-term solution for those who need to move more permanently. It also does not apply to those who move in advance of significant environmental harm, making it a poor fit for the kind of anticipatory, proactive movement that characterizes adaptive mobility in this book. There are, however, soft law frameworks that are more instructive and strengthen arguments for cooperation and assistance with adaptive mobility.[27] The Nansen Initiative, for example, was a state-led consultative process that culminated in the endorsement of a Protection Agenda on cross-border disaster and climate-related displacement by 109 states in 2015. Amongst other things, it advocates for the integration of human rights-based approaches to disaster risk reduction and adaptation, emphasizes proactive planning and measures, and seeks to expand the grounds to admit those forced across borders in the context of climate change and disasters.[28] Like other soft law frameworks, however, it offers guidance and does not create any new legal obligations.

III. Incorporating equity, differentiated responsibilities, and special circumstances

In addition to the integration of relevant rules, the operative principles of the climate regime shape adaptation obligations, both by providing guidance on action and as part of the interpretive context of climate treaties. As Chapter 5 establishes,

[25] Advisory Opinion in Case No 31, On the Request submitted to the Tribunal by the Commission of Small Island States of Climate Change and International Law (21 May 2024) ITLOS.

[26] This principle and the protection it may offer are discussed in Guy S Goodwin-Gill, 'Non-Refoulement, Temporary Refuge, and the "New" Asylum Seekers' in David James Cantor and Jean-François Durieux (eds), *Refuge from Inhumanity? War Refugees and International Humanitarian Law* (Brill 2014) 457–58.

[27] For example, the Sendai Framework calls for enhanced international cooperation to developing countries from developed countries; highlights small-island States, African countries, and Least Developed Countries (LDCs) as requiring support due to their high vulnerability; and provides concrete measures of support. United Nations Office for Disaster Risk Reduction (UNDRR), 'Sendai Framework for Disaster Risk Reduction 2015–2030' (2015) paras 8, 14, 18(f), 19, 28, 41; see also Office of the UN High Commissioner for Human Rights (OHCHR), 'The Slow Onset Effects of Climate Change and Human Rights Protection for Cross-Border Migrants' (22 March 2018) UN Doc A/HRC/37/CRP.4, paras 160–61.

[28] The Nansen Initiative, 'Agenda for the Protection of Cross-Border Displaced Persons in the Context of Disasters and Climate Change' vol 1 (2015) paras 20, 47, 51, 88.

the precautionary principle influences the timing of action on and support for adaptation. In addition to precaution, the principles of equity, CBDR-RC, and special circumstances are particularly important for adaptation obligations of assistance and cooperation. These principles inform obligations and strengthen arguments for timely, adequate support based on differentiation. They direct support for adaptation from developed countries that are responsible for and benefit from climate change towards developing countries and the vulnerable populations within them.

A. Equity and the CBDR-RC principle

The principles of the UNFCCC include guidance that parties should take action to protect the climate system 'for the benefit of present and future generations of humankind, on the basis of equity and in accordance with their common but differentiated responsibilities and respective capabilities'. Accordingly, developed country parties are instructed to 'take the lead in combating climate change and the adverse effects thereof'.[29]

Equity 'sits at the very heart of the international climate change regime'.[30] It incorporates broader concepts of fairness and justice into the regime, as well as arguments for burden-sharing and affirmative action as applied within and between states.[31] These arguments rest on the premise that states do not start on an equal footing. Accordingly, and as the UNFCCC instructs, developed countries must take the lead in climate action. Equity further argues for such action to extend beyond internal measures, to benefit those places and people most at risk. In doing so, it helps shift the impetus for assistance from altruism to obligation, operationalized in climate treaties through a clear entitlement: developing countries are owed support from developed countries. This shift has wider implications, contributing to the distributive justice that Philippe Cullet views as 'an essential part of international law and international environmental law in particular'.[32]

Equity can likewise introduce concern for the impacts to people into a regime that is primarily focused on state relations. The UNFCCC connects the principle with state action that benefits present and future generations, thereby invoking both intra- and intergenerational equity. The former, like other international

[29] UNFCCC, art 3(1).

[30] Cinnamon P Carlarne and JD Colavecchio, 'Balancing Equity and Effectiveness: The Paris Agreement & the Future of International Climate Change Law' [2019] New York University Environmental Law Journal 107, 116.

[31] Lavanya Rajamani, 'Integrating Human Rights in the Paris Climate Architecture: Contest, Context, and Consequence' (2019) 9 Climate Law 180, 199.

[32] See Philippe Cullet, 'Common but Differentiated Responsibilities' in Malgosia Fitzmaurice, Marcellinus Brus, and Panagiotis Merkouris (eds), *Research Handbook on International Environmental Law* (Edward Elgar 2021) 210.

environmental law principles (eg the polluter-pays and no-harm principles), considers the geographic distribution of environmental risks. The latter incorporates a temporal element, seeking to balance needs across generations.[33] The Paris Agreement explicitly references intergenerational equity in its preamble, calling for its consideration in action to address climate change. Inter- and intragenerational equity are linked, particularly for developing countries, as communities cannot be expected to fulfil obligations to future generations if they cannot meet basic needs today.[34] This makes assistance to developing countries important, not just to fulfil obligations to protect the interests and rights of people currently, but to better ensure adequate conditions for those who will follow.

The principle of CBDR-RC is a manifestation of equity, which is central to the climate regime, its obligations, and its continued development.[35] In addition to being included in the operative principles, CBDR-RC is reiterated in the commitments set out in the UNFCCC, including its adaptation obligations.[36] The principle is therefore foundational to differentiated obligations and actions in the regime. Differentiation provides clear guidance on assistance with adaptation, and it can distil what equity means between states into practical terms. However, where the emphasis on CBDR-RC should lie, if it includes corrective and distributive considerations, and whether and how differentiated obligations should shift are subject to dispute. Developing countries tend to focus on responsibilities, historical emissions, and the corrective aspect of the principle, while developed countries often emphasize capabilities so that other states will bear more responsibility as they develop.[37] Both perspectives feature in climate treaties, although the emphasis has changed over time. The UNFCCC divides parties into categories with different commitments. It keeps the onus of action on developed countries, indicating corrective and distributive elements. Developed countries have a duty to assist because of their capacity to act *and* their role in causing climate change. Developing country parties must adapt, but this is contingent on assistance from their developed country counterparts.[38] As developing country and LDC parties will be some of those most adversely affected but which have contributed the least to the causes of climate change, the operation of the principle is crucial to the implementation of adaptation obligations.

[33] Claire Molinari, 'Principle 3: From a Right to Development to Intergenerational Equity' in Jorge E Viñuales (ed), *The Rio Declaration on Environment and Development: A Commentary* (Oxford University Press 2015) 140.

[34] Edith Brown Weiss, 'Intergenerational Equity', *Max Planck Encyclopedia of Public International Law* (Oxford University Press 2013) para 11.

[35] See Lavanya Rajamani, 'Common but Differentiated Responsibilities' in Emanuela Orlando and Ludwig Krämer (eds), *Principles of Environmental Law*, vol VI (Edward Elgar 2018) 298.

[36] UNFCCC, art 4(1).

[37] See Bodansky, Brunnée, and Rajamani (n 17) 27; Carlarne and Colavecchio (n 30) 116–17.

[38] UNFCCC, art 4(7).

However, differentiation within the regime shifted from obligations based on category of state in the UNFCCC to a mixed approach in the Paris Agreement. In some areas, the Agreement seeks action by all parties, in other areas differentiated duties persist. Such a shift stems, in part, from the addition of 'in the light of different national circumstances' to the principle. Article 2 explains that the Agreement 'will be implemented to reflect equity and the principle of common but differentiated responsibilities and respective capabilities, in the light of different national circumstances'. The added language is also included in the preamble, the description of the parties' nationally determined contributions (NDCs), and discussion of mitigation and emission reductions.[39] It introduces an evolutive element to the principle: as conditions in a country shift, so too do its responsibilities. The expanded principle emerged as a compromise of sorts, the product of negotiations that balanced efforts to create commitments for all parties to mitigate, but that still accounted for the disparity in the causes and consequences of climate change and the capacity of countries to respond. Yet it may add little, as the qualifying language arguably reiterates the principle's accommodation of states' 'respective capabilities'.[40] Nonetheless, the additional emphasis on changing circumstances muddies the application of differentiation between states and calls into question the force of equity as an overarching organizing principle. At least for mitigation, all states are now expected to commit to action, although it is left to each country to determine its contribution.

For adaptation under climate treaties, differentiation remains critical and intact. Without the qualified principle that is incorporated into mitigation commitments, the UNFCCC's provisions continue to inform the scope and responsibility for adaptation under the Paris Agreement. The core differentiation for financial assistance also remains unchanged, with support owed from developed to developing countries for both mitigation and adaptation.[41] The contrast between these provisions and the article on mitigation evince an issue-based approach to differentiation in the Paris Agreement. The CBDR-RC principle still plays an important role across issues, but it is particularly pronounced for work on adaptation and finance. For example, the parties recognize that the implementation by developing countries of the framework for the GGA depends on 'the effective implementation by developed country Parties of means of implementation and support commitments'.[42] General questions remain about the principle's evolving status within the regime and its legal weight as a stand-alone principle.[43] But its use for the former is

[39] Paris Agreement, preamble, arts 2(2), 4(3), (4), (19).

[40] Lavanya Rajamani, 'Guiding Principles and General Obligation (Article 2.2 and Article 3)' in Daniel Klein and others (eds), *The Paris Agreement on Climate Change: Analysis and Commentary* (Oxford University Press 2017) 134.

[41] Paris Agreement, art 9(3).

[42] Decision 2/CMA.5 (n 23) para 32.

[43] For more, see Lavanya Rajamani, 'The Principle of Common but Differentiated Responsibilities and Respective Capabilities in the International Climate Change Regime' in Rosemary Lyster and

not based on the latter: whether it is a customary principle does not alter its role in the regime as a guiding principle that shapes interpretation and action.

Moreover, the legal status of the operative principles, while affecting arguments for their application, does not matter as much as the function they serve within the climate regime. These principles provide both interpretive guidance and normative bases to incorporate duties. For principles of differentiation, they move the need for support beyond reasons of reciprocity in state relations.[44] This is, in part, grounded in links to equity and justice that apply generally to adaptation and specifically to support for mobility. Concepts connected to differentiation—collective responsibility, ability to pay, and corrective and distributive justice—have been used to argue that developed countries are obligated to assist in mobility and relocation, and even to take in migrants.[45] Thus, differentiation is key to the analysis of adaptation obligations, and to ensuring adaptive mobility occurs and is supported by developed countries.

B. Special circumstances

The text of climate treaties also indicates that certain countries should be prioritized or are owed support due to their special circumstances. The operative principles set out in the UNFCCC, for example, call for full consideration of the 'specific needs and special circumstances of developing country Parties, especially those that are particularly vulnerable to the adverse effects of climate change, and of those Parties, especially developing country Parties, that would have to bear a disproportionate or abnormal burden under the Convention'.[46] The Paris Agreement reiterates this principle, its preamble recognizing 'the specific needs and special circumstances of developing country Parties, especially those that are particularly vulnerable to the adverse effects of climate change, as provided for in the Convention'. The special needs, situations, and circumstances of LDCs are further recognized with respect to funding, technology transfer, low greenhouse gas emissions development, and transparency and communications under the Agreement, as are the special circumstances of small island developing states for greenhouse gas emissions development and transparency and communications.[47]

Robert RM Verchick (eds), *Research Handbook on Climate Disaster Law: Barriers and Opportunities* (Edward Elgar 2018); Cullet, 'Common but Differentiated Responsibilities' (n 32) 219–20.

[44] See Philippe Cullet, 'Principle 7: Common but Differentiated Responsibilities' in *The Rio Declaration on Environment and Development* (n 33) 236.

[45] See eg Robyn Eckersley, 'The Common but Differentiated Responsibilities of States to Assist and Receive "Climate Refugees"' [2015] European Journal of Political Theory 481; Katrina M Wyman, 'Ethical Duties to Climate Migrants' in Benoît Mayer and François Crépeau (eds), *Research Handbook on Climate Change, Migration and the Law* (Edward Elgar 2017).

[46] UNFCCC, art 3(2).

[47] Paris Agreement, preamble, arts 4(6), 9(4), 11(1), 13(3).

The division between Annex I and II in the UNFCCC helps determine differentiation under the UNFCCC, with the Organization for Economic Cooperation and Development (OECD) members constituting Annex II and therefore obliged to assist with the costs of adaptation for developing country parties. Importantly, the adaptation assistance owed to particularly vulnerable countries in the UNFCCC is unqualified.[48] However, the distinction between developing and developed countries is not defined in the Paris Agreement, nor are those countries that are particularly vulnerable or with special circumstances identified, as the issue proved contentious in negotiations.[49] The Agreement leaves these terms general, aside from its references to LDCs and small island developing states, which are situated within the broader group of particularly vulnerable countries. For example, Article 9(4) is a key provision that helps specify where adaptation finance should be directed. It explains that the 'provision of scaled-up financial resources' should take into account 'the priorities and needs of developing country Parties, especially those that are particularly vulnerable to the adverse effects of climate change and have significant capacity constraints, such as the least developed countries and small island developing States'. Likewise, capacity-building efforts under the Agreement are aimed at developing country parties, particularly those with the least capacity, that includes LDCs and small island developing states.[50]

Determining which countries have special circumstances is of practical importance, as it can help answer the question of which are particularly vulnerable and should be prioritized in access to adaptation financing. It offers a means to distinguish these countries from others with increasing capacity and greater resources to adapt.[51] In this way, the principle provides a means to create distinctions within the category of developing country parties. However, while there is consensus that vulnerability serves as a way to allocate financial assistance, the identification of particularly vulnerable countries remains unresolved.[52] As the text of the Paris Agreement indicates, at a minimum the principle refers to LDCs and small island developing states. The preamble of the UNFCCC also recognizes that 'low-lying and other small island countries, countries with low-lying coastal, arid and semi-arid areas or areas liable to floods, drought and desertification, and developing countries with fragile mountainous ecosystems are particularly vulnerable'. The geographic vulnerabilities described in the operative text of the UNFCCC may shed further light,[53] as do negotiations and Conference of the

[48] UNFCCC, art 4(4).

[49] Lavanya Rajamani and Emmanuel Guérin, 'Central Concepts in the Paris Agreement and How They Evolved' in *The Paris Agreement on Climate Change* (n 40) 86–87.

[50] Paris Agreement, art 11(1).

[51] As called for in Cullet, 'Common but Differentiated Responsibilities' (n 32).

[52] See Stacy-ann Robinson and others, 'Vulnerability-Based Allocations in Loss and Damage Finance' (2023) 13 Nature Climate Change 1055, 1056.

[53] UNFCCC, art 4(8); Chapter 2, III.

Parties (COP) decisions that identify developing countries in Africa as requiring prioritization.[54] The establishment of the Climate Vulnerable Forum in 2009 sought to highlight highly vulnerable countries and to broaden the focus to further include developing countries in parts of Latin America, the Middle East, and Asia.[55] In practice, financial and technical assistance has been provided broadly to developing countries and LDCs to support adaptation planning and the preparation of related submissions and reports to the UNFCCC secretariat. However, as such support tends to be dispersed through the financial mechanisms of the regime, it is difficult to argue that this evinces a definitive interpretation of 'particularly vulnerable' based on subsequent practice, as such practice must be attributable to the parties.[56]

Finally, in its discussion of loss and damage, the first Global Stocktake (GST) recognized the importance of both 'particularly vulnerable developing countries and segments of the population that are already vulnerable owing to geography, socioeconomic status, livelihood, gender, age, minority status, marginalization, displacement, or disability'. It also recognized national efforts that included 'anticipatory action and planning' and 'planning for displacement and planned relocation', and it called for improved coherence in efforts aimed at migration, displacement, and planned relocation.[57] Thus, in addition to being aimed at particularly vulnerable countries, action and support through new and existing funding arrangements now explicitly focus on impacts on people, through measures that contemplate human mobility. As agreed in 2023, funding should 'target people and communities in climate-vulnerable situations (including women, children, youth, Indigenous Peoples, and climate-induced migrants and refugees in developing countries that are particularly vulnerable to the adverse impacts of climate change).'[58] Critically for adaptive mobility, parties also agreed that climate finance institutions and funds should also 'promote the inclusion of climate-induced migrants and refugees in their funded activities'.[59]

[54] See eg Decision 1/CP.16, 'The Cancun Agreements: Outcome of the work of the Ad Hoc Working Group on Long-term Cooperative Action under the Convention' (15 March 2011) FCCC/CP/2010/7/Add.1, para 95; but see Nicholas Chan, '"Special Circumstances" and the Politics of Climate Vulnerability: African Agency in the UN Climate Change Negotiations' (2021) 56 Africa Spectrum 314.

[55] See Robinson and others (n 52) 1056; 'Membership' (*CVFV20.Org*) <https://cvfv20.org/membership/> accessed 24 October 2024.

[56] See Vienna Convention on the Law of Treaties (adopted 23 May 1969, entered into force 27 January 1980) 1155 UNTS 331 (VCLT) art 31(3)(b).

[57] Decision 1/CMA.5, 'Outcome of the First Global Stocktake' (13 December 2023) FCCC/PA/CMA/2023/16/Add.1, paras 122, 125, 131.

[58] Decision 5/CMA.5, 'Operationalization of the New Funding Arrangements, including a Fund, for responding to loss and damage referred to in Paragraphs 2–3 of Decisions 2/CP.27 and 2/CMA.4' (6 December 2023) FCCC/PA/CMA/2023/16/Add.1, Annex II, para 18.

[59] ibid Annex II, para 26.

IV. Concrete forms of support

Assistance with adaptation can take different forms. Within the climate regime, at a minimum, the assistance owed to developing countries is financial. As Chapter 2 outlines, such support arises from obligations to assist in adaptation set out in the UNFCCC and the Paris Agreement. Developed countries must assist in 'meeting costs of adaptation' for particularly vulnerable developing countries—including those discussed in the case studies—or the 'agreed full incremental costs' needed by developing countries more broadly.[60] The Paris Agreement requires developed countries to 'provide financial resources to developing country Parties', to scale up these resources, and to work to balance financial support between adaptation and mitigation.[61] Such support is also required for efforts that protect individuals, both by virtue of being targets of adaptation and through the integration of human rights, as Chapters 3 and 4 establish. Accordingly, developing states are owed financial support in covering the costs of adaptation planning and implementation, including for adaptive mobility. This is a straightforward way in which developed countries can meet their assistance obligations.[62] Developing countries can also request such support. Bangladesh, for example, intends to pursue more funding options for its efforts to prevent, address, and mitigate climate-induced displacement, including through the adaptation finance mechanisms in the climate regime.[63] However, there is a risk—and, indeed, likelihood—that without more accountability or recognition of assistance *as an obligation*, developed countries will provide superficial support rather than meaningful assistance. Such support must also be for anticipatory measures, which broadens the scope of funding for disasters and climate change. Current funding often occurs after the fact, rather than for *ex ante* measures that can help reduce harm before the worst impacts occur.[64]

In addition to financial assistance, developed country support could come in the form of information, assistance with risk assessments, or technological support. The UNFCCC obligates developed country parties to take all practicable steps to transfer technology and knowledge.[65] The Paris Agreement builds on this, creating obligations to 'strengthen cooperative action on technology development

[60] UNFCCC, art 4(3), (4).

[61] Paris Agreement, art 9(1), (3).

[62] See eg Frank Biermann and Ingrid Boas, 'Towards a Global Governance System to Protect Climate Migrants: Taking Stock' in *Research Handbook on Climate Change, Migration and the Law* (n 45) 417 (arguing sea level rise is undisputed as a climate impact and thus costs of associated mobility must be paid for by developed countries).

[63] Tasneem Siddiqui, Mohammad Towheedul Islam, and Zohra Akhter, 'National Strategy on the Management of Disaster and Climate Induced Internal Displacement (NSMDCIID)' (Bangladesh Ministry of Disaster Management and Relief 2015) 26.

[64] See Rosemary Lyster and Maxine Burkett, 'Climate-Induced Displacement and Climate Disaster Law: Barriers and Opportunities' in *Research Handbook on Climate Disaster Law* (n 43) 108.

[65] UNFCCC, art 4(3), (5).

and transfer' and to require that financial support 'shall be provided to developing country Parties' to implement their technology-related commitments, with support balanced between adaptation and mitigation.[66] Together with the finance provisions, these commitments make clear that developing countries are owed financial and technical assistance in their adaptation efforts.

The need to provide concrete support is also confirmed outside the climate regime. The CESCR has stated that developed countries' duties of international assistance and cooperation should include support for adaptation efforts 'particularly in developing countries, by facilitating transfers of green technologies, and by contributing to the Green Climate Fund'.[67] As discussed in Section II, assistance is required by an interpretation of adaptation obligations that integrates relevant duties to assist and cooperate from human rights law, to enable developing countries to meet their core rights obligations. Such support is 'particularly incumbent on States . . . in a position to assist'.[68] For vulnerable developing countries, assistance could come in multiple forms and at different times: for example, to provide support with shifts in agricultural policies or land-use practices that enables people to stay in place for as long as possible; to address climate impacts or improve urban living conditions through assistance in planning and with infrastructure; and to assist in adaptive mobility measures or relocation planning, implementation, and monitoring when foreseeable risks make it untenable for people to remain.

An assessment of risks and needs is a precursor to actions that affect people and their rights, to ensure that adaptation reduces vulnerability and builds resilience. These assessments require adequate funding and support. For adaptive mobility this translates into financial support or technical assistance with impact assessments for facilitated migration and planned relocation. Assessments can identify areas of risk, impacts on migrants and host communities, and measures that require continuing external support. Greater clarity about potential risks offers the opportunity for more comprehensive and coherent planning, which can better ensure access to essential resources and forecast future support needs. The integration of human rights principles and rules, and the recognition of rights-related risks, can further shape concrete adaptation measures, helping to narrow the focus or balance competing interests. A rights-based approach also requires inclusive and meaningful participation in decision-making, to better ensure that adaptation responds to input from affected persons and complements local coping strategies.[69]

Furthermore, the success of adaptive mobility measures will be dictated by the conditions in places where people move. Support should accordingly extend to

[66] Paris Agreement, art 10(2), (6); see Chapter 2, IV.

[67] CESCR, '2018 Statement' (n 18) para 6.

[68] CESCR, 'General Comment No 8' (n 15) para 38.

[69] See Margaretha Wewerinke-Singh and Tess Van Geelen, 'Protection of Climate Displaced Persons under International Law: A Case Study from Mataso Island, Vanuatu' (2018) 19 Melbourne Journal of International Law 1, 31.

common destinations and places of resettlement or planned relocation, which will likely face challenges in integrating and accommodating migrants while improving adaptive capacity for all residents. As the case studies discuss, migrants frequently move to urban areas in search of new livelihood opportunities, and cities are expected to continue to grow. Yet there are systemic obstacles in these settings, that range from a lack of adequate infrastructure and difficulty accessing social services to discrimination and land tenure issues. These obstacles create and worsen socio-economic vulnerabilities. Climate impacts will further strain these areas, directly and via increased populations. Assistance with urban planning, cooperation across cities, and more formal collaborative efforts can help address known and expected challenges. As with broader adaptive efforts, support could also come through assistance in the review of relevant laws and policies or sharing best practices.[70]

The timeliness of adaption support is also important. Chapter 5 analyses the central role the precautionary principle plays in shaping adaptation measures, to better ensure that they anticipate and address significant harm before it occurs. To facilitate this, there must be a means to efficiently access finance such that support is not so slow that it undermines any efforts to undertake proactive, precautionary adaptive measures. Support must not be a one-off, and it should be available throughout the process of adaptation. Precaution provides a means to argue for such support, and for early assistance with measures that protect human rights. Mariya Gromilova advocates for a reading of climate principles alongside human rights norms: human rights law provides a 'baseline against which to measure the extent and scope of assistance' required under the climate change regime, while the precautionary principle means that 'there is no need to wait for human rights violations' before providing such assistance.[71] Thus, assistance can and should come before severe harm occurs, to support more anticipatory adaptation measures that avoid rights violations. Additionally, long-term impacts that allow for 'climate foresight' should, and as argued in Chapter 4 must, be considered in adaptation planning, policies, and funding.[72] In the Solomon Islands, for example, town planners from Australia helped develop a range of adaptation strategies that included a detailed plan for a potential planned relocation. The development of these strategies resulted in identification of various challenges—employment continuity,

[70] The C40, for example, is a global network of mayors from nearly 100 cities that offers a space for collaboration, shared learning, and planning to build equitable, resilient communities and take action on climate change. 'About C40' (*C40 Cities*) <www.c40.org/about-c40/> accessed 25 October 2024.

[71] Mariya Gromilova, 'Finding Opportunities to Combat the Climate Change Migration Crisis: The Potential of the "Adaptation Approach"' (2016) 33 Pace Environmental Law Review 105, 153.

[72] Such foresight helps prioritize adaption measures with longer time horizons as opposed to short-term coping strategies that lack enduring effects. See Michele Klein Solomon and Koko Warner, 'Protection of Persons Displaced as a Result of Climate Change: Existing Tools and Emerging Frameworks' in Michael B Gerrard and Gregory E Wannier (eds), *Threatened Island Nations: Legal Implications of Rising Seas and a Changing Climate* (Cambridge University Press 2013) 291.

provision of certain services, and loss of current assets—which all need to be addressed.[73] Similar assistance that is funded and facilitated by developed states is a means to satisfy adaptation obligations.

Developed countries could also assist with gathering, understanding, and distilling information needed by the public to participate in decision-making. This assistance helps ensure that developing countries are able to protect and fulfil procedural rights and satisfies parties' obligations to cooperate. The UNFCCC includes an obligation to cooperate in 'education, training and public awareness related to climate change'; and the Paris Agreement 'to cooperate in taking measures, as appropriate, to enhance climate change education, training, public awareness, public participation and public access to information'.[74] And while 'as appropriate' introduces discretion into an obligation, when linked to an example—facilitated migration or planned relocation—it takes on a specific context that makes mobility, and participation and education about it, appropriate and necessary.

Finally, satisfying adaptation obligations could be accomplished through support for international mobility and admittance of migrants into other states. Given the dominant security narrative attached to climate change and displacement, any argument that countries are required to admit migrants is likely to be met with staunch opposition.[75] Likewise, the legal basis for such a stand-alone duty is contestable: while general rights-based assistance obligations are relevant, there is no general right to enter a country that is not one's own. Nonetheless, some scholarship argues that a shared duty to receive international migrants exists, one that stems from states' contributions to the causes of climate change. This case has been made for small island states and those who must permanently relocate, as further discussed in Chapter 8.[76] Arguments for concrete support provided in this book give more force to an emerging duty to admit, but it will require more to make such a duty cognizable. Significant changes to the way international migration is viewed and governed would be necessary, as the status quo currently inhibits movement and limits the benefits that can be offered by mobility.[77] Assistance and cooperation would also be needed to enable a safe and legal means to migrate across borders. The forms of support needed vary depending on the countries involved, but may generally include agreements on international relocation, collaborative planning and adoption of consistent policies, and assistance for destination countries. Addressing precarious and abusive labour markets would offer more

[73] Ezekiel Simperingham, 'State Responsibility to Prevent Climate Displacement: The Importance of Housing, Land and Property Rights' in Dimitra Manou and others (eds), *Climate Change, Migration and Human Rights: Law and Policy Perspectives* (Routledge 2017) 92.

[74] UNFCCC, art 4(1)(i); Paris Agreement, art 12; see also Chapter 5.

[75] See Introduction Chapter; Chapter 1.

[76] See eg Eckersley (n 45).

[77] Chapter 1, IV.

reliable livelihood opportunities elsewhere and improve outcomes for migrants,[78] as would efforts to build networks and improve access to remittances and social services—schools, health facilities, and documentation.[79]

In summary, concrete forms of assistance and cooperation could include:

- financial, technical, or technological support for adaptive mobility measures at all stages of the process, including risk assessments, planning, implementation, and monitoring
- assistance in building capacity to enable decision-making on and implementation of adaptive mobility measures
- support and assistance with formulating relevant laws and policies
- support for inclusive participatory processes of decision-making on adaptive mobility, including planned relocation
- assistance with securing or provision of land elsewhere
- facilitating transfer of remittances
- coordination of cross-border mobility through *inter alia* admittance into another state, expansion of visas, labour migration programmes, or free movement agreements
- regional or bilateral agreements to enable cooperation and support for adaptive mobility[80]

This list is not exhaustive, but it illustrates some forms of support, with concrete measures discussed further in the case studies.

V. Barriers, open questions, and ways forward

Despite the importance of assistance and cooperation, and the book's arguments that they are legally required, there are barriers that must be resolved before any meaningful action can be taken to support mobility measures in developing countries. This book does not deeply engage with some of these, including questions of political will or about the general efficacy and enforcement of international environmental obligations. Nor does it canvas all potential financial mechanisms for the delivery of adaptation support. However, there are several specific barriers that warrant mentioning, as they provide areas of targeted inquiry and can also

[78] See François Crépeau, 'Towards a Mobile and Diverse World: "Facilitating Mobility" as a Central Objective of the Global Compact on Migration' (2018) 30 International Journal of Refugee Law 650, 656.

[79] Jon Barnett and Michael Webber, 'Accommodating Migration to Promote Adaptation to Climate Change' (The World Bank 2010) WPS5270 22–24.

[80] See Lauren Nishimura, 'Adaptation Obligations and Adaptive Mobility' (2022) 69 Forced Migration Review 39.

be partially addressed through the integration of human rights. There are, similarly, open questions that remain on the operationalization of support, with the interpretive tools used in this book able to help answer some, but not all, of these questions.

Foremost, there is currently a gap in adaptation funding. Although developed country parties are obligated to help meet the costs of adaptation for developing countries, their financial commitments to date fall far short of the estimated amount needed for adaptation. In 2009, at the COP in Copenhagen, developed countries committed to a goal of collectively mobilizing USD 100 billion annually by 2020 to address the needs of developing countries.[81] The decision adopting the Paris Agreement extended this goal to 2025 and linked it to Article 9(3) of the Agreement, which calls for developed countries to 'continue to take the lead in mobilizing climate finance from a wide variety of sources' and declares that '[s]uch mobilization of climate finance should represent a progression beyond previous efforts'.[82] The Special Rapporteur on human rights and the environment reminded developed countries of their commitment, calling on them 'to provide at least $100 billion annually to assist vulnerable States . . . in coping with the impacts of a problem that they did little to create'. He cited the process of relocation as an example of measures that require financial assistance.[83] The OECD has been tracking progress to reach this USD 100 billion goal since 2015. Its analysis for the years from 2013 through 2020, for example, found that while climate finance is increasing, it has yet to reach the goal. It estimates that for 2020, the climate finance provided for developing countries was USD 83.3 billion. Most of this financing targeted mitigation rather than adaptation, although the relative and total amounts for adaptation are on the rise.[84] Pledges have also been made to the Adaptation Fund, the Green Climate Fund (GCF), the LDC Fund (LDCF) and the Special Climate Change Fund (SCCF), but they still do not come close to meeting the goal.[85]

The amount pledged, even if met, would still fall short of current and projected needs. Estimates of adaptation costs range from USD 215 to 387 billion annually by 2030 and 500 billion by 2050.[86] These cost estimates are significantly more

[81] Decision 2/CP.15, 'Copenhagen Accord' (20 March 2010) FCCC/CP/2009/11/Add.1.

[82] Decision 1/CP.21, 'Adoption of the Paris Agreement' (29 January 2016) FCCC/CP/2015/10/Add.1, para 53; Paris Agreement, art 9(3).

[83] 'Statement of United Nations Special Rapporteur David R. Boyd on the Conclusion of His Mission to Fiji' (OHCHR 2018) <www.ohchr.org/EN/NewsEvents/Pages/DisplayNews.aspx?NewsID=23964&LangID=E> accessed 23 September 2022.

[84] See OECD, 'Aggregate Trends of Climate Finance Provided and Mobilised by Developed Countries in 2013–2020' (2022) <www.oecd.org/content/dam/oecd/en/publications/reports/2022/07/aggregate-trends-of-climate-finance-provided-and-mobilised-by-developed-countries-in-2013-2020_d1d90868/d28f963c-en.pdf/> accessed 22 September 2022.

[85] Decision 1/CMA.5 (n 57) para 79; see also Decision 5/CMP.17 'Fourth review of the Adaptation Fund' (6 November 2022) FCCC/KP/CMP/2022/9/Add.1, paras 8–11, 14.

[86] Decision 1/CMA.5 (n 57) para 68; Adaptation Committee, '25 Years of Adaptation under the UNFCCC: Report of the Adaptation Committee' (UNFCCC 2019) 9.

than the support available or promised, creating a shortfall that has been recognized within the climate change regime. The first GST under the Paris Agreement noted the widening financial gap, the inadequate levels of support for adaptation in developing countries, and the IPCC finding that adaptation financing would need to increase 'manyfold' to meet needs.[87] Similar challenges are recognized in enhancing technical assistance and funding the technology mechanism under the Paris Agreement.[88] Alongside the adaptation finance gap, the GST acknowledged that current levels of technology development and transfer for adaptation are insufficient to respond to climate impacts in developing countries.[89]

These gaps mean that developed states' adaptation obligations are already being violated, and they will continue to be violated until more funding and support is provided. The need to meet and increase existing goals for adaptation funding is frequently recognized within the COP and by governments. In 2021, for example, the UK Presidency of the COP invited ministers from Canada and Germany to develop a plan laying out how and when developed country parties could meet their annual USD 100 billion goal. The plan was 'delivered on behalf of all developed countries', and it includes guiding principles for collective action that highlight the need to increase finance for adaptation, strengthen the financial mechanism of the climate regime, and address barriers to accessing climate finance, amongst other things.[90] It does not, however, include any concrete commitments, and it was not officially adopted at the COP. Yet the COP has recognized the need to increase adaptation finance. The Glasgow Climate Pact, adopted in November 2021, urged developed country parties to 'significantly scale up' and at least double their collective financial support for adaptation, to meet needs of developing country parties and to achieve the balance between mitigation and adaptation in the scaled-up resources called for in the Paris Agreement.[91] The Pact also stressed 'the urgency of enhancing ambition and action' on adaptation.[92]

This call for urgent action on and support for adaptation continues to be repeated in subsequent climate regime outputs, as the parties work towards transitioning from adaptation planning to implementation. The framework for the GGA aims to enhance adaptation action and support. It recognizes 'the importance of the timeliness and predictability of adaptation finance and the need to accelerate efforts to enhance access to adaptation finance, with a focus on facilitating direct

[87] Decision 1/CMA.5 (n 57) paras 16(b), 81.
[88] Decision 14/CMA.5, 'Enhancing climate technology development and transfer to support implementation of the Paris Agreement' (11 December 2023) FCCC/PA/CMA/2023/16/Add.2, para 17.
[89] Decision 1/CMA.5 (n 57) para 81.
[90] See 'Climate Finance Delivery Plan: Meeting the US$100 Billion Goal' (2021) <https://ukcop26.org/wp-content/uploads/2021/10/Climate-Finance-Delivery-Plan-1.pdf> accessed 23 September 2022.
[91] See Decision 1/CMA.3, 'Glasgow Climate Pact' (8 March 2022) FCCC/PA/CMA/2021/10/Add.1, paras 15, 18; Paris Agreement, art 9(4).
[92] Decision 1/CMA.3, para 5.

access by harmonizing and simplifying access procedures'.[93] The need for such a focus arises from difficulty accessing climate funds and long-standing questions about the perceived equity, fairness, and processes involved in decision-making for these funds.[94] The Global Environment Facility (GEF), for example, manages the LDCF and SCCF, both of which are sources of adaptation funding. Criticisms of these financial mechanisms include their inefficiency and inaccessibility to the most vulnerable states, those that should be prioritized.[95] Adaptation funding is unevenly distributed, and it has not always reached those places and countries that need it most and are owed support. The Pacific Islands, for example, are facing significant impacts but receive only a small share of global funds.[96] In addition, most climate-related resources allocated under the GEF have supported mitigation projects.[97] This exacerbates access issues for developing countries, highlighting the importance of targeted adaptation funds and the need for more direct access modalities.

Confusion also surrounds what is meant by incremental costs, the need for these to be agreed upon, and the measures included in the costs that developed countries must help meet in Article 4(3) of the UNFCCC. For example, under the GEF guidelines, funding is provided to meet incremental costs for 'measures to achieve agreed global environmental benefits' beyond an established baseline.[98] This requirement introduces uncertainty about the meaning of baseline and incremental costs in the context of adaptation and the need to show global benefits.[99] Fortunately, both the LDCF and SCCF, while managed by the GEF, do not require global environmental benefits for project funding. Yet these and other adaptation funds, including the GCF, continue to grapple with issues of incremental costs and when to cover full costs of adaptation. Full costs should be covered for particularly vulnerable developing countries, in accordance with the commitments created in the UNFCCC. Article 4(4) does not reference incremental costs, instead creating an obligation for developed countries to assist particularly vulnerable developing countries in meeting the costs of adaptation. The Paris Agreement reiterated the commitment to assist particularly vulnerable developing countries 'to meet the

93 Decision 2/CMA.5 (n 23) paras 7, 25.

94 See eg IPCC (n 4) 413.

95 See Rosemary Lyster, 'Climate Justice, Adaptation and the Paris Agreement: A Recipe for Disasters?' (2017) 26 Environmental Politics 438, 452–53.

96 Carol Farbotko and others, 'Transformative Mobilities in the Pacific: Promoting Adaptation and Development in a Changing Climate' (2018) 5 Asia & The Pacific Policy Studies 393, 399.

97 Laurence Boisson de Chazournes, 'Financial Assistance' in Lavanya Rajamani and Jacqueline Peel (eds), *The Oxford Handbook of International Environmental Law* (Oxford University Press 2021) 947.

98 GEF, 'Operational Guidelines for the Application of the Incremental Cost Principle' (2007) GEF Policy Paper <www.thegef.org/sites/default/files/documents/OPERATIONAL.GUIDELINES.FOR_.THE_.APPLICATION.OF_.THE_.INCREMENTAL.COST_.PRINCIPLE_0_0_0_0.pdf> accessed 17 May 2024.

99 MJ Mace, 'Adaptation under the UN Framework Convention on Climate Change: The International Legal Framework' in W Neil Adger and others (eds), *Fairness in Adaptation to Climate Change* (The MIT Press 2006) 63–64.

costs of adaptation' through its mechanisms, and without any caveats about agreed or incremental costs.[100] This should provide a source of adaptation financing with less constraint, as evinced by the Adaptation Fund and its full cost approach.[101]

Further questions surround the nascent loss and damage funding arrangement, established in 2022.[102] It is not yet clear what the relationship of such funding will be with other finance mechanisms, including those that fund adaptation. As Chapter 2 discusses, there is overlap and interaction between loss and damage and adaptation efforts. Given their complementarity, there is a risk that developed countries might seek to discharge their obligations to provide adaptation assistance through voluntary contributions to a loss and damage fund. It is also unclear how measures related to climate mobility will be funded. Frank Biermann and Ingrid Boas argue for a specific fund for migrant protection and resettlement, in part because it would remove migration issues from competing with other adaptation measures.[103] Climate displacement is a focus within the loss and damage workstream, and decisions on the loss and damage fund state that the fund will provide financing for relocation and migration. However, parties and relevant institutions are also recommended to develop and implement 'additional funding arrangements' to address gaps in the eligibility for, access to, and adequacy of finance, with displacement, relocation, and migration specifically mentioned as requiring such additional funding.[104]

More broadly, the nature of climate finance is an obstacle to adequate support. International assistance tends to be viewed as aid rather than obligation, reinforcing the perception that contributing to adaptation funds is voluntary. Money is often lent rather than given: more than two-thirds of climate financing is provided in the form of loans, with less than 20 per cent of global adaptation financing coming in the form of grants.[105] While public financing for adaptation in low-income countries is offered more often via grants, the amount low-income countries received for adaptation overall is far less than higher-income countries.[106] The state of climate finance represents a failure to meet both legal and moral

[100] Paris Agreement, art 6(4), (6).

[101] See Adaptation Fund Board, 'Operational Policies and Guidelines for Parties to Access Resources from the Adaptation Fund (OPG)' (2022) para 15.

[102] Decision 2/CP.27, 'Funding arrangements for responding to loss and damage associated with the adverse effects of climate change, including a focus on addressing loss and damage' (20 November 2022) FCCC/CP/2022/10/Add.1.

[103] Biermann and Boas (n 62) 416–17; but see Sheila C McAnaney, 'Sinking Islands? Formulating a Realistic Solution to Climate Change Displacement' (2012) 87 NYU Law Review 1172, 1200–01 (limitations of a new fund).

[104] Decision 5/CMA.5 (n 58) Annex I, para 6; Annex II, para 17.

[105] Action Aid, 'The Vicious Cycle: Connections Between the Debt Crisis and Climate Crisis' (2023) Policy Brief <https://actionaid.org/publications/2023/vicious-cycle> accessed 9 September 2024; Global Center on Adaptation and Climate Policy Initiative, 'State and Trends in Climate Adaptation Finance 2024' (2024).

[106] OECD, 'Scaling Up Adaptation Finance in Developing Countries: Challenges and Opportunities for International Providers' (2023) 36.

obligations. It disregards the commitments made to provide financial assistance to developing countries. It ignores the ecological debt owed to these countries for the historical and continued extraction of their resources. And it perpetuates a cycle of vulnerability and financial debt many countries are locked in, which is rooted in colonialism, slavery, and exploitation that helped entrench global systems—energy, economic, and political—and that was worsened by the Covid-19 pandemic. Scholars and advocates seeking to address the damage created by these systems invoke a range of reparative actions, from debt cancellation to global tax reform.[107] Developed countries meeting and scaling up their financial commitments is also among these demands, a demand that arguments for clear legal adaptation obligations to assist and cooperate bolsters. Likewise, principles of equity, CBDR-RC, and special circumstances offer an additional basis to address the historical causes and consequences of climate change, placing the duty squarely on the developed countries that have long benefited from a fossil fuel-driven global economy. Thus, while the book focuses on international law and adaptive mobility, the normative and legal arguments that underpin its approach to adaptation and assistance can accompany, interact with, and may inform larger projects for climate reparations.[108]

Other structural and systemic barriers—increasingly recognized in discussions of adaptation—go beyond solutions rooted in international law. Indeed, some scholars point out that law itself entrenches barriers, by replicating injustices and privileging corporations or those in power.[109] This can be true, alongside the need for and benefit of finding new ways to use existing legal tools. The latter is a premise of this book, which addresses the need to clarify adaptation obligations and integrate human rights, while not discounting more radical changes that could prove essential in the absence of urgent climate action.[110] Furthermore, integrating human rights, and duties of assistance and cooperation, strengthens arguments that developed countries are obliged to support developing countries in adaptive efforts to protect people and their rights. This integration is itself important, given the resistance from developed countries to duties of support arising

[107] See Olúfẹ́mi O Táíwò and Patrick Bigger, 'Debt Justice for Climate Reparations' (Climate and Community Project 2022) <www.climateandcommunity.org/debt-justice-for-climate-reparations> accessed 9 September 2024; Ketaki Zodgekar and others, 'A Dangerous Debt-Climate Nexus: In a Warming World, Technocratic Fixes Are Inadequate Responses to Vulnerability. Achieving Climate and Economic Justice in the Caribbean Calls for Reparative Solutions' (2023) 55 NACLA Report on the Americas 319.

[108] See eg Maxine Burkett, 'Climate Reparations' (2009) 10 Melbourne Journal of International Law 509; Carmen G Gonzalez, 'Migration as Reparation: Climate Change and the Disruption of Borders' (2020) 66 Loyola Law Review 401; Olúfẹ́mi O Táíwò, *Reconsidering Reparations* (Oxford University Press 2022).

[109] See eg Maxine Burkett, 'Behind the Veil: Climate Migration, Regime Shift, and a New Theory of Justice' (2018) 53 Harvard Civil Rights-Civil Liberties Law Review 445.

[110] Discussed in the concluding chapter.

from the UN Charter and the ICESCR.[111] The cooperative framework of the climate regime and its obligations to assist anchor legal requirements to provide such support. However, whether developing countries are obligated to seek assistance is an open question. Section III establishes that in some circumstances states should request international assistance, and non-binding instruments call for a state duty to seek assistance where, for example, it lacks the capacity to adequately respond to a disaster.[112] Capacity is certain to be tested by worsening climate impacts, but the status of an *obligation* to request assistance is not settled. Regardless, a state must proactively seek funding in order to access the support available through the financial mechanisms of the climate regime.

More concretely, there is a risk that the assistance that is provided will not reach the communities and individuals in need of support. Multiple levels of coordination are required for international support to reach local communities. This accords with the recognition in the Paris Agreement that adaptation is global challenge 'with local, subnational, national, regional, and international dimensions' that requires international cooperation and assistance.[113] Yet such support must first be available, and without qualifications. Thus, in addition to the need to substantially increase the amount of finance and to provide more direct access to funds, inequities in decision-making and distribution processes must be addressed. For mobility, inequities in negotiating power could mean that developing countries are unable to retain full control over decisions about migrants or borders. This could force countries into making concessions to secure funding. Assistance could, for example, become a means to justify intervention by developed states in the domestic policies of developing countries and lead to further management and deterrence of international mobility.[114] Inequities also affect individual migrants. The responsibility and resources needed for migration fall on the migrant and their networks. Without support from states, and without anchoring action and assistance in legal obligations, this will remain the case, even as migration is increasingly viewed as an adaptive solution. Thus, the individual migrant becomes responsible for their own resilience, shifting the burden of adaptation outcomes to the people and communities facing the greatest harm from climate change, rather than the countries contributing most to its causes.[115]

[111] As discussed in John H Knox, 'Human Rights Principles and Climate Change' in Cinnamon P Carlarne, Kevin R Gray, and Richard Tarasofsky (eds), *The Oxford Handbook of International Climate Change Law* (Oxford University Press 2016) 227; Matthew Craven, 'The Violence of Dispossession: Extra-Territoriality and Economic, Social, and Cultural Rights' in Mashood Baderin and Robert McCorquodale (eds), *Economic, Social, and Cultural Rights in Action* (Oxford University Press 2007) 77.

[112] International Law Commission (ILC), 'Draft Articles on the Protection of Persons in the Event of Disasters' (2016) UN Doc A/71/10, arts 11–12.

[113] Paris Agreement, art 7(2), (6).

[114] Benoît Mayer, *The Concept of Climate Migration: Advocacy and Its Prospects* (Edward Elgar 2016) 252, 302.

[115] See Silja Klepp and Libertad Chavez-Rodriguez, 'Governing Climate Change: The Power of Adaptation Discourses, Policies, and Practices' in Silja Klepp and Libertad Chavez-Rodriguez (eds),

Likewise, the tension between state and locally led adaptation must be managed. There will be trade-offs, but adaptation should not lead to state decisions that override local plans and preferences. The interaction between these levels of governance underscores the importance of proactive, anticipatory planning and the inclusive participation of affected persons. Participation from both communities and governmental actors is essential to bridging the 'political gap' that can arise from state-based governance and that can deepen antagonisms towards those already marginalized.[116] Contextual differences must also be interrogated, to understand how rights are understood and used, not just how they could be called upon.[117] Similarly, adaptive mobility should be a part of a larger adaptation programme, but it must be driven by contextual conditions, foreseeable risks, and participatory processes that put people at the centre of decision-making. Mobility is therefore not necessarily the first or only option, but it emerges as part of a continuum of measures and efforts to adapt to changing conditions.

International climate finance mechanisms are beginning to take on board some of the inequities and obstacles that hamper access to funding. Steps have been taken to integrate human rights safeguards into the policies of adaption funding sources within the regime, including the GCF and Adaptation Fund. These are two of the Paris Agreement's primary funding mechanisms, tasked with support for LDCs and developing country parties in the creation and implementation of National Adaptation Plans (NAPs).[118] They are also potential sources to help pay for the costs of mobility-related initiatives.[119] Both funds have an 'Environmental and Social Policy', which require that all funded projects and programmes respect human rights (Adaptation Fund) or be designed or implemented in a manner that promotes, protects, and fulfils universal respect for and observance of human rights (GCF). Both include assessments that screen for social, environmental, and/or human rights impacts, and that consider how marginalized or vulnerable groups are affected.[120] Developments within climate finance mechanisms further

A Critical Approach to Climate Change Adaptation: Discourses, Policies, and Practices (Routledge 2018) 18; Lauren Nishimura, '"Climate Change Migrants": Impediments to a Protection Framework and the Need to Incorporate Migration into Climate Change Adaptation Strategies' (2015) 27 International Journal of Refugee Law 107, 119–20.

[116] See Maxine Burkett, 'Lessons from Contemporary Resettlement in the South Pacific' (2015) 68 Journal of International Affairs 75, 82–83.

[117] See Marcus Hedahl, 'Directional Climate Justice: The Normative Relationship between Moral Claim Rights and Directed Obligations' (2014) 5 Journal of Human Rights and the Environment 35.

[118] See Decision 1/CP.21 (n 82) paras 47, 55, 59–60; Decision 13/CMA.1, 'Matters relating to the Adaptation Fund' (2 December 2018) FCCC/PA/CMA/2018/3/Add.2.

[119] See Ilona Millar and Kylie Wilson, 'Towards a Climate Change Displacement Facility' in *Research Handbook on Climate Change, Migration and the Law* (n 45) 434; International Organization for Migration (IOM), 'IOM Contributions to Global Climate Negotiations: 22nd Conference of Parties to the United Nations Framework Convention on Climate Change (UNFCCC)' (2016) 5.

[120] See eg Adaptation Fund Board, 'Environmental and Social Policy' (Adaptation Fund 2016); Green Climate Fund Board, 'Environmental and Social Policy' (2018) Decision B.19/10, para (b), Annex X.

improve the prospects for adaptation funding: GCF projects are now country-led, an improvement from a system that initially allowed donor governments to dictate terms for accessing funds. Additionally, the GCF must now allocate half of its funds to adaptation, and half of that to particularly vulnerable countries. To that end, the COP approved funding through the GCF for NAPs of up to USD 3 million per country.[121] For human mobility, there are 'encouraging signs' that financial support is growing, given its incorporation into some of the GCF's recent projects.[122]

However, questions remain about how to allocate the costs of adaptive mobility. Differentiation within the climate change regime dictates that developed countries contribute to adaptation funds and thus provides a starting point to distribute the costs associated with planning for and facilitating adaptive mobility. Support is connected to both capacity and responsibility, in part through the CBDR-RC principle. The principle is, as discussed in Section III, subject to development and debate, raising practical questions about how much each developed country owes and whether amounts should be based on current ability to pay, historical contributions to the causes of climate change, how much a state has benefited from emissions, or some combination of these considerations. These questions do not undermine the core legal arguments in the book, and they leave room for both legal and political solutions. Principles of equity and non-discrimination, from the climate regime and human rights law, could help resolve allocation issues at the intergovernmental level. As Dinah Shelton argues, 'any allocation of benefits and burdens that makes vulnerable populations worse off, even if the harm is felt outside the boundaries of the state, cannot be regarded as equitable or in conformity with international human rights law'.[123] Developing states may therefore call upon developed states to refrain from climate action that creates extraterritorial harm and to proactively assist in adaptation, including in measures to facilitate adaptive mobility.

VI. Conclusion

Adaptation obligations to assist and cooperate add a collective dimension to support, which can help shift the burden of adaptation measures and adaptive mobility. These obligations ensure that adaptation is not an effort that states must undertake on their own. At a minimum, they create a duty for developed countries to help pay

[121] Decision 10/CP.22, 'Report of the Green Climate Fund to the Conference of the Parties and guidance to the Green Climate Fund' (31 January 2017) FCCC/CP/2016/10/Add.1.

[122] Task Force on Displacement, 'Report of the Task Force on Displacement, Recommendations' (2018) para 81.

[123] Dinah Shelton, 'Equitable Utilization of the Atmosphere: A Rights-Based Approach to Climate Change' in Stephen Humphreys (ed), *Human Rights and Climate Change* (Cambridge University Press 2010) 124.

for the costs of adaptation in developing countries. Bolstered by the integration of human rights law and the operative principles of the regime, this chapter argues that these obligations can direct support towards adaptation efforts that protect people, including adaptive mobility measures. Such an interpretation requires support for those countries, and the vulnerable people in them, that face some of the greatest harms but have contributed the least to climate change. Understanding adaptation obligations in this way affords arguments to enhance and broaden the scope of responsibility for climate-related mobility. It can strengthen calls for a range of concrete forms of support: from funding and technical assistance to the admittance of migrants. Without adequate assistance—which can facilitate proper planning and access to resources—governments are more likely to employ ad hoc responses that fail to protect people or their rights. Adaptation requires significant capital and resources, although inaction will likely prove more costly. Yet substantial gaps in finance and barriers to support exist. This chapter identified some of these, while also arguing that clearer pathways for support are necessary, possible, and begin with developed countries meeting their obligations. Wealthy countries must contribute more funding and assistance. They must satisfy their commitments in good faith and provide support that is continuous, timely, and accessible. Doing so plays a crucial role in whether and how adaptive mobility measures are undertaken, with adequate support better enabling anticipatory action and adaptive mobility.

7
Low-lying Areas: Internal Mobility in Bangladesh

I. Introduction

The previous chapters established the legal basis to argue that states must plan for and facilitate adaptive mobility, arguing for the integration of relevant human rights duties and an interpretation of adaptation obligations that includes the operative principles in climate treaties. It also established the legal obligations of developed countries to support developing countries in their adaptation efforts. The next three chapters will illustrate these arguments in three different contexts. These case studies explore circumstances that give rise to obligations to enable adaptive mobility, through an interpretation of adaptation obligations that integrates relevant human rights and the precautionary principle. Each chapter considers vulnerable developing countries entitled to support from developed countries within the climate regime. Each also focuses on vulnerable individuals and communities whose subsistence and survival are put at risk by the impacts of climate change. These are the individuals that should be prioritized, as a matter of international law and climate justice, and as argued for through the interpretive process that underpins adaptive mobility in this book.

This chapter focuses on Bangladesh. Like the case studies that follow, it first draws on empirical studies and research to highlight a range of climate change impacts. Impacts often affect multiple resources, creating overlapping and interconnected risks that together could trigger the need for adaptive action. Bangladesh is a highly vulnerable country, with much of its population living in low-lying areas where people are expected to face significant and foreseeable risks that require adaptation. The chapter next considers different forms of mobility, including past and predicted movement. Many people already migrate within Bangladesh, and the government is preparing for a significant increase in internal displacement as climate impacts worsen. The chapter contrasts this displacement with approaches that require anticipatory, planned adaptive measures, including on mobility. It then turns to the international assistance and cooperation owed to Bangladesh by developed countries, exploring the support needed to ensure that adaptive mobility is possible and those affected lead the process.

Climate Change, Human Rights, and Adaptive Mobility. Lauren Sakae Nishimura, Oxford University Press.
 DOI: 10.1093/9780198930068.003.0008

Alongside the subsequent case studies, this chapter makes the case for the integration of human rights into adaptation early and often, in planning, implementation, and monitoring. This is not a panacea for the many and complex issues surrounding climate adaptation. Nor does it mean that adaptive mobility measures are required in all circumstances. States retain discretion in selecting appropriate measures. However, the action needed to satisfy adaptation obligations, including facilitating migration or planned relocation, offers a means to help ensure the rights of some of those most vulnerable to climate impacts. Together with adequate support, such action can lead to better outcomes for people, which helps avoid precarious, forced movement and allows people to move with more choice and access to rights and resources.

II. Climate impacts, vulnerabilities, and risks to people and their rights

Bangladesh is perennially considered one of the most climate-vulnerable countries in the world, its ranking a product of recent and long-term risks.[1] It is a Least Developed Country (LDC) facing a wide range of adverse effects from climate change: sea level rise, flooding, more intense and frequent storms, riverbank and coastal erosion, saline intrusion, and severe droughts. These effects are increasingly foreseeable and predicted to occur sooner and with more severity as mitigation goals are not met.[2]

Bangladesh has long dealt with flooding and extreme weather events. It contains areas that are low-lying, prone to natural disasters, and with fragile ecosystems—all of which are identified by the United Nations Framework Convention on Climate Change (UNFCCC) as requiring special consideration. Eighty per cent of Bangladesh is low-lying coast or river floodplains, and 30 to 70 per cent of the country is flooded each year.[3] Sea level rise poses an increasing risk to individuals and communities. Almost half of Bangladesh's population live at 5 metres or less

[1] See eg David Eckstein, Vera Künzel, and Laura Schäfer, 'Global Climate Risk Index 2021: Who Suffers Most From Extreme Weather Events? Weather-Related Loss Events in 2019 and 2000 to 2019' (Germanwatch eV 2021); Rajib Shaw, Yong Luo, and Tae Sung Cheong, 'Asia' in HO Pörtner and others (eds), *Climate Change 2022: Impacts, Adaptation and Vulnerability. Contribution of Working Group II to the Sixth Assessment Report of the Intergovernmental Panel on Climate Change* (Cambridge University Press) 1467.

[2] See Joern Birkmann and others, 'Poverty, Livelihoods and Sustainable Development' in *Climate Change 2022: Impacts, Adaptation and Vulnerability* (n 1) 1235; Ove Hoegh-Guldberg and others, 'Impacts of 1.5°C Global Warming on Natural and Human Systems' in V Masson-Delmotte and others (eds), *Global Warming of 1.5°C. An IPCC Special Report on the impacts of global warming of 1.5°C above pre-industrial levels and related global greenhouse gas emission pathways, in the context of strengthening the global response to the threat of climate change, sustainable development, and efforts to eradicate poverty* (Cambridge University Press 2018) 203, 231.

[3] Saleemul Huq and Jessica Ayers, 'Climate Change Impacts and Responses in Bangladesh' (2008) IP/A/CLIM/NT/2007-09 2.

above sea level.[4] Even with large, immediate cuts to global greenhouse gas emissions, the population exposed to risks from rising sea levels is expected to double by 2100, with higher emission scenarios creating significantly greater risks.[5] Yet the impacts of climate change do not occur in isolation, often interacting with or exacerbating other hazards. The combination of sea level rise and changes to sea surface temperatures can increase the frequency and severity of tropical cyclones, cause shifts in seasonal monsoons, and inundate land, threatening Bangladesh with more harm than any other non-island nation.[6] Sea level rise also interacts with tidal and storm surges to increase coastal flooding and erosion.[7] More prolonged and severe flooding will further reduce access to land and freshwater. Large portions of the country's coastal areas are at risk from these impacts. The combination of slow-onset climate events with storm surges and other sudden-onset events threatens up to 30 per cent of Bangladesh's coastal area in the coming decades.[8]

Sea level rise will likewise worsen saltwater intrusion, which currently reaches up to 100 kilometres inland and reduces access to freshwater. Twenty million people in Bangladesh already experience salinity in their drinking water, and 26 million people are exposed to very high salinity in groundwater.[9] If temperatures continue to rise at their current pace, salinization will reach over 40 per cent of the country's Ganges-Brahmaputra Delta.[10] Human activity—such as dams, river diversions, embankments, and over-abstraction—exacerbates impacts, including salinization, and increases the risks of downstream flooding.[11] Extreme rainfall, expected to become more frequent, can also lead to more surface flooding, soil erosion, and waterlogging. In 2017, for example, floods impacted 220,000 hectares

[4] Displacement Solutions, 'Climate Change Displaced Persons and Housing, Land and Property Rights: Preliminary Strategies for Rights-Based Planning and Programming to Resolve Climate-Induced Displacement' (2010) 22 <https://globalprotectioncluster.org/sites/default/files/2023-04/Climate_Change_Displaced_Persons_2010_EN.pdf> accessed 30 May 2024; Gordon McGranahan, Deborah Balk, and Bridget Anderson, 'The Rising Tide: Assessing the Risks of Climate Change and Human Settlements in Low Elevation Coastal Zones' (2007) 19 Environment and Urbanization 17, 26.

[5] Ben Strauss and Scott Kulp, 'Bangladesh and the Surging Sea: A Vulnerability Assessment with Projections for Sea Level Rise and Coastal Flood Risk' (Climate Central 2017) 9–11.

[6] Kanta Kumari Rigaud and others, 'Groundswell: Preparing for Internal Climate Migration' (World Bank 2018) 146.

[7] Ebru Kirezci and others, 'Projections of Global-Scale Extreme Sea Levels and Resulting Episodic Coastal Flooding over the 21st Century' (2020) 10 Scientific Reports 11629.

[8] Jane McAdam, *Climate Change, Forced Migration, and International Law* (Oxford University Press 2012) 163; Ministry of Environment, Forest and Climate Change (MoEFCC), Government of the People's Republic of Bangladesh, 'Third National Communication of Bangladesh' (2018) 191–93 (identifying areas at risk of salinization).

[9] Mohammad Shamsudduha and others, 'Multi-Hazard Groundwater Risks to Water Supply from Shallow Depths: Challenges to Achieving the Sustainable Development Goals in Bangladesh' (2020) 12 Exposure and Health 657, 664; Potsdam Institute for Climate Impact Research and Climate Analytics, 'Turn Down the Heat: Climate Extremes, Regional Impacts, and the Case for Resilience' (World Bank 2013) 109.

[10] Hoegh-Guldberg and others (n 2) 232–33.

[11] See Rigaud and others (n 6) 146; MA Caretta and others, 'Water' in HO Pörtner and others (eds), *Climate Change 2022: Impacts, Adaptation and Vulnerability* (n 1) 611.

of paddy crop and contributed to a 30 per cent increase in rice prices, with climate change doubling the likelihood of such flooding.[12]

These adverse effects put people's health, livelihoods, and the enjoyment of related rights to life and to an adequate standard of living at risk. The Committee on Economic, Cultural and Social Rights (CESCR) expressed concern about the impacts of climate change on the enjoyment of rights in Bangladesh, despite the country's 'negligible contribution' to the causes of climate change.[13] Impacts include those that affect health: salinization increases risk of hypertension from salt consumption, and the incidence and areas of exposure to climate related water- and vector-borne diseases is expected to increase, causing more people to suffer from yellow fever, cholera, dengue, diarrhoea, and malaria.[14] Furthermore, climate change can create stress, anxiety, and mental health impacts, arising from both the damage caused and worry about future prospects.[15] Impacts to livelihoods are a further concern. Many in the country are dependent on land and natural resources for their food and livelihoods, with up to two-thirds of people involved in farming activities in some capacity and nearly half the labour force employed through agriculture.[16] The adverse effects of climate change threaten subsistence by disrupting water supplies, food production, and resource quality and availability. Salinization, for example, affects food sources and limits the availability of freshwater fish in coastal and delta areas, where fisheries provide between 50 to 60 percent of the animal protein for people's diets.[17]

Similarly, extreme events like cyclones and storm surges combine with other climate impacts to hamper livelihoods and put lives at risk. For example, shrimp farmers and fry catchers are often detrimentally affected by damage to coastal aquaculture from these events, with farmers observing that rising temperatures and changes to rainfall patterns affect production and lower the growth rate of shrimp.[18] Individuals and communities have learned from past events and adapted their homes and livelihoods. However, the pace and scale of climate impacts will test people's traditional coping strategies.[19] Individuals are already responding to extreme events, salinization, and riverbank erosion through recourse to prohibited livelihood activities. A study

[12] Caretta and others (n 11) 585; Ruksana H Rimi and others, 'Risks of Pre-Monsoon Extreme Rainfall Events of Bangladesh: Is Anthropogenic Climate Change Playing a Role?' (2019) 100 Bulletin of the American Meteorological Society S1.

[13] CESCR, 'Concluding observations on the initial report of Bangladesh' (18 April 2018) UN Doc E/C.12/BGD/CO/1, para 13.

[14] See Shaw, Luo, and Cheong (n 1) 1508; Ministry of Environment and Forests, Bangladesh, 'NAPA: Updated Version' (2009) 33; Third National Communication (n 8) 159–63.

[15] Birkmann and others (n 2) 1209.

[16] Third National Communication (n 8) 182; Ministry of Environment and Forest Government of the People's Republic of Bangladesh, 'National Adaptation Programme of Action (NAPA)' (2005) 23.

[17] Valentina Lauria and others, 'Importance of Fisheries for Food Security across Three Climate Change Vulnerable Deltas' (2018) 640–41 Science of The Total Environment 1566; Shaw, Luo, and Cheong (n 1) 1491.

[18] See Md Adibul Islam, Md Shariful Islam, and Md Abdul Wahab, 'Impacts of Climate Change on Shrimp Farming in the South-West Coastal Region of Bangladesh' (2016) 3 Research in Agriculture Livestock and Fisheries 227.

[19] McAdam (n 8) 162.

of the impact of environmental stressors on livelihoods found that when adaptation strategies and support are limited, many people turn to 'illegal livelihoods' banned by the government. In coastal Bangladesh, these livelihood strategies include use of fine mesh nets to collect shrimp fry and logging in the Sundarbans. These strategies are often deployed by poor and highly vulnerable individuals, with law enforcement measures only worsening vulnerability.[20] Yet there is evidence that some coastal communities are diversifying their livelihoods in the aftermath of cyclones, as well as self-organizing to build capacity and prepare for future extreme weather events. Social capital and networks played a key role, and findings from these cases highlight the importance of community participation and knowledge sharing.[21]

Consideration of community dynamics further requires recognition of the disparate distribution of risks. The government of Bangladesh recognizes the need for adaptation measures and institutional support, to sustain livelihoods and improve health, well-being, and access to food and drinking water.[22] In doing so, it analyses vulnerability as this book does, by considering risks that arise from geography and social identity. It identifies those experiencing poverty, women, children, the elderly, and persons with disabilities as more likely to suffer a disproportionate share of harm from climate change.[23] Women and children have historically experienced greater harm from extreme events. Inequities are being replicated in existing policies and local adaptation measures, which can overlook gender imbalances and reinforce marginalization and disparities in food insecurity for women.[24] Furthermore, where short-term gains are prioritized over long-term prospects, environmental change leads to worsening conditions for agriculture, income, and health—particularly for women.[25]

Conditions in urban settings can also heighten vulnerability. Bangladesh has one of the highest population densities in the world and one of the highest growth rates for urban populations.[26] Its densely populated areas are prone to flooding, erosion, and salinization. Poverty increases vulnerability, including in urban areas.

[20] Istiakh Ahmed and others, 'Climate Change, Environmental Stress and Loss of Livelihoods Can Push People towards Illegal Activities: A Case Study from Coastal Bangladesh' (2019) 11 Climate and Development 907.

[21] See M Salim Uddin and others, '"Disasters Threaten Livelihoods, and People Cope, Adapt and Make Transformational Changes": Community Resilience and Livelihoods Reconstruction in Coastal Communities of Bangladesh' (2021) 63 International Journal of Disaster Risk Reduction 102444.

[22] MoEFCC, 'National Adaptation Plan of Bangladesh (2023–2050)' (Bangladesh NAP) (2022) 69–79.

[23] ibid 42–8; NAPA 2009 (n 14) 33–34; Government of the People's Republic of Bangladesh, MoEF, 'Bangladesh Climate Change Strategy and Action Plan' (BCCSAP) (2009) 13.

[24] See Camille Parmesan and others, 'Terrestrial and Freshwater Ecosystems and Their Services' in *Climate Change 2022: Impacts, Adaptation and Vulnerability* (n 1) 784, 793; Mostafa AR Hossain and others, 'Impacts and Responses to Environmental Change in Coastal Livelihoods of South-West Bangladesh' (2018) 637–38 Science of The Total Environment 954.

[25] See Intergovernmental Panel on Climate Change (IPCC), *Managing the Risks of Extreme Events and Disasters to Advance Climate Change Adaptation*, A Special Report of Working Groups I and II of the IPCC (Field CB and others (eds), Cambridge University Press 2012) 452.

[26] Rigaud and others (n 6) 145.

Those living in slums and informal settlements, many of whom have migrated from rural areas, are highly vulnerable. They face greater risk of disease and harm to their health from climate impacts.[27] Bangladesh is also home to a large Rohingya refugee population, who are made more vulnerable due to their living conditions and exposure to environmental hazards. Precarious living conditions contribute to ongoing risks to people's lives, livelihoods, and basic rights. Indeed, current living conditions for urban migrants and refugees often violate human rights: many reside in inadequate housing and lack access to essential resources like drinking water, healthcare, and education.[28]

III. Current and predicted mobility

Climate and other environmental impacts have long played a role in human mobility, contributing to new and shifting patterns of movement in Bangladesh. Such movement tends to occur within the country, which will continue to be the case in the context of climate change.[29] Globally, up to 80 per cent of all human mobility is within countries.[30] Likewise, mobility as an adaptation strategy will primarily be internal, particularly in the absence of international cooperation and anticipatory, proactive planning and action.

Bangladesh is experiencing high levels of internal mobility. Internal migrants amount to as much as 10 per cent of the population in the country.[31] The Internal Displacement Monitoring Centre (IDMC) reports that between 2008 and 2023, 18.8 million people have been internally displaced, more than a million people on average annually.[32] Nearly all of these are weather-related displacements, and this figure is likely an underestimate as it does not capture the contribution of most slow-onset events and impacts. Furthermore, as Chapter 1 discussed, the effects of climate change—particularly slow-onset events—rarely operate as the exclusive driver of human mobility. The adverse impacts that arise from disasters almost always result from a confluence of social, economic, and environmental factors. Yet

[27] Sabina Faiz Rashid, Showkat Gani, and Malabika Sarker, 'Urban Poverty, Climate Change and Health Risks for Slum Dwellers in Bangladesh' in Rajib Shaw, Fuad Mallick, and Aminul Islam (eds), *Climate Change Adaptation Actions in Bangladesh* (Springer 2013) 294, 317.

[28] See Rabab Fatima, Anita Jawadurovna Wadud, and Sabira Coelho, 'Human Rights, Climate Change, Environmental Degradation and Migration: A New Paradigm' (2014) International Organization for Migration (IOM) and Migration Policy Institute (MPI) Issue in Brief.

[29] Chapter 1, III.

[30] W Neil Adger and others, 'Human Security' in Christopher B Field and others (eds), *Climate Change 2014: Impacts, Adaptation, and Vulnerability. Part A: Global and Sectoral Aspects. Contribution of Working Group II to the Fifth Assessment Report of the Intergovernmental Panel on Climate Change* (Cambridge University Press 2014) 767.

[31] Amina Maharjan and others, 'Migration and Household Adaptation in Climate-Sensitive Hotspots in South Asia' (2020) 6 Current Climate Change Reports 1.

[32] 'Bangladesh' (IDMC) <www.internal-displacement.org/countries/bangladesh/> accessed 30 May 2024.

the latter play an important and increasingly central role in human movement. In coastal areas, for example, sudden-onset events like cyclones drive large-scale displacement, with other climate events, including sea level rise and salinization, creating long-term risks. Out-migration, weak growth, and population losses in these areas are, in part, attributed to the interaction of these increasing risks.[33] More generally, people use temporary, seasonal, and circular migration as a livelihood strategy or to cope with changing conditions, and they may move multiple times when faced with repeated environmental hazards.[34] There are well-established patterns of seasonal internal migration in northern Bangladesh, for example, that follow annual shifts from rainy to dry periods in agricultural areas.[35] Increases in temporary, short-distance migration out of rural areas also occur during droughts and periods of extreme heat.[36] Access to water and crop productivity will continue to influence migration out of rural areas, including rural-to-rural migration. In rice-growing areas in the northeast, out-migration is likely to follow loss of crop productivity, with movement toward rainfed and irrigated cropland areas in the Ganges River Basin where water availability is expected to improve.[37] Movement is also predicted to become more permanent due to changes in flooding and risks from sea level rise, with slow-onset events contributing to internal movement over longer distances.[38]

Much of the internal mobility in Bangladesh is rural-to-urban movement, a trend that is anticipated to grow as climate impacts intensify. The rural poor are more frequently displaced, generally over the short term in search of labour, but this can be followed by more permanent migration.[39] People often leave environmentally fragile areas and head to cities, where their stay becomes long-term or permanent. Global predictions estimate up to 60 per cent of people could live in urban areas by 2030 and 70 per cent by 2050; without proper planning most could be living in slums and working in informal employment sectors.[40] The majority of internal migration is to cities in Bangladesh, which has steadily increased in recent decades. A large portion of this increase is movement towards the garment industry in Dhaka and Chittagong.[41] Yet movement from rural areas to

[33] See Matthew Walsham, 'Assessing the Evidence: Environment, Climate Change and Migration in Bangladesh' (IOM 2010); Richard Marshall and Shibaab Rahman, 'Internal Migration in Bangladesh: Character, Drivers and Policy Issues' (UNDP 2013) 20.

[34] See McAdam (n 8) 167–69.

[35] Marshall and Rahman (n 33) 8.

[36] G Cissé and others, 'Health, Wellbeing and the Changing Structure of Communities' in *Climate Change 2022: Impacts, Adaptation and Vulnerability* (n 1) 1083, 1085.

[37] Rigaud and others (n 6) 157.

[38] D Dodman and others, 'Cities, Settlements and Key Infrastructure' in *Climate Change 2022: Impacts, Adaptation and Vulnerability* (n 1) 930.

[39] See Cissé and others (n 36) 1084.

[40] Adger and others (n 30) 767; Rigaud and others (n 6) 35.

[41] Kristin Dreiling and others, 'People on the Move: Strengthening Adaptation Responses to Support Human Movement in a Changing Climate' (United States Agency for International Development (USAID) 2021) Technical Report 46.

urban centres is primarily driven by push factors, including poverty and livelihood constraints associated with climate and environmental impacts, rather than pull factors and perceived opportunities.[42] On average, around 2,000 people move to Dhaka daily.[43] Dhaka is Bangladesh's capital and largest city, and its slum population doubled in size from 1996 to 2006 following significant rural-to-urban migration.[44] A reported 70 per cent of those in these slums moved after experiencing some form of environmental hardship.[45] Displacement is leading and will continue to lead to growth in Dhaka and Bangladesh's other urban areas, with many vulnerable migrants ending up in slums. Conditions in slums heighten vulnerability and make it more likely that residents will face difficulties accessing basic resources and enjoying their fundamental human rights. The lack of public services is a pressing policy concern within the country, as is the rapid growth of urban areas.[46]

Migration requires financial and social capital. Those without adequate resources may not move or may undertake precarious journeys. Women and children, for example, may be unable to move because of constraints to capital and social barriers. Poverty can likewise create or increase vulnerabilities and limit mobility. An individual's vulnerability can, in turn, compound risks during migration and upon arrival. It can also affect where people move, leading some of the most vulnerable migrants to end up living in conditions that exacerbate economic and social marginalization. These conditions mean that individuals internally displaced from elsewhere in Bangladesh, often to urban areas, can experience losses of identity and belonging, unable to return home despite longing to move.[47] Indeed, migrants in the region have faced multiple crises from climate impacts and the Covid-19 pandemic, although adaptation policies that include social services and account for a wider range of factors contributing to mobility and immobility can reduce vulnerabilities.[48]

[42] Abul Barkat, Asmar Osman, and SK Ali Ahmed, 'Bangladesh Migration Governance Framework' (MiGOF) (IOM 2020) 56.

[43] Sadiq Khan and Mohammad Atiqul Islam, 'World Leaders Must Prepare for the Climate Migration Challenge' (*Mayors Migration Council*, 17 December 2021) <https://mayorsmigrationcouncil.org/news/climate-migration-oped-migrants-day/> accessed 31 May 2024.

[44] Ministry of Environment and Forests, Government of the People's Republic of Bangladesh, 'Second National Communication of Bangladesh' (2012) 175.

[45] See Deepti Mahajan, 'No Land's Man: Migration in a Changing Climate', *On the Move: Migration Challenges in the Indian Ocean Littoral* (Stimson Center 2010) 9 (citing IOM); Mi Zhou and Dorien Braam, 'Community Resilience and Disaster-Related Displacement in South Asia' (Norwegian Refugee Council (NRC) 2015) Nansen Initiative Research Paper (noting no statistics but that those in Dhaka slums claim the environment negatively affects their lives).

[46] See Section IV; Dreiling and others (n 41) 18, 46.

[47] See Sonja Ayeb-Karlsson, Dominic Kniveton, and Terry Cannon, 'Trapped in the Prison of the Mind: Notions of Climate-Induced (Im)Mobility Decision-Making and Wellbeing from an Urban Informal Settlement in Bangladesh' (2020) 6 Palgrave Communications 1; Birkmann and others (n 2) 1208, 1219; Dodman and others (n 38) 930.

[48] See Georgina Cundill and others, 'Toward a Climate Mobilities Research Agenda: Intersectionality, Immobility, and Policy Responses' (2021) 69 Global Environmental Change 102315; Shaw, Luo, and Cheong (n 1) 1470; Section IV.

While most analysis of climate in Bangladesh highlights the internal mobility that is occurring and will continue to occur, international labour migration is likely to be undertaken by those who have the resources and connections. Both internal and international labour migration have long been used as a means to diversify or supplement income sources in South Asia. Migrants, often young men, send remittances home that can help reduce environmental and non-climatic risks.[49] Labour migrants from Bangladesh often travel to Gulf States and Malaysia. India is also a destination for migrant workers, and for international migrants in general from Bangladesh.[50] Movement from Bangladesh to India has occurred following previous sudden-onset events; these, along with other past migrations, account for between 12 and 17 million people who have moved from Bangladesh to India since the 1950s.[51]

Bangladesh is also home to nearly a million stateless Rohingya refugees from Myanmar. Bangladesh is not party to the 1951 Refugee Convention and does not have a legally binding framework for refugees.[52] It maintains that stay for refugees is temporary and on humanitarian grounds. The vast majority live in the world's largest network of refugee camps in Cox's Bazar, although more than 30,000 refugees have been relocated to the island of Bhasan Char since 2021.[53] The conditions in the camps have been criticized, both for conditions that violate the right to an adequate standard of living and for constraining freedom of movement. Government authorities have closed community-led schools, destroyed shops, and restricted travel within and between camps.[54] These camps are also highly exposed to environmental hazards and vulnerable to the adverse effects of climate change. Floods, landslides, and the risk of extreme weather

[49] Maharjan and others (n 31).

[50] See Deutsche Gesellschaft für Internationale Zusammenarbeit (GIZ) and International Labour Organization (ILO), 'Labour Market Trends Analysis and Labour Migration from South Asia to Gulf Cooperation Council Countries, India and Malaysia' (GIZ and ILO 2015) 16, 18; MiGOF (n 42) 38. However, most of these destination countries are not parties to the International Convention on the Protection of the Rights of All Migrant Workers and Members of Their Families (ICRMW), which provides rights protections for labour migrants. UN Treaty Collection, 'Status of the ICRMW' <https://treaties.un.org/Pages/ViewDetails.aspx?chapter=4&clang=_en&mtdsg_no=IV-13&src=TREATY> accessed 30 May 2024.

[51] The Nansen Initiative, 'Climate Change, Disasters, and Human Mobility in South Asia and Indian Ocean' (2015) Nansen Initiative Background Paper 14, 16.

[52] However, one indicator of the government's goal of engaging with partners on migration is to ratify the Refugee Convention. MiGOF (n 42) 24.

[53] The government of Bangladesh and UN High Commissioner for Refugees (UNHCR) jointly issued documentation to refugees in Cox's Bazar and signed a Memorandum of Understanding to establish a framework and joint commitment to ensure access to protection and essential services in Bhasan Char. UNHCR, 'Bangladesh' (*Global Focus*) <https://reporting.unhcr.org/operational/operations/bangladesh> accessed 30 May 2024.

[54] UN Human Rights Council (UNHRC), 'Report of the Special Rapporteur on the Situation of Human Rights in Myanmar, Thomas H Andrews' (13 June 2022) UN Doc A/HRC/49/76, Annex II (Bangladesh mission); Human Rights Watch, 'Bangladesh: New Restrictions on Rohingya Camps' (4 April 2022) <www.hrw.org/news/2022/04/04/bangladesh-new-restrictions-rohingya-camps> accessed 30 May 2024.

events, storms, and tidal surges are influencing mobility within the camps and beyond. One study found that environmental conditions are, at a minimum, an indirect driver of migration within and out of the camps, interacting with concerns about security, conflict, and better access to basic resources and livelihoods. Shelter is often inadequate to protect against weather and sudden-onset events, with reports of extreme heat and precipitation affecting living conditions, livelihoods, and health. Climate impacts can also exacerbate intersectional risks women and girls face in these camps. Given restrictions on movement, however, it is likely that many refugees may be unable to move, despite a desire to leave.[55]

Without proper planning and adaptive action that ensures rights in anticipation of foreseeable risks, the status quo will persist. Those vulnerable will be left worse off, and the displacement that has occurred in the past will likely be replicated. For example, case studies of displacement in southwest and coastal Bangladesh, due largely to coastal flooding, erosion, and storm surges, found that with few other options many were forced to a nearby narrow embankment. This left almost 90 per cent of those displaced without access to livelihood opportunities and with little assistance from the government.[56] These kinds of precarious displacements will increase in magnitude and scope without concerted climate action—putting the country at greater risk than nearly any other for widespread population movement—with estimates of climate change contributing to the displacement of 6.5 million and another 20 to 35 million expected to be forced from their homes by mid-century.[57] Estimates for the impact of slow-onset events and their influence on displacement range from 1.9 million to 19.9 million internal migrants by 2050, depending on the level of continued greenhouse gas emissions and investment in development pathways.[58] Absent intervention, the government of Bangladesh estimates that by 2050 one in every seven people will be displaced due to climate impacts, comprising nearly half of the projected internal migrants of the South Asian region as a whole.[59]

55 See Jessica Marsh and Mohammad Aziz, 'Impact of Climate Change on the Migration and Displacement Dynamics of Rohingya Refugees' (Danish Refugee Council 2024).

56 UN Habitat, 'Forced Evictions, Global Crisis, Global Solutions' (2011) 56; Displacement Solutions (n 4) 4–6.

57 Second National Communication (n 44) 175; UN Habitat (n 56) 56; Displacement Solutions (n 4) 16, 22. While this provides a rough picture, sweeping estimates can be inaccurate or justify harmful measures as discussed in Chapter 1, III.

58 Rigaud and others (n 6) 148.

59 Tasneem Siddiqui, Mohammad Towheedul Islam, and Zohra Akhter, 'National Strategy on the Management of Disaster and Climate Induced Internal Displacement (NSMDCIID)' (Bangladesh Ministry of Disaster Management and Relief 2015) 4; Bangladesh NAP (n 22) 47 citing Viviane Clement and others, 'Groundswell Part 2: Acting on Internal Climate Migration' (World Bank 2021).

IV. Integration of human rights and adaptive mobility

Displacement is not inevitable. As this book argues, if Bangladesh interprets and applies its obligations to act on adaptation in a way that integrates human rights duties, then measures should be taken that boost resilience and both enable people to remain in place and to plan for adaptive mobility. Bangladesh is party to eight of the core international human rights treaties, and its constitution includes obligations to guarantee human rights and to protect and improve the environment for present and future generations.[60] Accordingly, when taking climate action Bangladesh must ensure human rights, based on its domestic and international human rights obligations and as the preamble of the Paris Agreement reminds parties.[61] It must also take action because these human rights duties are relevant and shape obligations that trigger adaptation. Adaptation measures that focus on short-term risks can offer essential resources to meet immediate needs. In coastal Bangladesh, for example, initial adaptation measures include raising houses above coastal flood levels or harvesting rainwater for drinking where salinization is prevalent.[62] Yet these kinds of measures may make it more likely that people stay in increasingly hazardous areas, leading to maladaptation if not combined with more sustainable solutions that address longer-term risks. As Section II establishes, more of Bangladesh's land is predicted to become incompatible with continued access to food, water, and housing over time. Anticipatory action is needed now, to better address these access issues in the future. Adaptation plans and policies must therefore account for current conditions and longer-term foreseeable changes, particularly given that return will not be possible to certain places, as floods take over land and coastal areas further erode.

Bangladesh adaptation assessment and planning began two decades ago. It has created multiple relevant planning documents and policies, including its National Adaptation Programme of Action (NAPA) (2005 and 2009), National Adaptation Plan (NAP) (2022), Bangladesh Climate Change Strategy and Action Plan (2009), Bangladesh Climate and Gender Action Plan (2013, updated 2024), Bangladesh Delta Plan 2100 (2018), updated Standing Order on Disaster (2019), and three National Communications to the UNFCCC. Together, they establish that adaptation programmes should recognize the inequalities and societal constraints that make some people more vulnerable than others. The government repeatedly recognizes the 'special needs' of marginalized and vulnerable groups,

[60] See The Constitution of the People's Republic of Bangladesh (4 November 1972) preamble, arts 11, 18A.

[61] See Chapter 3, III. The preamble acknowledges states should 'respect, promote and consider' their human rights obligations when taking action to address climate change.

[62] Michele Klein Solomon and Koko Warner, 'Protection of Persons Displaced as a Result of Climate Change: Existing Tools and Emerging Frameworks' in Michael B Gerrard and Gregory E Wannier (eds), *Threatened Island Nations: Legal Implications of Rising Seas and a Changing Climate* (Cambridge University Press 2013) 291–92.

for adaptation planning and action and for responses to 'displacement issues'; it includes in these groups women, children, the elderly, Indigenous people, and persons with disabilities.[63] Bangladesh's Climate Change and Gender Action Plan, for example, recognizes that women 'often lead the way in adapting to climate change impacts'. The Plan, in part, aims to take action to support women who have migrated due to climate change through livelihood activities in rural and urban areas.[64]

Overall, Bangladesh's adaptation planning considers 'climate migrants' as a vulnerable group in need of adaptive strategies, particularly in urban settings. Migration is not generally viewed as an adaptive strategy, but rather a problem the government seeks to solve. Consequently, planning aims to address mobility after the fact or pursues measures that 'discourage mobility'. Bangladesh's NAP, for example, includes activities to develop an inventory of potential and existing climate migrants, offer livelihood training to discourage mobility, improve migrants' financial conditions, and provide training for coping with new environments. It also claims it will reduce the vulnerability of 15 million climate migrants if implemented.[65] Yet the NAP's discussion of proactive mobility measures is limited to a few mentions of planned relocation for settlements at high risk of cyclones, floods, and erosion, and it notes that some coastal communities may need to be relocated before 2050.[66] The only detailed discussion of mobility is a project to house 'climate refugees' by constructing 139 five-storey buildings that can 'rehabilitate 4,409 climate refugee families' in Cox's Bazar, where the Rohingya refugee camps are also located. This adaptive intervention is envisioned for 'affected vulnerable landless and homeless families', meaning support will occur only after displacement and significant upheaval to their lives.[67]

The NAP includes short-, medium-, and long-term adaptation strategies and actions, aligned with the nature of adaptation as a process that must consider longer time scales to address changing conditions. Its broader perspective can likewise encompass the evolving measures needed to provide basic resources—adequate food, water, housing, and healthcare—as required by the integration of human rights duties. Bangladesh's national adaptation goals aim to ensure these resources, incorporated into strategies that seek to develop climate-resilient agriculture for food, nutrition, and livelihood security and climate-smart cities, amongst others. It plans to accomplish its adaptation goals through livelihood diversification, changes in agricultural practices, building climate-resilient housing and shelters, and construction and rehabilitation of infrastructure and flood and drainage management

[63] See n 23; NSMDCIID (n 59) 12.
[64] Ministry of Environment and Forest, 'Bangladesh: Climate Change and Gender Action Plan' (2013) 22, 44.
[65] Bangladesh NAP (n 22) 11, 68, 77, 95, 99, 149–50, 162.
[66] ibid 80, 100.
[67] ibid ii, 5.

measures.[68] Little mention is made of human rights in its adaptation strategies, although other related documents and policies discuss rights concerns. For example, the country's Third National Communication to the UNFCCC mentions barriers to human rights access that women can face due to violence, and discrimination in land and asset ownership, employment, and participation in decision-making. It also highlights the protection of the rights of children in discussing adaptation and education.[69]

Human rights take a more central role in relevant policy frameworks on human mobility. In 2015, the government recognized that 'a clear gap' in its policy framework existed surrounding climate mobility, as its NAPA and initial climate planning documents only make cursory mention of climate-related migration and displacement.[70] It has attempted to fill this gap with its National Strategy on the Management of Disaster and Climate Induced Internal Displacement (NSMDCIID), which looks to proactive and human rights-based strategies for all stages of displacement. It is Bangladesh's first policy to directly explore the connection between the environment, climate change, and human mobility. While the NSMDCIID includes adaptation and disaster risk reduction (DRR), it focuses primarily on managing and responding to internal displacement.[71] Accordingly, like the subsequent NAP and climate analyses, mobility is viewed as a negative consequence of climate change and not an adaptive strategy that could minimize risks to people and their rights. Yet the NSMDCIID establishes the government's willingness to incorporate human rights into its climate strategies, particularly related to mobility. Indeed, Bangladesh identified the same rights as this book—to life, housing, health, food, and water—as essential to the management of mobility.[72]

More recently, in 2020, Bangladesh adopted a Migration Governance Framework (MiGOF), based on an internationally agreed MiGOF template approved by member states of the International Organization for Migration (IOM). The MiGOF seeks to set out a clear, comprehensive vision for migration governance and guide policy and relevant activities of government actors. It covers both international and internal migration and aims to ensure safe migration, aligned with the Global Compact for Safe, Orderly and Regular Migration (GCM). Bangladesh's MiGOF targets multiple forms of mobility in a whole government approach, and it includes several goals relevant to climate change. First, in the short term, it contemplates development of a national strategy on addressing migratory movement caused by climate change. Second, in the medium term, it seeks to govern climate change-induced migration and reintegration of displaced persons.

[68] ibid 72–79; NSMDCIID (n 59) 14–17.
[69] Third National Communication (n 8) 28, 179–81.
[70] NSMDCIID (n 59) 5; Second National Communication (n 44) 175 (brief discussion of link between disasters and migration).
[71] NSMDCIID (n 59) 5–6.
[72] Siddiqui, Islam, and Akhter (n 59) 12, 18.

Finally, its overall aim includes respecting human rights, and it incorporates a principle of adherence to international standards and the fulfilment of migrants' rights as guidance.[73]

The MiGOF was followed by two documents: a National Strategy on Internal Displacement Management in 2021 (Strategy) and the National Action Plan (2022–2042) to Implement the National Strategy on Internal Displacement Management (Action Plan) in 2022. Climate and disaster displacement are the focus of the Strategy and the Action Plan, both of which take a rights-based approach and offer suggestions to cover three phases of displacement: prevention and measures before displacement, protection during displacement, and durable solutions after displacement. Both engage more than prior migration strategies with adaptation measures, as a key means to reduce vulnerability and prevent displacement. Notably, the Strategy underscores the need to prepare 'vulnerable people for potential migration and/or relocation/resettlement in a context when local adaptation and prevention is no longer a viable option'. And it identifies the facilitation of temporary or circular international labour migration as an adaptive measure that could help avoid displacement.[74] The Action Plan builds on this, developed through a participatory process with local policymakers and representatives from communities identified as vulnerable. Its self-described 'explicitly rights-based, globally appropriate, and locally applicable' approach offers 'rights-based and community-centric solutions'.[75]

The Action Plan provides a promising example of how a government might incorporate more anticipatory action to address displacement. It makes a direct, detailed connection between human rights, adaptation, and preventing displacement, and several of its key targets contemplate mobility as an adaptive action. These targets primarily focus on international labour mobility, with governmental activities aimed at encouraging remittances, creating financial products for labour migrants, and enabling skill development and livelihood training. There is, however, a reference to potential agreements with labour-receiving countries, and pursuit of technical assistance to do so from the IOM. Planned relocation is also discussed for people already displaced, although there is mention of its use to help avoid displacement.[76] Together, the MiGOF, Strategy, and Action Plan signal progress in coordinating policy efforts to manage migration, and in recognition of the role adaptation plays in reducing vulnerability and preventing displacement. If implemented, the measures they include will be critical to managing the risks from

[73] MiGOF (n 42) 2, 6, 11, 19, 30.

[74] Ministry of Disaster Management and Relief (MoDMR), Government of Bangladesh, 'National Strategy on Internal Displacement Management' (2021) 11, 18.

[75] MoDMR, 'National Action Plan (2022–2042) to Implement the National Strategy on Internal Displacement' (2022) 1–2.

[76] ibid 3, 17, 26–28, 38–39.

climate impacts in Bangladesh.[77] Yet obstacles to coordination exist, reportedly from fragmentation across agencies.[78] Moreover, aside from the notable exceptions in these documents, relevant strategies and adaptation plans do not deeply engage with mobility *as* a form of adaptation and instead consider adaptive interventions to prevent displacement or to address risks after people have moved. The framing of mobility as a problem persists, as evident in the 2022 Mujib Climate Prosperity Plan's aim to reduce climate-induced migration, and to reach 'zero migration' by 2041.[79]

Thus, there remains a need for more concrete anticipatory action to plan for and facilitate migration. This may take several forms. The government could, for example, focus on internal options: it could facilitate migration by first and foremost recognizing mobility as an adaptation strategy. It could then seek to ensure safe, rights-protective conditions for migrants during transit and in their destinations. This would broaden the current focus from protecting rights during and after displacement to ensuring rights for all forms of mobility, including adaptive mobility.[80] Doing so will require the government to account for the discrimination migrants might face. Bangladesh has repeatedly recognized the disproportionate impact climate change has on certain groups, but it must move beyond recognition to proactively pursue planning processes and measures that reduce barriers to access and enable inclusive participation. It must continue to connect national plans with local impacts, a connection not always made in prior adaptation planning. Such a disconnect, where local contexts and practices are not considered, can lead to maladaptation.[81]

To address potential constraints on resources and livelihood opportunities in places of resettlement or planned relocation, adaptation plans and policies should consider migrants and receiving communities. Efforts to build communities that are physically and socially resilient will be needed, to offer more sustainable outcomes and to help ensure that mobility is adaptive when undertaken. Accordingly, and consistent with an approach that integrates human rights, state and local governments must work together to ensure access to essential resources,

[77] See Tasneem Siddiqui, Md Ekhtekharul Islam, and Tamim Billah, 'Policy Architecture to Address Disaster and Climate Change Induced Displacement in Bangladesh' (Researching Internal Displacement 2023) <https://researchinginternaldisplacement.org/short_pieces/policy-architecture-to-address-disaster-and-climate-change-induced-displacement-in-bangladesh/> accessed 6 December 2024.

[78] See Serena Tagliacozzo, Sonja Ayeb-Karlsson, and Bayes Ahmed, 'An Integrated Governance Framework to Map Out and Act on the Interrelationships between Human Mobility and Disaster Risk' (2023) 50 Journal of Ethnic and Migration Studies 1277, 1289.

[79] MoEFCC, 'Mujib Climate Prosperity Plan 2022–2041' (2022).

[80] See Action Plan (n 75).

[81] See HM Tuihedur Rahman and Gordon M Hickey, 'What Does Autonomous Adaptation to Climate Change Have to Teach Public Policy and Planning About Avoiding the Risks of Maladaptation in Bangladesh?' (2019) 7 Frontiers in Environmental Science 2; Dodman and others (n 38) 964.

health services, and employment opportunities.[82] Policies, laws, and institutional coordination should reflect the need to provide these necessities, while aligning with an anticipatory, rights-based approach to human mobility. Such an approach could include the transformation of cities and towns into 'migrant-friendly' spaces. Researchers have argued that this option makes sense for a populous country like Bangladesh, with few areas that are not 'vulnerable hotspots'.[83] Creation of these migrant-friendly spaces would facilitate a form of adaptive mobility that allows individuals to choose to move to these locations. Something similar is also proposed in the Action Plan, which calls for the creation of peri-urban areas and urban growth centres, the latter through public-private partnership at regional and local levels. The Action Plan notes that these centres will reduce pressure from displacement on mega cities like Dhaka and Chittagong.[84]

In addition, some community-based planned relocation is likely necessary. Planned relocation is the coordinated movement of people away from disaster or risk-prone areas. Unlike temporary evacuation, it is considered a permanent solution. It is typically considered a measure of last resort, a view shared in the Action Plan.[85] Planned relocation can be used in several ways in the context of climate change. The first use is as a preventative measure, to move people away from a hazardous location; the second is to enable a durable solution within a country for those already displaced; and the third, considered exceptional, is international relocation where large portions of a state are not fit for habitation.[86] Planned relocation may also be considered to make room for mitigation or adaptation projects, as has occurred for large-scale development projects. Bangladesh's current policies and strategies recognize planned relocation as both a potential adaptation strategy and a durable solution for those displaced, a view which is supported by community-based initiatives in parts of the country.[87] Discussion of planned relocations are thin on detail in the government's existing plans, and will require much more foresight, assessment, and effort before they are an option in earnest. The significant costs and losses they entail should not be understated. They are often

[82] States must provide access to basic services and essential resources as required by human rights law and the Guiding Principles. See Chapter 4; United Nations Commission on Human Rights (UNCHR), 'Guiding Principles on Internal Displacement' (11 February 1998) UN Doc E/CN.4/1998/53/Add.2, principle 18(2).

[83] Mizan R Khan and others, 'High-Density Population and Displacement in Bangladesh' (2021) 372 Science 1290.

[84] MoDMR, Government of Bangladesh (n 74) 49–50.

[85] ibid 17 (planned relocation 'as a strategy of last resort for avoiding displacement').

[86] See The Nansen Initiative, 'Agenda for the Protection of Cross-Border Displaced Persons in the Context of Disasters and Climate Change' vol 1 (2015) para 96; Jane McAdam and Elizabeth Ferris, 'Planned Relocations in the Context of Climate Change: Unpacking the Legal and Conceptual Issues' (2015) 4 Cambridge Journal of International and Comparative Law 137, 139.

[87] See Action Plan (n 75) 17, 38–39; Bangladesh NAP (n 22) 80, 100; NSMDCIID (n 59) 11, 21, 24; Ezekiel Simperingham, 'State Responsibility to Prevent Climate Displacement: The Importance of Housing, Land and Property Rights' in Dimitra Manou and others (eds), *Climate Change, Migration and Human Rights: Law and Policy Perspectives* (Routledge 2017) 91.

'politically and emotionally charged, may not necessarily be seen as desirable by exposed populations and are most successful when approached proactive[ly] and strategically to avoid increasing the socioeconomic vulnerability of those who are relocated'.[88]

Valuable insights can be found in past relocations, exposing the risks of forcible population movements and processes that are not inclusive. In Bangladesh, the government's planned relocation of people displaced by riverbank erosion in the southeastern part of the country, as well as the relocation of Rohingya refugees to Bhasan Char, have shown the harm and rights violations that can arise, in part because of a lack of consultation and consent from those affected.[89] To avoid replicating these outcomes, any future planned relocation should comply with international laws and standards. They must therefore be participatory and inclusive. They should involve affected people in planning, implementation, and monitoring; account for barriers to participation; include adequate compensation; and offer lasting access to essential resources and livelihoods.[90] To do so, those facilitating planning processes should first consider how decisions will be made, who is making them, and what meaningful participation entails. These processes should aim for informed consent, while also providing agency and options to enable direct involvement and control.[91]

Proper planning is also required to prepare for the unavoidable growth of urban areas, particularly as climate impacts put greater constraints on rural and coastal communities. Bangladesh illustrates a context to explore how the integration of human rights can shape planning that better accommodates the movement to urban centres that is already underway. Accomplishing this would mean a change from current mobility patterns, marked by migration into informal urban settlements that often leaves migrants worse off.[92] The outcomes of urban migration as an adaptation strategy are mixed: in many cases it leads to greater insecurity and poverty; yet it can still serve as a tool to diversify income and provide remittances.[93] Governments tend to characterize the spontaneous movement of people from rural to urban areas as something to deter, rather than a positive adaptation

[88] Cissé and others (n 36) 1118 (citations omitted).

[89] See Khan and others (n 83).

[90] For further discussion of planned relocation, see Chapter 8, IV; see also Cissé and others (n 36) 1118; Anne Tadgell, Brent Doberstein, and Linda Mortsch, 'Principles for Climate-Related Resettlement of Informal Settlements in Less Developed Nations: A Review of Resettlement Literature and Institutional Guidelines' (2018) 10 Climate and Development 102.

[91] See Chapter 5, III; Erica Bower, Rachel Harrington-Abrams, and Betsy Priem, 'Complicating "Community" Engagement: Reckoning with an Elusive Concept in Climate-Related Planned Relocation' (2024) 88 Global Environmental Change 102913.

[92] See Section III.

[93] Walsham (n 33) 25; Office of the UN High Commissioner for Human Rights (OHCHR), 'The Slow Onset Effects of Climate Change and Human Rights Protection for Cross-Border Migrants' (22 March 2018) UN Doc A/HRC/37/CRP.4, para 87.

strategy.[94] For Bangladesh, internal migration is viewed as key driver of urbanization that is both an obstacle to and component of national development.[95] Adaptation strategies that anticipate and accept continued urban migration are more likely to contribute to development gains. By foregrounding people and their rights, and by shifting the focus towards long-term resilience, these strategies can also improve outcomes for migrants.

More specifically, if the government facilitates continued movement to large urban centres like Dhaka and Chittagong—which will occur absent intervention—it must do so in a way that does not put people in precarious positions or further perpetuate the growth of slums. The government has acknowledged the need to improve conditions in urban slums, provide security of tenure, and shift mobility away from precarious areas. It noted that urban planning does not adequately account for climate change, nor has internal migration been the focus for those government actors addressing climate impacts.[96] None of the existing social safety programmes target internal mobility, and the benefits they offer do not travel with migrants. The lack of coordination or clear institutional management of internal mobility is a further barrier, as is the lack of focus on improving conditions so that 'migration is a choice rather than a necessity'.[97] But Bangladesh recognizes these obstacles, and it is contemplating ways forward in planning documents. For example, its MiGOF identifies making social benefits portable as a key element to advance the socioeconomic well-being of migrants and improve migration governance.[98] Similarly, its most recent climate communication recognizes the importance of urban development planning, which must account for 'the risks, vulnerabilities and adverse impacts of climate change'. The communication explains that planning processes should emphasise those most vulnerable, namely the urban poor, and ensure that decentralized, holistic development occurs to encourage people to stay in urban centres other than Dhaka and Chittagong.[99] Given that both cities are expected to be subject to sea level rise, increased storm surges, or flooding from nearby rivers—and thus be places of out- as well as in-migration—Bangladesh plans to focus on the growth of smaller urban areas to reduce pressure on already overwhelmed infrastructure in larger cities. This aligns with the suggested development of migrant-friendly cities and towns, as well as the proposed peri-urban areas in the Action Plan.[100] The NSMDCIID proposes a

[94] Koko Warner and others, 'National Adaptation Plans and Human Mobility' (2015) 49 Forced Migration Review 8.

[95] Marshall and Rahman (n 33) 5; MiGOF (n 42) 43.

[96] See NSMDCIID (n 59) 23; MiGOF (n 42) 58; Second National Communication (n 44) 175.

[97] MiGOF (n 42) 58; see also Tagliacozzo, Ayeb-Karlsson and Ahmed (n 78) 1289.

[98] MiGOF (n 42) 10.

[99] Third National Communication (n 8) 203; see also General Economics Division, Bangladesh Planning Commission, 'Making Vision 2041 a Reality: Perspective Plan of Bangladesh 2021–2041' (February 2020) 45.

[100] See nn 83, 84 and accompanying text.

range of efforts in these alternative areas, which include generating new employment opportunities; providing affordable housing and educational opportunities; improving transit to allow for longer commutes; and ensuring access to healthcare, water, sanitation, and electricity.[101] The Action Plan, similarly, calls for creation of 'low-cost housing, educational institutions, healthcare facilities, access to safe drinking water, sanitation and electricity'.[102] If implemented, many of these measures would satisfy adaptation obligations to address risks to rights in urban areas. They would also be a significant undertaking, one that is likely to require financial and technical assistance.

Furthermore, adaptation obligations that integrate relevant human rights must incorporate procedural guarantees in the implementation of adaptation, including adaptive mobility measures. The CESCR's concluding observations to Bangladesh confirm the need for climate action to be 'formulated and implemented on the basis of human rights and with the meaningful participation of affected communities and civil society'.[103] The latter includes community-based groups, which have been key actors in efforts to organize and address displacement due to coastal flooding. For example, around 200 community-based groups formed the Association of Climate Refugees to work on climate-related mobility issues and empower those at risk in Bangladesh.[104] The importance of inclusive participation to migration outcomes is apparent in Bangladesh, evident in its absence in past planned relocation and decisions on mobility.[105] The government recognizes rights to access information and to participate in decision-making in general, as well as the need to carry out assessments to understand vulnerability for climate-related mobility specifically.[106] It must do more, however, to ensure procedural rights in practice. To do so, it must prioritize individuals and groups that face barriers to participating in decision-making, who are also often constrained in their mobility. Women, for example, face obstacles to meaningful participation due to social inequalities that, along with the gendered division of labour and perceived lack of agency, can limit mobility.[107] Resettlement or planned relocation without the participation and against the will of those affected risks loss of livelihood, cultural expression and practices, and displacement while undermining the legitimacy of climate measures.[108] Moreover, as Jane McAdam observed in her analysis of Bangladesh, keys to successful adaptation are the acceptance of those who must adapt, their perception of effective strategies, and measures that allow for a continued life with dignity.[109]

101 NSMDCIID (n 59) 17.
102 Action Plan (n 75) 50.
103 CESCR (n 13) para 14. See also Chapter 5.
104 UN Habitat (n 56) 56. However, this group no longer appears to be active.
105 See Khan and others (n 83).
106 See Action Plan (n 75); NSMDCIID (n 59); Bangladesh NAP (n 22) 118.
107 Shaw, Luo, and Cheong (n 1) 1514.
108 Adger and others (n 30) 779; Alice Kaswan, 'Adaptation Justice' in Daniel A Farber and Marjan Peeters (eds), *Climate Change Law*, vol I (Edward Elgar 2016) 607.
109 McAdam (n 8) 174–75 (citations omitted).

Finally, while internal mobility is and likely will continue to be the predominant form of movement in Bangladesh, international movement is also a potential adaptive strategy. In general, international labour migration could be scaled up to provide opportunities for migrants to diversify income and boost adaptive capacity at home via remittances. Remittances already generate significant economic benefits, and they are an adaptation strategy that Bangladesh has promoted to transfer risks and shift livelihood opportunities.[110] However, for international mobility to truly be adaptive, more emphasis on living conditions and legal protections is needed to foster positive integration abroad. Support is needed to enable movement as well, given the significant costs and resources often required to move across borders. Bangladesh's strategies incorporating labour migration recognize that constraints need to be addressed systematically, while also highlighting the importance of reducing the costs of migration and that loans will likely be required to facilitate international mobility.[111] However, scholars criticize labour migration and the reliance on remittances, which keep the onus of adaptation on migrants and maintains an unfair economic status quo.[112]

In addition to existing labour mobility arrangements, other state-facilitated programmes are proposed to enable international migration. These include the expansion of regional labour markets, bilateral or regional mobility arrangements, seasonal international migration, and the creation of temporary or circular migration programmes with other countries.[113] But to improve resilience and reduce vulnerability, these measures—like internal mobility options—must include participatory processes and ensure human rights. Domestic practices provide a starting point to meet international commitments and the legal duties owed to those living within Bangladesh's borders. The government's treatment of immigrants and refugees must therefore change, as housing or relocating people in environmentally fragile locations clearly runs afoul of any approach that integrates human rights. Such a change should, for example, translate into better engagement with and measures for Rohingya refugees, to ensure their access to rights and resources. In addition, the conditions of those who remain must be factored into adaptation and mobility policies. Migration can leave those who stay more vulnerable to future risks from disasters.[114]

[110] See Action Plan (n 75) 26; NSMDCIID (n 59) 15, 24; Bangladesh Planning Commission (n 99) 45.

[111] Bangladesh Planning Commission (n 99) 45; NSMDCIID (n 59) 23.

[112] See Carol Farbotko and others, 'Transformative Mobilities in the Pacific: Promoting Adaptation and Development in a Changing Climate' (2018) 5 Asia & the Pacific Policy Studies 393, 399.

[113] See Action Plan (n 75) 26–29; MiGOF (n 42) 10, 24, 26–30; NSMDCIID (n 59) 15; McAdam (n 8) 209–10; The Nansen Initiative, 'Climate Change, Disasters, and Human Mobility in South Asia and Indian Ocean' (2015) Outcome Report 9–12.

[114] See Michelle Yonetani, 'Mapping the Baseline: To What Extent Are Displacement and Other Forms of Human Mobility Integrated in National and Regional Disaster Risk Reduction Strategies?' (Platform on Disaster Displacement (PDD) 2018) 32.

V. International assistance and cooperation

To meet its adaptation goals, Bangladesh recognizes that it will need assistance in its efforts. The government highlights the importance of financial and technical support from developed countries, calling for compensation for climate impacts and for developed countries to meet their promise of enhanced climate finance.[115] For mobility, it sees a need to '[s]eek international cooperation and support in case internal relocation is not possible' in international forums and through diplomacy.[116] The CESCR also encouraged Bangladesh to strengthen its international cooperation to help mobilize the 'financial and technological support to which it is entitled' for climate action.[117] As an LDC and particularly vulnerable country, Bangladesh is owed special consideration and support from developed countries under the UNFCCC. Such international support is vital to efforts to prevent displacement, to allow for in situ adaptation, and to put concrete plans into action for adaptive mobility. For the latter, costs may include those needed for urban infrastructure rehabilitation or mobility programmes, to assess and monitor relevant mobility measures, or to implement planned relocation.

To date, Bangladesh has received adaptation funding from the Least Developed Countries Fund (LDCF), Adaptation Fund, Green Climate Fund (GCF), and the Global Environment Facility (GEF). In 2010, it also established the Bangladesh Climate Change Resilient Fund (BCCRF) and the Bangladesh Climate Change Trust Fund (BCCTF). The latter is resourced from its own budget, the former derives funds from developed countries, but both are intended to help implement the country's climate action plan (BCCSAP). The BCCRF is implemented by the government, with the World Bank providing technical and secretariat support. As of 2018, the BCCRF had dispersed over USD 130 million to address climate change in the country, with another USD 400 million allocated to the BCCTF.[118] In its NAP, however, the government highlighted the significant gap between the international financing provided and the current and projected needs. Estimated costs for adaptation to tropical cyclones alone is over USD 55 billion by 2050, with another USD 26 billion for inland monsoons and over USD 23 billion to fund its adaptation projects under the Bangladesh Delta Plan 2100.[119] Ad hoc funding may also come from loss and damage funds. One non-profit organization, Young Power in Social Action (YSPA), has provided new homes for families displaced from their homes along the Chittagong coast due to erosion. YSPA's work is funded from the

[115] Bangladesh NAP (n 22) 2; MiGOF (n 42) 13; BCCSAP (n 23) 31.

[116] Action Plan (n 75) 11.

[117] CESCR (n 13) para 14.

[118] See Third National Communication (n 8) XIV, 206, 230; SM Munjurul Hannan Khan, Saleemul Huq, and Md Shamsuddoha, 'The Bangladesh National Climate Funds: A Brief History and Description of the Bangladesh Climate Change Trust Fund and the Bangladesh Climate Change Resilience Fund' (International Institute for Environment and Development 2011).

[119] Bangladesh NAP (n 22) 110–11.

Scottish government, the first country to earmark funding for climate change loss and damage.[120]

The timing of support is also crucial to adaptation outcomes.[121] For example, evidence from the use of forecast-based-financing (FbF) in the country shows that it led to better access to essential resources, including food and water, and a means for safer evacuation. FbF provides funding when forecasts for an extreme event reach a set threshold. It allows for anticipatory action and helps prevent families from accruing additional debts from high-interest loans. FbF was used in 2017, in advance of flooding in the Brahmaputra River Basin in northern Bangladesh, and in 2020, before Cyclone Amphan. In both cases, assistance in advance of an extreme event reduced harm and led to better outcomes for recipients compared to similarly situated people.[122] The need for early support is also evident on a larger scale, to help initiate ambitious adaptation plans in Bangladesh. So too is the benefit of funding that is consistent, continuous—a requirement of the Paris Agreement—and that allows for longer-term action. Bangladesh has argued that adaptation financing must be substantially scaled up so that it can develop and implement adaptation programmes, which include measures related to internal and international migration, planned relocation, or 'climate refugees'. It has also argued for developed states to accept 'climate refugees'.[123]

In addition to financial assistance, Bangladesh plans to pursue bilateral agreements and to engage with regional cooperative processes.[124] These kinds of cooperation are necessary to afford migrants legal rights and means to live abroad. Bilateral, multilateral, or regional agreements can provide pathways for international migrants, while also satisfying adaptation obligations to assist and cooperate. More concretely, cooperative efforts could include labour agreements, free movement agreements, liberalization of visas, or providing land for relocation, all of which would need to integrate procedural and substantive human rights protections.[125] Easing barriers to or facilitating remittances are further forms of

[120] Sadiqur Rahman, 'How Bangladesh Is Supporting Climate Refugees' *BBC* (8 December 2023) <www.bbc.com/future/article/20231206-how-bangladesh-is-supporting-climate-refugees> accessed 31 May 2024.

[121] Chapter 6, IV.

[122] See Clemens Gros and others, 'Household-Level Effects of Providing Forecast-Based Cash in Anticipation of Extreme Weather Events: Quasi-Experimental Evidence from Humanitarian Interventions in the 2017 Floods in Bangladesh' (2019) 41 International Journal of Disaster Risk Reduction 101275; Ezekiel Simperingham, 'Climate Displacement and Community-Led Solutions: Perspectives from the Asia-Pacific Region' in Simon Behrman and Avidan Kent (eds), *Climate Refugees: Global, Local and Critical Approaches* (Cambridge University Press 2022) 228–29; IFRC, 'Acting Early to Protect Bangladesh's Vulnerable Population from Cyclone Amphan' (*Forecast-based Financing*) <www.forecast-based-financing.org/2020/05/19/cyclone-amphan-is-approaching-bangladesh/> accessed 31 May 2024.

[123] Ad-hoc Working Group on Long-term Cooperative Action, 'Ideas and Proposals on the Elements Contained in Paragraph 1 of the Bali Action Plan: Submissions from Parties (Part I)' (19 May 2009) FCCC/AWGLCA/2009/MISC.4 (Part I) 27.

[124] MiGOF (n 42) 25–26.

[125] Discussed in Chapters 9, V, 10, V.

assistance that can quickly benefit receiving communities. More generally, coordination will be needed to ensure procedural rights and the participation of affected people in home and destination countries.

There are also complexities and limits to cooperation and support that must be navigated, particularly with neighbouring countries. India and Bangladesh, for example, share ecosystems that cross borders, and both countries consistently rank amongst the most climate-vulnerable.[126] Yet their political and economic differences will factor into how support and cooperation unfold between the two countries. India considers itself a developing country for the purposes of adaptation funding and support, and thus could also be entitled to assistance in adaptation efforts.[127] Its status as a developing country would also mean that it does not owe financial assistance to Bangladesh, at least not by virtue of any commitment made in climate treaties. But joint access to the climate regime's financial mechanisms could enable adaptation projects to be undertaken cooperatively, through measures that affect both countries. Likewise, mobility between the two countries has a long history, which will continue to be influenced by environmental factors. India's diversion of water from rivers that flow into Bangladesh is a long-standing source of tension between the two countries that has contributed to migration within and across borders.[128] Any future rerouting and shifts in water access could lead to more migration. The two countries also share the Sundarbans, the largest mangrove forest in the world, where sea level rise and extreme weather events are shaping mobility within and between the countries.[129] Under climate treaties, each country has independent obligations to undertake anticipatory adaptation and to ensure human rights, and both have cooperative obligations. At a minimum, these obligations should lead to policies to coordinate any adaptation with cross-border impact, including on mobility. Bangladesh's leadership in international policy efforts evinces a willingness to cooperate on human mobility issues: it pushed for inclusion of migration in the Sustainable Development Goals, advocated for the GCM, and served as Chair of the Platform on Disaster Displacement.

[126] See Architesh Panda, 'Climate Induced Migration and Interdependent Vulnerabilities between Bangladesh and India' in Bishawjit Mallick and Benjamin Etzold (eds), *Environment Migration and Adaptation: Evidence and Politics of Climate Change in Bangladesh* (A H Development Publishing House 2015).

[127] See eg Government of India, 'Submission of India for Consideration by the Standing Committee on Elements in the Development of the Further Guidelines for the Fifth Review of the Financial Mechanism' (2013) Submission to the Conference of the Parties (COP); Ministry of Finance, Government of India, 'Climate Summit for Enhanced Action: A Financial Perspective from India' (2019).

[128] See eg Ataur Rahman and others, 'A Review and Analysis of Water Research, Development, and Management in Bangladesh' (2022) 14 Water 1834, 9; Li Xu and James S Famiglietti, 'Global Patterns of Water-Driven Human Migration' (2023) 10 WIREs Water e1647, 16.

[129] See eg MA Caretta and others, 'Water' in *Climate Change 2022: Impacts, Adaptation and Vulnerability* (n 1) 588–90, 611.

VI. Conclusion

Bangladesh's current plans and policies signal that it prioritizes both climate adaptation and mobility, and that it is already taking action to adapt. The government retains discretion in how to satisfy its adaptation obligations, which is consistent with the Paris Agreement's language that adaptation must be implemented 'as appropriate'.[130] However, the predicted impacts in certain vulnerable areas will render in situ adaptation efforts insufficient and, at some point, mobility will be the most 'appropriate' response to protect against the foreseeable risks to people's lives, health, and access to essential resources. Addressing these risks will require more anticipatory action, and a reframing of mobility as a potential solution rather than a problem to address only after people move. The country's current preparation for increased displacement is an important step, but its aim to eliminate climate-related mobility going forward belies the reality of the risks climate impacts pose. To ensure rights are not violated and obligations to adapt are met, Bangladesh should instead continue to pursue plans and measures to build resilient urban centres and facilitate mobility to these locations. Urban migration is and will continue to be the predominant form of mobility in the country, and investing in adequate social services and access to resources will benefit both current residents and new migrants. At the same time, some communities in coastal or heavily flooded areas may need more permanent resettlement. Longer-term plans should account for this need and put measures in place to ensure rights-based transitions.

Bangladesh also illustrates the importance of connections between national planning and local practice, which further requires the inclusive participation of affected individuals and communities. These connections are crucial to address different local needs and to adequately plan for a wide range of adaptation measures, from agricultural practices and livelihood diversification to internal adaptive mobility and planned relocation. Likewise, a balance of strategies over different time scales will be required to ensure that existing indebtedness, precarity, and risks are not exacerbated or deferred. If it succeeds in striking a balance while responding to community preferences, Bangladesh is far more likely to meet its adaptive goals. Yet it cannot do so without the assistance and cooperation it is owed. As this chapter shows, regional and international support will play a critical role in adaptation, and in whether and how mobility is undertaken.

[130] Paris Agreement (adopted 12 December 2015, entered into force 4 November 2016) 3156 UNTS 79, art 7(9); Chapter 2, IV.

8
Small Island States: Choosing to Stay and Planning to Move in the Pacific

I. Introduction

This chapter provides a case study of the Pacific Islands. Along with chapters that focus on Bangladesh and the Sahel, it shows the need for more proactive, anticipatory approaches to mobility. However, what that mobility can and should look like differs depending on the context. Small island countries, including those in the Pacific, have contributed the least to the causes of climate change but are some of the first to be facing widespread threats to cultural resources, livelihoods, homes, and habitable land. Not all countries and communities are responding to these risks in the same way, with different perceptions of and prospects for mobility in the region. These differences highlight the tension that must be managed between choosing to stay and planning to move, and the need to consider and facilitate both.

Like the other case studies, this chapter begins by highlighting the impacts from climate change and the risks they create for people and the enjoyment of their human rights. It describes the confluence of sea level rise and other projected climate impacts in the Pacific, the threats these impacts exacerbate, and the foreseeable harms that can result. The chapter turns next to current and predicted human mobility in the Pacific Islands and the lessons that can be learned from past planned relocations. It then sets out the ways in which the integration of human rights into the interpretation of adaptation obligations can lead to action on adaptive mobility, while recognizing there are different perspectives and approaches to human movement. There is a unifying desire to be able to stay in place across the region, to maintain connections to culture, home, and land. Pacific Island countries, however, have approached the issue of human mobility and climate change in different ways. Some, like Fiji and Kiribati, contemplated planned relocation or migration early on in their adaptation planning and policies. Other countries, like Tuvalu and the Republic of the Marshall Islands (RMI), focused on adaptation measures that enable people to stay, only more recently recognizing mobility as an adaptive measure of last resort. Furthermore, the strength of state governance and the influence of customary land tenure systems are factors that affect adaptation and planned relocation. These differences in approaches, capacity, and influences underscore the importance of balancing complex and divergent interests in adaptation strategies that plan for mobility, while prioritizing action that enables

Climate Change, Human Rights, and Adaptive Mobility. Lauren Sakae Nishimura, Oxford University Press.
 DOI: 10.1093/9780198930068.003.0009

people to remain in their homes for as long as possible. The chapter closes with a discussion of international assistance and cooperation. Although technical and financial assistance are the baseline of support required, other forms of cooperation can satisfy adaptation obligations. This chapter considers the additional measures needed to facilitate international adaptive mobility, exploring arguments for duties to admit migrants and the need to consider affected people in communities of origin and destination.

II. Climate impacts, vulnerabilities, and risks to people and their rights

The Pacific Island region includes 22 states and territories, most of which are considered small island developing states (SIDS). Small island countries like those in the Pacific emit less than 1 per cent of the world's greenhouse gases, yet they will suffer some of the most significant impacts. These include sea level rise, increasing temperatures, more frequent and intense tropical storms, flooding, salinization, storm surges, changes in precipitation patterns, ocean acidification and warming, and coral bleaching. Impacts are already occurring and increasingly threaten ecosystems and the people living in the region.[1]

Rising seas pose a grave threat in the Pacific Islands, with many communities living at or near sea level. Although sea level is often discussed as global mean rise, the level of increase is not uniform across regions. Between 1993 and 2009, small island states in the western Pacific experienced sea level rise four times the global average.[2] Rising sea levels can have significant impacts on these states due to their low elevations. The Pacific Islands contain more land in low-elevation coastal zones (LECZs)—contiguous coastal areas that are less than 10 metres above sea level—than any other region in the world. The whole island of Tuvalu is less than 5 metres above sea level and, along with RMI, over 90 per cent of its population lives in an LECZ.[3] In Solomon Islands and Vanuatu, over 60 per cent of people live within 1 kilometre of the coast. More than half of the infrastructure in many islands in the Pacific is within 500 metres of the coast, and for Kiribati, Tuvalu, and RMI, more than 95 per cent of infrastructure is within LECZs. Given the proximity

[1] Michelle Mycoo and others, 'Small Islands' in HO Pörtner and others (eds), *Climate Change 2022: Impacts, Adaptation and Vulnerability. Contribution of Working Group II to the Sixth Assessment Report of the Intergovernmental Panel on Climate Change* (Cambridge University Press 2022) 2045.

[2] Leonard A Nurse and others, 'Small Islands' in Vincente R Barros and others (eds), *Climate Change 2014: Impacts, Adaptation, and Vulnerability. Part B: Regional Aspects. Contribution of Working Group II to the Fifth Assessment Report of the Intergovernmental Panel on Climate Change* (Cambridge University Press 2014) 1619.

[3] Sixteen per cent of the region's land is in LECZs. Gordon McGranahan, Deborah Balk, and Bridget Anderson, 'The Rising Tide: Assessing the Risks of Climate Change and Human Settlements in Low Elevation Coastal Zones' (2007) 19 Environment and Urbanization 17, 24, 26.

of most infrastructure to the coast, even islands with higher elevations like those in Fiji face risks from sea level rise.[4]

Tropical storms are causing significant harm in the region, and the number of intense cyclones is expected to grow. These storms are affecting people's livelihoods, homes, and infrastructure. For example, a 2015 cyclone in Vanuatu led to damages in the agricultural sector estimated at USD 56.5 million, over 60 per cent of the country's GDP, and destruction from a 2016 cyclone exceeded 20 per cent of Fiji's current GDP.[5] Storms also combine with sea level rise, extreme ocean waves, and phases of El Niño or La Niña to contribute to increased coastal flooding, saltwater intrusion, erosion, and loss of land.[6] Increasing salinity can reduce water supplies and hamper food production, destroying agricultural areas and crops. Droughts are further impacting some islands, as rising temperatures and changes to precipitation patterns affect rainwater and harm the ecosystems needed for livelihoods and survival. Such impacts place increasing strain on already limited freshwater resources, further reducing access to potable water.[7]

Projected impacts will continue to alter ecosystems. Across all warming scenarios air temperatures will rise, and coastal storm surges and flooding are expected to significantly increase in frequency and extent. Some low-lying islands and atolls are likely to be flooded during every high tide or storm event, inundating water supplies with saltwater and substantially constraining access to freshwater.[8] Climate events will further reduce biodiversity and threaten endemic species in the region. Fish catch could be cut by more than half by 2100, relative to two to four decades ago.[9] Coral bleaching will also continue. Even if the temperature target for the Paris Agreement is met, and emission reductions hold warming to 1.5°C, an estimated 70 to 90 per cent of reefs will be lost, and losses will increase to 99 per cent if the world warms to 2°C or more above the pre-industrial levels.[10]

Together, these impacts make the Pacific Islands one of the regions confronting the greatest threats from climate change, and its inhabitants some of the first to experience significant harm. Those living in states composed almost entirely of atoll islands, including Kiribati, Tuvalu, and RMI, are particularly at risk. Likewise,

[4] Neil L Andrew and others, 'Coastal Proximity of Populations in 22 Pacific Island Countries and Territories' (2019) 14 PLOS ONE e0223249; Lalit Kumar and Subhashni Taylor, 'Exposure of Coastal Built Assets in the South Pacific to Climate Risks' (2015) 5 Nature Climate Change 992.

[5] Mycoo and others (n 1) 2045, 2048, 2064.

[6] ibid 2095.

[7] ibid 2065; Office of the UN High Commissioner for Human Rights (OHCHR), 'The Slow Onset Effects of Climate Change and Human Rights Protection for Cross-Border Migrants' (22 March 2018) UN Doc A/HRC/37/CRP.4, para 111.

[8] Mycoo and others (n 1) 2045–46, 2095.

[9] ibid 2046.

[10] Ove Hoegh-Guldberg and others, 'Impacts of 1.5°C Global Warming on Natural and Human Systems', *Global Warming of 1.5°C. An IPCC Special Report on the impacts of global warming of 1.5°C above pre-industrial levels and related global greenhouse gas emission pathways, in the context of strengthening the global response to the threat of climate change, sustainable development, and efforts to eradicate poverty* (Cambridge University Press 2018) 8.

communities in riverine, delta, and coastal areas of islands, along with those in urban areas, are more vulnerable.[11] As in other contexts, vulnerability can be compounded by socioeconomic conditions and existing marginalization. In the Pacific Islands, climate impacts exacerbate vulnerability for women who experience gender inequality, gender-based violence, and reduced employment opportunities. These experiences, along with barriers to political participation, worsen poverty, health problems, and access to housing. In Tuvalu, RMI, Vanuatu, and Tonga, for example, women rarely own or are legally unable to own land independently. Hardships for women can in turn impede the development and nutrition of children.[12]

Overall, those living in the Pacific Islands face certain key risks from climate impacts. The Intergovernmental Panel on Climate Change (IPCC) identifies eight for small islands: loss of marine and coastal biodiversity and ecosystem services, submergence of reef islands, loss of terrestrial biodiversity and ecosystem services, water insecurity, destruction of settlement and infrastructure, degradation of human health and well-being, economic decline and livelihood failure, and loss of cultural resources and heritage. These risks interact to create direct, indirect, and cascading effects. They combine to contribute to an additional, overarching risk of reduced habitability, as islands become increasingly unable to sustain people and their communities.[13] Thus the prospect of uninhabitable land, long raised as a possibility in the Pacific Islands, remains a pressing concern. As a result, cultural resources, practices, and identity are threatened, making multiple forms of loss likely. Disconnection from land can lead to potential loss of nationality or citizenship, as well as what Kate Wilkinson Cross and Pefi Kingi call 'cultural statelessness', felt by those moving from their homelands regardless of whether they cross state borders.[14] Studies in Tuvalu, Fiji, and other Pacific Islands also show that concern about cultural losses can lead to anxiety, mental health impacts, and sadness among Indigenous peoples and local communities.[15] These losses are not, however, experienced in the same way across the region. Context affects vulnerability, and existing conditions compound risks and harms from climate change. These conditions include demographic pressures, development challenges, governance limitations, and issues with services like sanitation,

[11] John Campbell and Olivia Warrick, 'Climate Change and Migration Issues in the Pacific' (UN Economic and Social Commission for Asia and the Pacific 2014) 28–29.

[12] World Bank Group, 'Systematic Country Diagnostic for the Eight Small Pacific Island Countries: Priorities for Ending Poverty and Boosting Shared Prosperity' (2016) 22–23, 26, 28, 57, 89.

[13] See Mycoo and others (n 1) 2069, 2073.

[14] Kate Wilkinson Cross and Pefi Kingi, '"*Fonua*" Cultural Statelessness in the Pacific and the Effects of Climate Change' in Tendayi Bloom and Lindsey N Kingston (eds), *Statelessness, Governance, and the Problem of Citizenship* (Manchester University Press 2021); see also Michelle Foster and others, 'The Future of Nationality in the Pacific: Preventing Statelessness and Nationality Loss in the Context of Climate Change' (2022) (discussing legal risks to nationality for cross-border migrants).

[15] MA Caretta and others, 'Water' in *Climate Change 2022: Impacts, Adaptation and Vulnerability* (n 1) 594, 634 (citations omitted).

education, and healthcare.[16] Climate impacts alongside contextual conditions can mean that some people will encounter threats to their food, water, livelihoods, and homes before others, influencing whether and when they move. For example, those living in remote atoll islands may need to move sooner, due to low-lying locations, reliance on reef ecosystems for food and livelihoods, risks to housing and infrastructure, and limited in situ adaptation options.[17] Yet uninhabitability is a contested concept: even in atolls, what makes a place habitable is not reducible to universally defined criteria but is also historically and culturally specific.[18]

III. Current and predicted mobility

Small islands, including those in the Pacific, are experiencing disproportionately high rates of displacement caused by extreme weather events.[19] As elsewhere, past displacement following disasters or sudden-onset events has been primarily internal. People often move temporarily, although protracted displacement occurs and is identified as an area in need of further study. Mobility from rural to more central urban islands of a country has also been attributed to a mixture of climate change and socioeconomic factors; these factors include access to adequate food, education, and employment, as well as pressure from overpopulation.[20] States recognize this mobility in their adaptation plans and submissions to the climate regime. For example, Solomon Islands recognized displacement will be caused by rising seas, Tuvalu cited lack of food access as a cause of internal migration, and Fiji highlighted increased rural-to-urban migration.[21]

When people move internationally, it tends to be within the region or to Australia or New Zealand. Climate impacts are not currently seen as a major contributor to such movement, which is primarily related to economic factors, education, and family reunification. It is expected that the drivers of mobility will shift in the coming decades, however, as sea level rise and other climate events heighten exposure to risks and make it increasingly difficult to sustain livelihoods

[16] See The Nansen Initiative, 'Human Mobility, Natural Disasters and Climate Change in the Pacific' (2013) Background Paper 9; Ilan Kelman, 'No Change from Climate Change: Vulnerability and Small Island Developing States' (2014) 180 The Geographical Journal 120.

[17] See eg Mycoo and others (n 1) 2046, 2096.

[18] Carol Farbotko and John Campbell, 'Who Defines Atoll "Uninhabitability"?' (2022) 138 Environmental Science & Policy 182.

[19] See G Cissé and others, 'Health, Wellbeing and the Changing Structure of Communities' in *Climate Change 2022: Impacts, Adaptation and Vulnerability* (n 1) 1045.

[20] Viviane Clement and others, 'Groundswell Part 2: Acting on Internal Climate Migration' (World Bank 2021) 230; see also The Nansen Initiative, 'Background Paper' (n 16) 8–10.

[21] Ministry of Environment, Conservation and Meteorology, 'Solomon Islands NAPA' (2008) 40; Ministry of Natural Resources, Environment, Agriculture and Lands, 'Tuvalu's NAPA' (2007) 6; Government of the Republic of Fiji, 'Republic of Fiji NAP: A Pathway towards Climate Resilience' (2018) 75. Some Pacific Island countries have developed policies to govern internal mobility, discussed in Section IV.

and infrastructure.[22] Environmental conditions may also play a role in decisions not to return to small island states after moving abroad.[23]

Immobility, whether by choice, compulsion, or some degree of both, is also increasingly recognized in the Pacific Islands. The connection between environmental change and so-called 'trapped' populations is often made, which rests on the notion that some people may not be able to move from places of environmental risk due to a lack of resources. The United Kingdom's Foresight Report introduced the idea that poorer households are likely to be 'trapped' in conditions that prevent their movement away from places where they are vulnerable to environmental change. The report also acknowledged that people may choose to stay, contrasting such a choice with those 'trapped' individuals who are forced to remain.[24] As with migration, however, choice and compulsion are not a tidy dichotomy, and immobility is therefore described as a continuum from people who are unable to move to people who choose not to move due to attachments to land and home.[25] The desire to stay takes a central role in the Pacific, and its normative articulation as a right reflects the unity between people and place, albeit expressed in different terms across the region.[26] Economic factors and livelihoods are important, but so too are access to places of social and cultural importance. This aligns with the subjective nature of decisions on mobility, which are influenced by perceived risks, preferences, networks, and social conditions. Greater nuance therefore demands more critical thinking about involuntary immobility, as well as its use to justify planned relocation.[27]

Nonetheless, there is an expectation that planned relocations will become increasingly necessary in some low-lying coastal areas, particularly under scenarios with higher degrees of warming.[28] The region has a long history of community relocations. Some involve relocating to places that are less exposed to environmental hazards, often to nearby sites that are within customary land boundaries. Less commonly, they involve moving over greater distances, including across borders.[29] Planned relocations are generally costly, require time and planning, and involve multiple actors. Lessons can be learned from past relocations, which demonstrate the risks people face when adequate planning and participation are not

[22] Cissé and others (n 19) 1100 (citations omitted).

[23] See eg Kees van der Geest and others, 'Climate Change, Ecosystem Services and Migration in the Marshall Islands: Are They Related?' (2020) 161 Climatic Change 109.

[24] Foresight, 'Migration and Global Environmental Change' (UK Government Office for Science 2011) Final Project Report 14, 67, 104–05, 127.

[25] See Cissé and others (n 19) 1085.

[26] Discussed further in Section IV; Farbotko and Campbell (n 18).

[27] See Sonja Ayeb-Karlsson, Andrew W Baldwin, and Dominic Kniveton, 'Who Is the Climate-Induced Trapped Figure?' (2022) 13 WIREs Climate Change e803.

[28] Cissé and others (n 19) 1117–18.

[29] See John Campbell, 'Climate-Induced Community Relocation in the Pacific: The Meaning and Importance of Land' in Jane McAdam (ed), *Climate Change and Displacement: Multidisciplinary Perspectives* (Hart 2010) 60.

incorporated into the process. Furthermore, when human rights are not taken into account, communities are often left poorer or worse off in relocation sites.[30] These problems will be replicated without the integration of relevant rights, or if affected people do not participate in, shape, and lead relocation processes. The impacts on the land and resources of people living in areas of relocation should also be considered, as must their participation. Participation is a right for all, not just those in power or with leadership roles. Participatory processes must therefore be meaningful rather than a formality.[31] They must be inclusive, to account for the differences in needs and preferences across individuals and communities. Likewise, communities are not homogenous, and a contextual understanding of who makes decisions and constitutes the community is critical to addressing inequities in decision-making and ensuring broader engagement.[32] Analyses of several colonial-era international relocations highlight the need for consultation and participation, as well as the importance of perceived choice and agency. These past experiences help establish informed consent as a prerequisite for planned relocation; without providing agency to those who move, as well as a commitment to ensure rights, individuals are likely to lose both access to resources and cultural practices and identities. They are also unlikely to view relocation as a success when not provided some control over the process.[33] Yet beyond consent, people's preferences and self-determination are critical to the experience and outcome of relocations. Losses will be felt more acutely without such engagement, exacerbating the disconnect between homes left and new destinations.

More recent relocation experiences highlight the importance of continued connections, despite the unavoidability of losses. Planned relocation to move people away from environmental or geophysical hazards has occurred in a number of countries in the Pacific, including Fiji, Solomon Islands, Vanuatu, Samoa, and Papua New Guinea.[34] Collective ownership and cultural practices

[30] See generally UN High Commissioner for Refugees (UNHCR), 'Planned Relocations, Disasters and Climate Change: Consolidating Good Practices and Preparing for the Future' (2014) paras 34–48; Jane McAdam, 'Relocation and Resettlement from Colonisation to Climate Change: The Perennial Solution to "Danger Zones"' (2015) 3 London Review of International Law 93.

[31] See Chapter 5, II–III; see also Committee on Economic, Social and Cultural Rights (CESCR), 'General Comment No 14: The Right to the Highest Attainable Standard of Health (Art 12)' (11 August 2000) UN Doc E/C.12/2000/4, paras 11, 54; CESCR, 'General Comment No 15: The Right to Water (Arts 11 and 12 of the Covenant)' (20 January 2003) UN Doc E/C.12/2002/11, paras 24, 48.

[32] Erica Bower, Rachel Harrington-Abrams, and Betsy Priem, 'Complicating "Community" Engagement: Reckoning with an Elusive Concept in Climate-Related Planned Relocation' (2024) 88 Global Environmental Change 102913.

[33] See Jane McAdam, 'Historical Cross-Border Relocations in the Pacific: Lessons for Planned Relocations in the Context of Climate Change' (2014) 49 The Journal of Pacific History 301; Jane McAdam and Elizabeth Ferris, 'Planned Relocations in the Context of Climate Change: Unpacking the Legal and Conceptual Issues' (2015) 4 Cambridge Journal of International and Comparative Law 137, 144; Tammy Tabe, 'Climate Change Migration and Displacement: Learning from Past Relocations in the Pacific' (2019) 8 Social Sciences 1, 10–13.

[34] Erica Bower and Sanjula Weerasinghe, 'Planned Relocation in the Pacific: A Regional Snapshot' (Platform on Disaster Displacement (PDD) 2021).

tied to land, oceans, and ecosystems are integral to the identity and well-being of many in the Pacific Islands. Relocating can sever these connections to land and home, which can lead to cultural, spiritual, and physical losses.[35] Closer proximity to places of origin may, however, help maintain connections. Several planned relocations that occurred over short distances allowed some of those relocated to continue to pursue similar livelihoods. These relocations often involved Indigenous communities, and they tended to be to single destination sites within countries.[36] Some communities also reported improvements to well-being. This is the case for several communities in Fiji, for example, where an improved sense of community, safety from climate hazards, and overall cohesion were outcomes of village relocations. These communities did not encounter issues in acquiring new land, as they relocated nearby or within their customary land areas, retained agency through decision-making and social networks, and were supported by the government.[37] Overall, past relocations indicate that contextual factors and processes shape outcomes. These include the location of relocation sites and distance from places of origin, ability to maintain cultural ties, planning and implementation processes, the role played by affected people in decision-making, and available resources and financing. The timing and duration of relocation processes, as well as measures to ensure social cohesion, also influence outcomes.[38] Differences in perceptions and preferences between states, communities, and individuals further affect how and whether mobility is undertaken, although common experiences and colonial legacies contribute to shared concerns and conditions. Indeed, colonialism helped determine where people currently reside, including in places of heightened exposure to environmental risks.[39]

[35] See Karen E McNamara, Ross Westoby, and Alvin Chandra, 'Exploring Climate-Driven Non-Economic Loss and Damage in the Pacific Islands' (2021) 50 Current Opinion in Environmental Sustainability 1, 3–5.

[36] See Bower and Weerasinghe (n 34).

[37] See Annah E Piggott-McKellar and others, 'Moving People in a Changing Climate: Lessons from Two Case Studies in Fiji' (2019) 8 Social Sciences 133; Celia McMichael, Manasa Katonivualiku, and Teresia Powell, 'Planned Relocation and Everyday Agency in Low-Lying Coastal Villages in Fiji' (2019) 185 The Geographical Journal 325.

[38] Erica R Bower and others, 'Enabling Pathways for Sustainable Livelihoods in Planned Relocation' (2023) 13 Nature Climate Change 919.

[39] See eg Rebecca Monson and Daniel Fitzpatrick, 'Negotiating Relocation in a Weak State: Land Tenure and Adaptation to Sea-Level Rise in Solomon Islands' in Susanna Price and Jane Singer (eds), *Global Implications of Development, Disasters and Climate Change* (Routledge 2015); Rebecca Monson and others, 'Planning for Climate Mobility: Developing a Relational Approach to Planned Community Relocation in Solomon Islands' in Andreas Neef, Natasha Pauli, and Bukola Salami (eds), *De Gruyter Handbook of Climate Migration and Climate Mobility Justice* (De Gruyter 2024) 387; Solomon Islands Government, 'Planned Relocation Guidelines 2022' (International Organization for Migration (IOM) 2022) 5; Piérick CM Martin and others, 'Responding to Multiple Climate-Linked Stressors in a Remote Island Context: The Example of Yadua Island, Fiji' (2018) 21 Climate Risk Management 7, 12–13.

IV. Integration of human rights and adaptive mobility

Measures to facilitate migration are not always required by adaptation obligations that integrate human rights duties. Nor is mobility necessary or wanted in all circumstances of foreseeable risk in the Pacific. A range of adaptation measures are being explored, from constructing sea walls to providing alternative livelihood options; strategies are also being developed to evaluate and respond to future risks. Some of these strategies focus on mobility, others are aimed at planning for, preventing, and responding to displacement in the region.[40] Human rights can both drive and shape these efforts. Without access to resources underpinning the enjoyment of rights, and to adequate housing, food, and water in particular, people are more prone to displacement; people living in more exposed areas are likewise more likely to be displaced. But not all displacement is unavoidable. The slow-onset nature of sea level rise and the foreseeability of other expected impacts in the region make anticipatory action possible, including mobility measures that help prevent or address displacement.[41] Preventive, anticipatory action accords with adaptation obligations that incorporate precaution and require states to take adequate measures to prepare for and avoid risks before they lead to significant harm. Integrating relevant human rights duties into the interpretation of adaptation obligations turns adaptive measures towards the protection of people and their rights. States in the Pacific are bound by these duties by virtue of customary or treaty-based international law. While a handful of countries in the region are not party to the International Covenant on Economic, Social and Cultural Rights (ICESCR) or the International Covenant on Civil and Political Rights (ICCPR)—Kiribati, Tuvalu, and Vanuatu, for example—they must comply with customary international law and are parties to other human rights treaties that include the rights that are a focus of this book.[42]

The Pacific Islands are often presented as urgent cases, where predicted climate impacts make it likely that coastal communities will need to relocate. This sense of urgency is bolstered by a narrative often attached to the region of impending loss of territory and disappearing small island states. While this narrative has been called into question, as has the notion of habitability, climate change will create significant risks to ecosystems and people in the region that will make it difficult

[40] See eg Internal Displacement Monitoring Centre (IDMC), 'Project Brief: Pacific Response to Disaster Displacement' (2019).

[41] See International Law Association (ILA), 'Sydney Declaration of Principles on the Protection of Persons Displaced in the Context of Sea-Level Rise' (2018) Resolution 6/2018 <https://disasterdisplacement.org/wp-content/uploads/2019/06/ILA-Sydney-Declaration.pdf> accessed 14 September 2019.

[42] All are parties to the Convention on the Rights of the Child (CRC), Convention on the Elimination of All Forms of Discrimination Against Women (CEDAW), Convention against Torture and Other Cruel, Inhuman or Degrading Treatment or Punishment (CAT), and the Convention on the Rights on Persons with Disabilities. Vanuatu is also party to the ICCPR.

to remain in certain highly exposed places.[43] Consequently, there will be a need for some communities to relocate, with more significant international movement likely to occur over time.[44] As the IPCC recognizes, 'for some populations, migration away from their homeland may become the only viable response'.[45] This kind of relocation is distinct in its collective nature, the foresight that predictions provide about places and people at risk, and the inability of migrants to return home.[46] Before such international relocation occurs, however, a range of other adaptive options can and should be pursued, including internal planned relocations. Thus, the region illustrates the importance of timing, the need to plan for mobility, and the shifting strategies that adapting adequately to climate change will entail.

Internal relocations are already being undertaken or contemplated by communities and governments in the region. The permanent relocation of thousands of Carteret Islanders is underway, from atoll islands to mainland Bougainville. Families in the Carteret Islands are confronting a confluence of environmental and economic conditions that both drive and constrain migration.[47] However, access to land and funding have hampered community-based efforts. Internal relocations in Fiji are moving people to higher ground as sea level rise, erosion, and storm surges encroach on coastal communities. Likewise, in Solomon Islands, some communities in low-lying areas facing coastal erosion are relocating internally. Fiji's relocations involve government actors and support, while relocations in Solomon Islands were initially on an ad hoc basis, with decisions to move made by households that predate a national policy on planned relocation.[48] Customary landholders are likely to play an important role in Solomon Islands, where 87 per cent of land is held under customary land tenure and relatively weak state governance on land issues has been observed.[49]

Nevertheless, the existence of dedicated policies on internal mobility signals increasing state attention to mobility as an adaptation strategy, including in

[43] See Section II; Farbotko and Campbell (n 18).

[44] See eg *Teitiota v New Zealand*, 'Views Adopted by the Committee under Article 5(4) of the Optional Protocol, Concerning Communication No 2728/2016' (Human Rights Committee (HRC), 23 September 2020) UN Doc CCPR/C/127/D/2728/2016, para 9.11; Jane McAdam and others, 'International Law and Sea-Level Rise: Forced Migration and Human Rights' (Fridtjof Nansen Institute 2016) para 50.

[45] Nerilie Abram and others, 'Framing and Context of the Report' in HO Pörtner and others (eds), *IPCC Special Report on the Ocean and Cryosphere in a Changing Climate* (Cambridge University Press 2019) 114.

[46] See Maxine Burkett, 'The Nation Ex-Situ: On Climate Change, Deterritorialized Nationhood and the Post-Climate Era' (2011) 2 Climate Law 345, 351.

[47] See John Connell, 'Last Days in the Carteret Islands? Climate Change, Livelihoods and Migration on Coral Atolls' (2016) 57 Asia Pacific Viewpoint 3.

[48] See Bower and Weerasinghe (n 34); Carol Farbotko and others, 'Transformative Mobilities in the Pacific: Promoting Adaptation and Development in a Changing Climate' (2018) 5 Asia & The Pacific Policy Studies 393, 396–97.

[49] Monson and Fitzpatrick (n 39).

Solomon Islands. Some policies contemplate internal planned relocation as a form of adaptation, as is the case for Solomon Islands and Fiji.[50] Vanuatu's policy on climate change and disaster-induced displacement recognizes internal relocation, or what it sees as 'sustainable integration' elsewhere, as a durable solution for displaced persons. The policy declares that mobility requires safeguards to protect and assist communities and to ensure human rights.[51] The acknowledgement that human rights should be considered and protected is a baseline for state policies and processes on mobility and relocation. This is the minimum that is required to meet rights obligations and provide the resources necessary for people to enjoy rights and a dignified life.

Human rights can also play a central role in compelling state action on planned relocation and in guidance on how it can be accomplished. The interpretation of adaptation obligations, which all countries in the region are bound by, must integrate relevant human rights. Integrating positive duties related to rights to life and to an adequate standard of living, for example, includes the need to address societal conditions that put lives and access to resources at risk and to protect people from arbitrary displacement.[52] This integration should lead to action to facilitate migration or relocation away from areas in the Pacific Islands where rights can no longer be protected in place. In some circumstances, it may require movement of whole communities, as is occurring or being considered in several countries. The obligations in climate treaties provide a legal basis to argue that such planning for and facilitating action on mobility as an adaptive measure is required, and that it must incorporate human rights duties and standards. These obligations must also include a precautionary perspective, to shape action in advance of impacts that would otherwise result in dire situations and rights violations.

However, state measures to enable adaptive mobility in the region raise several tensions that must be managed. The first is the need to plan for relocation as part of a process of adaptation, while recognizing that planned relocation is generally a measure of last resort. Relocation is included in a number of disaster risk reduction (DRR) strategies as a preventive measure to address disasters and climate change, one which requires ample timing and implementation.[53] Some scholars go so far as to characterize strategies to prepare for and anticipate climate change in the Pacific as 'essentially translat[ing] to planning for community relocation'

[50] See Ministry of Economy, Republic of Fiji, 'Planned Relocation Guidelines: A Framework to Undertake Climate Change Related Relocation' (2018); Solomon Islands Planned Relocation Guidelines (n 39).

[51] Government of Vanuatu, 'National Policy on Climate Change and Disaster-Induced Displacement' (2018).

[52] See Chapter 4, III.

[53] For examples of relocation in DRR strategies globally, see Michelle Yonetani, 'Mapping the Baseline: To What Extent Are Displacement and Other Forms of Human Mobility Integrated in National and Regional Disaster Risk Reduction Strategies?' (PDD 2018) 34–35.

and rehabilitation.[54] Policymakers recognize that relocation requires significant advanced planning, including long-term land use plans, as well as consultation with those affected and time to enact a holistic approach that is more than 'moving houses, [but is rather] about moving lives'.[55] Whether as a strategy to move people from hazardous areas or a durable solution for those displaced, relocations can be transformative and adaptive.[56] Yet past relocations indicate that outcomes are largely determined by the level of institutional and governmental support and community perception, agency, and leadership.[57] Previous relocations and mobility further evince that, regardless of the outcome, affected people will be forced to make difficult choices and grapple with losses arising from disconnection to homelands.

There are ongoing efforts to develop and implement specific guidelines on planned relocation by countries in the region, which are an important first step towards anticipatory, rights-based approaches to relocation. Yet these guidelines recognize planned relocation as a measure of last resort. Solomon Islands and Fiji, for example, have developed guidelines on planned relocation, both of which consider the planned relocation of communities as a strategy of last resort, to be undertaken only after other feasible adaptation measures have been exhausted. Both guidelines also explain that relocation should contribute to 'migration as adaptation', as part of broader adaptation efforts to ensure livelihoods and to adapt according to the specific needs of those affected by relocations.[58] Solomon Islands adopted its guidelines in 2022, recognizing the urgent need for adaptation and the '[c]limate-driven' relocations underway and likely required in the future. These guidelines are a self-declared 'living document', meant to evolve with changing needs. They draw on guidance and best practices developed internationally, highlighting that human rights are central to any relocation, must be ensured, and necessitate ongoing monitoring of relocation processes and outcomes. They further recognise the government's obligation to take preventative action to uphold rights despite constraints and the need for external support.[59]

[54] Dalila Gharbaoui and Julia Blocher, 'The Reason Land Matters: Relocation as Adaptation to Climate Change in Fiji Islands' in Andrea Milan and others (eds), *Migration, Risk Management and Climate Change: Evidence and Policy Responses*, vol 6 (Springer 2016) 153.

[55] See Karen E McNamara and Hélène Jacot Des Combes, 'Planning for Community Relocations Due to Climate Change in Fiji' (2015) 6 International Journal of Disaster Risk Science 315, 317 (quoting Fijian climate policy officer); Government of the Republic of the Marshall Islands, 'Initial Communication under the UNFCCC' (2000) 52–53.

[56] Chapter 7, IV, nn 85–87 and accompanying text.

[57] See Section III, nn 30–35 and accompanying text.

[58] Fiji Planned Relocation Guidelines (n 50) 1, 8, 11; Solomon Islands Planned Relocation Guidelines (n 39) 4, 8, 15.

[59] Solomon Islands Planned Relocation Guidelines (n 39) 5, 6, 14, 22; informed by Brookings Institution, Georgetown University, and UNHCR, 'Guidance on Protecting People from Disasters and Environmental Change through Planned Relocation' (2015); Georgetown University, IOM, and UNHCR, 'A Toolbox: Planning Relocations to Protect People from Disasters and Environmental Change' (2017).

Fiji is explicit in its acknowledgement of the connection between human mobility and climate change and its rights-based approach to planned relocation. Fiji was the first country to develop a specific policy and guidelines on planned relocation. It initially identified hundreds of coastal and riverbank communities put at risk by the impacts of climate change that will need to be relocated, with 48 it considers in 'urgent need of relocation' to be supported by government funding.[60] As part of its climate adaptation strategies, and to facilitate these relocations, Fiji developed guidelines on planned relocation in 2018. Its Standard Operating Procedures for Planned Relocation (SOP) followed in 2023, which lays out the processes for planned relocation.[61] These documents are referenced in domestic law: the Climate Change Act declares that decisions on planned relocation must be consistent with and follow the processes in the guidelines and in any SOP.[62] Fiji is a country where some forms of adaptive mobility are more appropriate than others in the region, as it contains mountainous and higher elevation terrain that make internal relocations possible. Its relocation guidance and regulations aim to ensure a 'gender responsive consultative and participatory process', community ownership of the relocation process, and a commitment to human rights while fulfilling the needs of those relocated.[63] These aims accord with Fiji's gender and human rights-based approach to adaptation more generally, as reflected in its National Adaptation Plan (NAP).[64] Like this book, Fiji's relocation guidelines directly link planning for and action on mobility to human rights and climate obligations. These guidelines view the Paris Agreement as bringing together rights and mobility, and as connecting a rights-based approach 'inexorably to climate change discourse'. They go on to explain that '[t]he Paris Agreement, together with the ICCPR and the ICESCR reflects the countries' rationale to relate climate change triggers to rights belonging to affected people', which requires that Fiji ensures their rights, meaningful engagement, and participation.[65] Moreover, as with adaptation obligations, the guidelines take a precautionary or 'pre-emptive approach' for all stages of the relocation process, and note that in implementing this process 'addressing human rights aspects is unavoidable and important in all . . . stages of movement', as 'relocated people carry their rights at all times'.[66] The Climate Change Act reiterates the adoption of approaches to planned relocation that are

[60] Republic of Fiji, '5-year & 20-year National Development Plan: Transforming Fiji' (2016) 102 <https://www.adb.org/sites/default/files/linked-documents/LD4%205yr%20and%2020yr%20DP%20Transforming%20Fiji.pdf> accessed 20 June 2024.

[61] Office of the Prime Minister, 'Standard Operating Procedures for Planned Relocation in the Republic of Fiji' (2023).

[62] Climate Change Act 2021, s 77(4)(d).

[63] Fiji Planned Relocation Guidelines (n 50) 1, 5, 8; see also SOP (n 61) s 2.1; Climate Change Act 2021, s 77(1)(j), (2), (3).

[64] Fiji NAP (n 21) 38–39.

[65] Fiji Planned Relocation Guidelines (n 50) 8; restated in SOP (n 61) s 2.2.1 (Box 1: Principles for Planned Relocation in Fiji).

[66] Fiji Planned Relocation Guidelines (n 50) 9.

'human rights based' and 'pre-emptive'.[67] Thus, explicit through these general terms and implicit in its approach to relocation is both Fiji's integration of human rights into adaptation strategies and its plans to take anticipatory action on mobility. In this way, Fiji's guidelines and law operationalize the interpretive approach advanced by this book.

The second tension that must be managed is between government action and community-led decision-making. Guidance on planned relocation emphasizes participatory and consultive processes driven by communities.[68] At the same time, states must protect rights and comply with their legal obligations, and their efforts to facilitate movement away from hazard-prone areas are one means to meet their commitments to people under their jurisdiction. Governments therefore enable and help manage relocations, but they must do so in collaboration with affected people to ensure community participation and ownership throughout the process. This can be a difficult balance to strike. When government actors fail to provide agency and control to affected people, communities are left worse off or perceive relocations to be unsuccessful.[69] There is a further risk that relocation can be used as a tool by states to accomplish other goals. States may, for example, use relocation as a pretext to take land for development or economic reasons, or they may facilitate movement to urban areas to consolidate labour in the interest of businesses or to achieve a political end. In these cases, people could be exposed to greater risks after relocation, despite purportedly moving to avoid harm. Moreover, moving people away from physical risks does not in itself address the reasons for their vulnerability to climate change. Without a focus on other social, economic, and cultural conditions, planned relocation is likely to exacerbate or create new vulnerabilities.[70]

One of the potential benefits of an interpretation of adaptation obligations that integrates human rights duties is that the participatory processes it prescribes can help prevent use of mobility to accomplish other goals. Participatory processes are more likely to lead to outcomes that respond to the needs and preferences of those affected and may help avoid illegal evictions or arbitrary displacement. More collaborative processes can better incorporate Indigenous knowledge and practices, long used to adapt to changing conditions—including through mobility—but often overlooked in policymaking.[71] Furthermore, and as discussed in Section II,

[67] Climate Change Act 2021, s 77(1)(j)(iii)–(iv).

[68] See Fiji Planned Relocation Guidelines (n 50) 3; Solomon Islands Planned Relocation Guidelines (n 39) 6, 8; UNHCR, 'Guidance on Planned Relocation' (n 59) paras 12, 26; 'The Peninsula Principles on Climate Displacement within States' (Displacement Solutions 2013) principle 10; International Federation of Red Cross and Red Crescent Societies (IFRC), 'Planned Relocation in the Context of Disasters and Climate Change: A Guide for Asia Pacific National Societies' (2021) 48.

[69] See n 33 and accompanying text.

[70] Carol Farbotko and others, 'Relocation Planning Must Address Voluntary Immobility' (2020) 10 Nature Climate Change 702.

[71] See Annah Piggott-McKellar, Celia McMichael, and Teresia Powell, 'Generational Retreat: Locally Driven Adaption to Coastal Hazard Risk in Two Indigenous Communities in Fiji' (2021) 21 Regional Environmental Change 1, 2; Jon Barnett and Celia McMichael, 'The Effects of Climate Change on the

broad engagement with those affected is critical to the outcomes of planned relocation. Inclusive planning processes can help identify preferences and constraints, inform and educate, and gather local knowledge. Participation is a right, which requires states to provide information and procedural protections. As Chapter 5 explains, although it is not required by international law, consent should still be an aim, and it can be better accomplished by decision-making processes that provide agency to those affected. Domestic law in Fiji, for example, calls for planned relocation 'only with the full free and prior informed consent of the communities, following inclusive and gender responsive consultation and participatory processes'.[72] Customary land tenure and traditional negotiation processes will also be central to any relocation, where 80 per cent of the land is under such tenure in most countries.[73] There is limited guidance on how to negotiate new land arrangements for customary landholders, which is critical to resolve in order to foster successful adaptive mobility and avoid displacement.[74] Security of land tenure is a basic requirement of the right to adequate housing that necessitates positive action by the state, particularly for those in areas highly exposed to climate impacts.[75] Local tenure provides certainty for migrants, and access to developed systems of land governance helps promote the sustainable use of land and longer-term access to basic resources.[76] In addition, while the legal duties that shape anticipatory action in this book are found in individual subsistence rights, the collective right to self-determination provides further grounds for communities to assert a right to decide how and whether to move, and for continued control over and ties to homelands. Outcomes are not a given, nor does the assertion of rights lead to a single or static option, but the process and the people deciding are paramount. Accordingly, an approach based on self-determination 'calls for consultative, inclusive, place-, and culture-specific processes that provide resources to support migrants' mobility decisions'.[77]

Geography and Timing of Human Mobility' (2018) 39 Population and Environment 339. Indigenous knowledge is itself evolving with changing conditions and refers to knowledge, decisions made locally, practice, worldviews, and connections to nature across generations. Piggott-McKellar, McMichael, and Powell 3.

[72] Climate Change Act 2021, s 77(1)(e); Chapter 5, III. Soft law and normative frameworks also require consent in situations that include planned relocation. Free, prior and informed consent is required, for example, by the UN Declaration on the Rights of Indigenous Peoples (UNDRIP). However, many Pacific countries have not endorsed the UNDRIP.

[73] Gharbaoui and Blocher (n 54) 150.

[74] See Vanuatu Policy on Climate Change and Disaster-Induced Displacement (n 51) 12.

[75] See Chapter 4, IV.C; CESCR, 'General Comment No 4: The Right to Adequate Housing (Art. 11 (1) of the Covenant)' (13 December 1991) UN Doc E/1992/23, para 8.

[76] Jon Barnett and Michael Webber, 'Accommodating Migration to Promote Adaptation to Climate Change' (The World Bank 2010) WPS5270, 9.

[77] Carmen G Gonzalez, 'Climate Change, Race, and Migration' (2020) 1 Journal of Law and Political Economy 109, 129; see Chapter 5, III.

The third tension that states must manage is the need to plan for mobility and the desire of people to stay. This builds on the other two tensions: planned relocation as both adaptive and a last resort, through processes that involve government action but are driven by community decision-making. These tensions are, in part, a reflection of history and culture. Pacific Islanders have deep connections to the ocean. The region has a history of ocean voyaging and navigation. Equally important are attachments to land and home, which are central to Indigenous identities and belonging.[78] Thus, 'mobility and stability' and the idea of both 'roots and routes' are important to people's experiences and identities in the Pacific.[79] However, the dynamism of these identities, informed by variations in historical contexts, contribute to different perspectives on climate change in the present. People's needs and preferences also vary. Likewise, planned relocation need not be a binary choice—between communities leaving or staying—as more gradual or incremental forms of mobility are possible.[80] The region therefore illustrates the multiple roles adaptation can play, differing views on mobility, and the range of strategies used across countries and communities when faced with increasing risks. These strategies are not fixed, and they are shifting alongside growing knowledge about—and rising severity of—climate impacts. Evidence suggests demographic distinctions as well, with different perspectives on climate change and mobility across generational lines.[81]

Well before mobility was on the agenda of the climate change regime, governments of the Pacific Islands actively lobbied the international community for migration opportunities. Some sought to alleviate demographic pressure, as was the case for Tuvalu in the 1980s and 1990s. But since the 2000s, emphasis on climate change and migration has risen steadily, often overlooking the other factors that contribute to human mobility.[82] Leaders of the Pacific Islands are at the forefront of advocacy on climate change. They have regularly identified the risks to human rights that climate impacts create, the threat to livelihoods and subsistence, and the need to protect climate migrants.[83] The turn towards climate change resulted in a dual focus on mobility as necessary adaptation and adaptation to remain in the

[78] See Carol Farbotko and Celia McMichael, 'Voluntary Immobility and Existential Security in a Changing Climate in the Pacific' (2019) 60 Asia Pacific Viewpoint 148, 151–52.

[79] See Campbell (n 29) 63; Burkett (n 46) 361–62.

[80] Piggott-McKellar, McMichael, and Powell (n 71) 11.

[81] See ibid (examples of 'generational retreat' in two villages in Fiji, where younger generations are initiating short-distance relocations outside of governmental processes); see also Nicole Herman-Mercer and others, 'Changing Times, Changing Stories: Generational Differences in Climate Change Perspectives from Four Remote Indigenous Communities in Subarctic Alaska' (2016) 21 Ecology and Society 28.

[82] Jane McAdam, *Climate Change, Forced Migration, and International Law* (Oxford University Press 2012) 31.

[83] See eg 'Malé Declaration on the Human Dimension of Global Climate Change' (14 November 2007); Pacific Island Forum, 'Forty-Eighth Pacific Islands Forum: Forum Communiqué' (8 September 2017) PIFS(17)9) para 52; Pacific Island Forum, 'Forty-Ninth Pacific Islands Forum: Forum Communiqué' (6 September 2018) PIFS(18)10).

Pacific Islands, both of which can protect human rights and help sustain communities. Most states also declare their desire to maintain sovereignty, and consultations in the region highlight both the need to plan for relocation well in advance and the desire to avoid it if possible.[84] This tension and varying strategies on mobility show that no single solution is appropriate across contexts.

The atoll states of Kiribati, Tuvalu, and RMI provide examples of divergent views on mobility. The low-lying nature and exposure of these atoll countries makes them particularly vulnerable to storm surges and sea level rise, which can lead to flooding, erosion, and destruction of agricultural lands. Kiribati was one of the first countries in the region to incorporate international mobility as part of its adaptation planning with its 'migration with dignity' policy. The policy is aimed at creating opportunities to migrate abroad now and in the future, and it highlights voluntary labour mobility, circular migration, and the benefits of remittances.[85] Kiribati purchased land from Fiji, with food security as the stated reason, but use of the land for relocation is also contemplated.[86] However, Kiribati's migration policy does not seek to preclude people from staying, and subsequent government emphasis is on in situ adaptation to make that possible.[87] Within the country, residents have different perspectives on what the aims of government policy should be: some wish to remain; others would prefer to migrate to New Zealand instead of Fiji due to work opportunities and social networks; and some analyses put development priorities ahead of mobility, although one need not come at the expense of the other.[88]

In contrast, Tuvalu showed an initial commitment to in situ adaptation rather than planning for mobility. Its National Adaptation Programme of Action (NAPA) considers migration and resettlement as a 'last resort to adaptation' to undertake only 'should the worse [sic] case scenario occur'.[89] This preference is echoed by residents of Tuvalu, some of whom are resolved to remain irrespective of long-term risks and rising sea levels. Such voluntary immobility flows from an attachment to land and cultural identity, an exercise of self-determination expressed in the desire to stay—and even face death—rather than be forced to leave.[90] Nonetheless, the

[84] See The Nansen Initiative, 'Human Mobility, Natural Disasters and Climate Change in the Pacific' (2013) Outcome Report.

[85] Discussed in 'Kiribati National Labour Migration Policy' (2015); Farbotko and others, 'Transformative Mobilities in the Pacific' (n 48) 397.

[86] See Elfriede Hermann and Wolfgang Kempf, 'Climate Change and the Imagining of Migration: Emerging Discourses on Kiribati's Land Purchase in Fiji' (2017) 29 The Contemporary Pacific 231, 234–35.

[87] See Government of Kiribati, 'Kiribati Climate Change Policy' (2019); Government of Kiribati, 'Kiribati Joint Implementation Plan for Climate Change and Disaster Risk Management (KJIP): 2019–2028' (2019).

[88] See Silja Klepp, 'Framing Climate Change Adaptation from a Pacific Island Perspective—The Anthropology of Emerging Legal Orders' (2018) 68 Sociologus 149, 160–61; Robert Oakes, 'Culture, Climate Change and Mobility Decisions in Pacific Small Island Developing States' (2019) 40 Population and Environment 480, 497.

[89] Tuvalu NAPA (n 21) 25.

[90] See Oakes (n 88) 484; Farbotko and McMichael (n 78) 154–55.

government's National Climate Change Policy now accepts 'the imminent possibility of the islands in Tuvalu becoming uninhabitable'. In doing so, it makes 'managed human mobility and protection of national sovereignty' a priority outcome, one that it recognizes as requiring a human rights-based approach and a respect for people's right to make their own decisions.[91] Its primary focus for adaptation remains on reducing vulnerability and enhancing resilience through measures to protect and rehabilitate coastal land, existing settlements, and infrastructure.[92] However, a 2023 bilateral treaty with Australia recognizes the need for cooperation on climate and security, and provides for 'mobility with dignity' from Tuvalu to Australia.[93]

Similarly, while RMI's government now recognizes relocation as a future adaptation measure, this was not always the case. The Marshallese root land at the heart of their existence and culture.[94] These kinds of connections compel people to stay despite deteriorating conditions, choosing immobility even in the face of extreme risks. Such a position was echoed in 2019 by Hilda Heine, the President of RMI at the time, when she declared that her government would not consider relocation in response to climate change. She acknowledged that some citizens asked the government to buy land offshore to relocate but said that national planning would focus on sustaining the country.[95] In 2021, foreign Minister Casten Nemra reiterated a focus on staying, explaining that it is 'not an option for us to relocate or become so-called climate change refugees' and that suggestions otherwise were 'degrading', as the islands are 'central to the identity of Marshallese'.[96] These firm positions, however, are mediated by more recent adaptation plans for the country. RMI's NAP, adopted in 2023, includes planned relocation and migration as adaptation in the country's medium-term strategies, to pursue between 2040 and 2070 after exhausting protective measures. Relocation is envisioned to areas of reclaimed or raised lands, with migration out of the country representing a shift in adaptation strategy, considered only after other internal relocations are completed.[97] This plan enables the focus of adaptation to remain on protection, while still planning for relocation and deferring decisions about mobility further in RMI's adaptation timeline. These decisions are to be guided by the NAP's stated

[91] Government of Tuvalu, Climate Change Department, 'Te Vaka Fenua o Tuvalu: National Climate Change Policy 2021–2030' (2021) 15.

[92] ibid 12–13.

[93] Australia-Tuvalu Falepili Union (signed 9 November 2023, entered into force 28 August March 2024).

[94] See Burkett (n 46) 360.

[95] 'Marshall Islands Rules out Relocation in Climate Fight' *RNZ* (4 March 2019) <www.rnz.co.nz/international/pacific-news/383854/marshall-islands-rules-out-relocation-in-climate-fight> accessed 20 June 2024.

[96] 'Climate-Affected Marshall Islands Don't Intend to Relocate' *RNZ* (23 April 2021) <www.rnz.co.nz/international/pacific-news/441080/climate-affected-marshall-islands-don-t-intend-to-relocate> accessed 20 June 2024.

[97] See 'Republic of the Marshall Islands: National Adaptation Plan (Pāpjelmae)' (2023) 24, 139, 147.

principles of self-determination, the right of citizens to determine responses to climate impacts, the right to remain in homelands, and the need for consultation with traditional landowners.[98]

Across these contexts, different approaches to adaptation show that it is possible to both plan for mobility and undertake measures that enable people to stay. Differences in the former may be a matter of timing and emphasis, while the clear preference is for the latter, as even well-planned, right-based migration will result in losses. Indeed, the loss of customary land—and its associated links to culture, individual and community identities, and livelihoods—is considered one of the most significant potential impacts of climate change in the region.[99] Moreover, the narrative of the 'disappearing state' may become a self-fulfilling prophecy due to changes in resources and reduced assistance, rather than as a result of physical impacts.[100] Reticence to relocate and scepticism about external involvement are also understandable given experiences of colonialism and forced movement. The RMI, for example, sees its history of displacement from nuclear weapons testing as influencing climate vulnerability and its plans for adaptation, with lessons to be learned from the trauma and environmental degradation people experienced.[101] There is further resistance by residents in the Pacific Islands to becoming or being labelled as 'climate refugees', framed as victims rather than agents defining solutions on their own terms.[102] This contributed to New Zealand's decision to drop a plan that would have provided a visa for people displaced by climate change; many Pacific Islanders prefer adaptation support first, followed by more extensive legal opportunities for migration.[103]

As is the case in Tuvalu and RMI, there is also a pushback against permanently moving more generally, with some voicing a decision to stay in the face of significant risks. Scholars argue that this 'voluntary immobility' must be included in planning for relocation.[104] This raises questions about the role of state government and individual rights to remain. Whether a person has a right to refuse to relocate, even if they may not survive, is an open legal question.[105] Recognition

[98] ibid 30–31.

[99] See Campbell and Warrick (n 11) 3, 10.

[100] McAdam, *Climate Change, Forced Migration, and International Law* (n 82) 120.

[101] RMI NAP (n 97) 19, 31.

[102] See n 96 and accompanying text; McAdam, *Climate Change, Forced Migration, and International Law* (n 82) 30, 40, 203; Robyn Eckersley, 'The Common but Differentiated Responsibilities of States to Assist and Receive "Climate Refugees"' (2015) 14 European Journal of Political Theory 481, 482 (using term but noting that leaders like the Kiribati president reject the refugee label); cf François Gemenne, 'The Refugees of the Anthropocene' in Benoît Mayer and François Crépeau (eds), *Research Handbook on Climate Change, Migration and the Law* (Edward Elgar 2017) 400 (using term to highlight the plight and responsibility for and towards those moving).

[103] See Helen Dempster and Kayly Ober, 'New Zealand's "Climate Refugee" Visas: Lessons for the Rest of the World' (*Center for Global Development*, 10 January 2020) <www.cgdev.org/blog/new-zealands-climate-refugee-visas-lessons-rest-world> accessed 15 August 2020.

[104] Farbotko and others, 'Relocation Planning Must Address Voluntary Immobility' (n 70).

[105] See Bruce Burson and others, 'The Duty to Move People Out of Harm's Way in the Context of Climate Change and Disasters' (2018) 37 Refugee Survey Quarterly 379, 397–98 (suggesting a

of a right to stay is emerging in literature on climate mobilities, with a legal basis articulated through rights to freedom of movement and choice of residence, rights to return—and remain—in one's country, rights to adequate housing and privacy, rights of Indigenous peoples, and the right to a healthy environment.[106] Some of these rights are not absolute; there are legal grounds for states to curtail freedom of movement and choice of residence, for example, on public health or security grounds. In doing so, states must show that restrictions are necessary, proportional to the risk, and that no reasonable alternative is possible.[107] No government in the region, however, is currently exploring forced relocation, and the overall desire is the opposite: to enable people to have the choice and dignity to stay, move, or pursue both. For example, the government of RMI stated that its 'priority is to ensure that our citizens can fulfil their right to remain in their home islands, and that it is their choice whether and when to migrate.'[108] Ensuring that choices remain with people, rather than external actors, embodies the aims of a collective right to self-determination. The right applies to adaptation to help shape decision-making and measures on migration and planned relocation.[109] Arguments to stay or go can therefore be grounded in rights to self-determination, security of tenure, access to resources, and to not be forcibly displaced. Central to both are processes that provide choice and recognition of the complexity that underlies responses to climate risks. Such a recognition extends considerations of mobility to include immobility and to understand that immobility, much like mobility, can involve some degree of choice and compulsion. The freedom to move includes the choice—and arguably the right—to stay, as much as constraints can limit the ability to move or contribute to forced movement. Adequate support to adapt in place, to facilitate migration, and to undertake planned relocation is critical to ensure that people have more choice and rights can be enjoyed. Likewise, an individual is not static in their preferences and the pressures they face. As Caroline Zickgraf argues, '(Im)mobility is dynamic not fixed: mobile people may become immobile and immobile

competent individual with full information may have such a right in the context of evacuations). Guidance on planned relocation also suggests a right to refuse, suggesting plans that accommodate those who choose not to take part in planned relocation. UNHCR, 'Guidance on Planned Relocation' (n 59) para 34.d.

[106] See Monica Iyer and Kerilyn Schewel, 'Articulating and Claiming the Right to Stay in the Context of Climate Change' (2024) 38 Georgetown Immigration Law Journal 207, 216–20; see also Farbotko and Campbell (n 18) 185; Taukiei Kitara, Samid Suliman, and Carol Farbotko, '*Fale Pili*: A Tuvaluan Perspective on Mobility Justice' (2024) 9 Applied Mobilities 233, 240.

[107] International Covenant on Civil and Political Rights, 16 December 1966 (UN Treaty Series, vol 999, 171) art 12(1), (3); HRC, 'General Comment No 27: Article 12 (Freedom of Movement)' (2 November 1999) UN Doc CCPR/C/21/Rev.1/Add.9, paras 11–18.

[108] Climate Change Directorate, Republic of the Marshall Islands, 'Adaptation Communication' (2020) 29.

[109] See Chapter 4, IV.D.

populations may become mobile; people may stay by choice but then eventually find themselves trapped.'[110]

There is, then, no single or simple perspective in the Pacific Islands on the role of mobility in confronting the threats posed by climate change. The range of perspectives and the shared view of relocation as a last resort, albeit one that requires anticipatory planning, cut against cursory suggestions like that of the Human Rights Committee (HRC) that moving entire populations is a feasible solution.[111] Such a suggestion further belies the history, colonialism, and racism that underlie climate risks and responses in the region.[112] Yet the HRC also concludes that there are human rights duties to provide timely adaptive action.[113] There are different ways to accomplish this. Integrating these human rights duties into adaptation obligations both compels and shapes the process of adaptation, which can include in situ adaptation and anticipatory adaptive mobility measures. Crucially, adaptation understood in this way incorporates procedural rights and inclusive engagement with affected individuals and communities, which can better ensure that Pacific Islanders are the arbiters of their future.

V. International assistance and cooperation

Whether countries in the region are undertaking adaptation in place or planning for adaptive mobility, support and cooperation are necessary. At a minimum, finance and technical assistance are due from developed countries under the commitments made in climate treaties. Certain forms of adaptive mobility, particularly international mobility, also require certain kinds of support to ensure rights and meet the needs of those affected. Whether through legal migration pathways or planned relocation, measures to facilitate international movement will require significant cooperation across states and work with communities in both countries. The scale of mobility currently differs, with internal planned relocations often aimed at communities or large portions of communities, while international mobility tends to take place on an individual or household basis. Nonetheless, the international movement of people, particularly as larger numbers of people move, can create different considerations for the ways procedural rights are guaranteed. Individuals and communities in places of origin, transit, and destination

[110] Caroline Zickgraf, 'Theorizing (Im)Mobility in the Face of Environmental Change' (2021) 21 Regional Environmental Change 126.

[111] *Teitiota v New Zealand* (n 44) para 9.12.

[112] See Dylan Asafo, 'The Racism in Climate Change Law: Critiquing the Law on Climate Change-Related Displacement with Critical Race Theory' (2022) 39 Berkeley Journal of International Law 249; Gonzalez (n 77).

[113] *Billy v Australia*, 'Views Adopted by the Committee under Article 5(4) of the Optional Protocol, Concerning Communication No 3624/2019' (HRC, 21 July 2022) UN Doc CCPR/C/135/D/3624/2019.

are entitled to information and to participate in decision-making. When the destination is across a state border, different actors and logistical considerations come into play. Cooperation and collaboration between both states is required to ensure participatory processes and that assistance reaches the communities for which it is intended. Likewise, efforts will be needed to address any obstacles that impede the participation of marginalized groups, including women and Indigenous peoples.

International adaptive mobility will involve additional hurdles. There must be agreement on multiple fronts: the timing and movement of people across borders, how to ensure access to basic resources and social services, and what durable solutions look like in a new state. Migrants often face discrimination and barriers to resources and services. Accounting for these risks throughout the process—and as argued for through a human rights-based approach—can help avoid some of the problems associated with international mobility. Such an approach includes integrating principles of non-discrimination and equity, which will likely require positive action to ensure international migrants are treated equally. There should also be means to protect against loss of nationality, retain citizenship, and access legal status and rights, which may require changes to domestic laws in multiple states.[114] Furthermore, in both places of origin and destination, obligations on adaptation that integrate human rights duties necessitate provision of essential goods and resources. States must, for example, observe protections required for adequate housing and must avoid forced evictions.[115] They also cannot derogate from minimum core obligations related to an adequate standard of living. Similarities between communities and social ties can help facilitate mobility processes[116] and support movement within the region or to places with established social networks.

In addition, the movement of people across borders presents opportunities to satisfy obligations of assistance beyond the provision of financing and technical support. States could, for example, offer migrants entry and enable their stay. Those at risk of losing their habitation, after exhausting other feasible adaptation options, are foreseeably some of the individuals and communities most vulnerable to climate impacts. Their legal entry and resettlement do not equate with entry for all migrants, or even all those adversely affected by climate change. Rather, it is one way that states can meet their adaptation obligations, and an avenue that must be pursued when other adaptation options cannot ensure access to basic resources and to the enjoyment of human rights. The admittance of migrants from small island developing states is a manifestation of assistance and cooperation that scholars like Robyn Eckersley advocate. Eckersley argues that while states have differentiated obligations to provide financial and technical assistance, allocated

[114] See Foster and others (n 14).

[115] Chapter 4, IV.D; CESCR, 'General Comment No 7: The Right to Adequate Housing (Art 11.1): Forced Evictions' (20 May 1997) UN Doc E/1998/22.

[116] Gharbaoui and Blocher (n 54) 163.

based on ability to pay, there is a shared responsibility to admit those forced to leave their home countries, as every country contributed to a collective failure to effectively address climate change. She argues that 'climate refugees' from small island states have the right to choose where to move, highlighting their need for permanent resettlement and that such a choice provides 'recognition of, and partial compensation for, the injustice and trauma of their loss and damage'. Adequate support is key to the political viability of her argument, to overcome resistance by states to such a right and to address concerns about the costs involved.[117] It is similarly crucial to measures aimed at enabling adaptive mobility. As this book explores, law can both facilitate and inhibit mobility. Law provides the authority to act, tools to adapt, and can create pathways for mobility away from areas where life and subsistence rights are at risk. It also provides a basis to argue that support for adaptive mobility—for migrants and states—is a legal obligation of developed countries. Yet it can likewise permit barriers to mobility, both in policy and practice, that contribute to past and present migration crises.[118] To mitigate or avoid such crises, and to continue to respond to and anticipate impacts from climate change, the law must itself be adaptive.[119] It must ensure appropriate structures and regulations are in place, which include those that apply to migration and planned relocation.

The climate change regime offers a starting place for international regulation. It can provide a forum to require increased finance, support in implementing adaptive mobility, and, more radically, ground an emerging duty to admit. Its principles and differentiated obligations can help determine who is responsible for assistance and to what degree, its financial mechanisms and institutional bodies can guide undertakings and offer spaces for dialogue and cooperation, and its general Conference of the Parties can provide oversight. All of this would, however, require a significant shift in financial contributions and action to ensure commitments are met. The need for increased funding for adaptation efforts has been called for by states in the Pacific Islands.[120] Governments are more likely to both plan for relocation within their own country or accept international migrants if provided with adequate support from other countries. Other countries are more apt to provide such support when the focus is on the collective and prospective aspects of adaptation obligations, which can better avoid obstacles created by blame or attribution.[121] At a local level, however, there is a need to ensure that any funding allocated for relocation or migration reaches the communities that it is intended to support. This

[117] See Eckersley (n 102) 495.

[118] As discussed in Chapter 1, IV.

[119] See eg Craig Anthony Arnold, 'Adaptive Law' in Rosemary Lyster and Robert RM Verchick (eds), *Research Handbook on Climate Disaster Law: Barriers and Opportunities* (Edward Elgar 2018).

[120] See eg National Communication, Marshall Islands (n 55) 1, 7–9.

[121] See eg Iris Marion Young, *Responsibility for Justice* (Oxford University Press 2011) 104–06 (setting out a social connection model of responsibility for wide-reaching injustices best addressed through collective action); Eckersley (n 102) 489 (citing Young).

will require addressing inequities in decision-making and distribution, and more opportunities for direct access to adaptation funding sources.[122]

Furthermore, once states make the connection between adaptation obligations and human rights integration, it gives them flexibility in exploring how to meet their commitments. This allows states both needing and providing support to look to guidance grounded in human rights. Instruments that create legal obligations, as well as those based in best practices, can therefore inform interpretation and decision-making. Guidelines and guidance on planned relocation can provide greater detail on how to formulate laws, policies, and programmes to implement adaptive planned relocation.[123] They discuss, for example, assessment of risks, maintaining standards of living for both relocated and receiving communities, and regulations in national laws that provide accountability mechanisms and adequate support.[124] A toolbox on relocation offers checklists for governments and detailed guidance on a legal framework; information, participation, and consultation; the needs of affected people; land issues; and monitoring, evaluation, and accountability.[125] Relevant guidance further includes the Nansen Protection Agenda, the culmination of a multi-year consultative process endorsed by 109 states in 2015. The Protection Agenda advocates for the integration of rights-based approaches to DRR, adaptation, and cross-border displacement. It calls for proactive planning and measures, managed mobility, and the admission of those facing climate impacts or disasters.[126] It proposes humanitarian protection measures to facilitate admission, such as the expansion of visa categories or free movement agreements. Thus, the Protection Agenda can and should be instructive to states. Yet it and other non-binding guidance are not an alternative to legal commitments. Instead, their suggestions can help flesh out content or provide direction on how to satisfy legal obligations, obligations that this book argues can drive and shape anticipatory adaptation and measures to enable adaptive mobility.

Cooperation and assistance are already apparent in the region, and the suggestions made in rights-based guidance are being taken up in government policies. Fiji, for example, has recognized the high probability of international mobility and the need to integrate migrants into host states across the Pacific through collaboration.[127] The Fijian government has agreed to accept migrants from Kiribati and other Pacific Islands, and some communities have gone so far as to declare

[122] Discussed further in Chapters 6, V; 9, IV.

[123] See eg United Nations Commission on Human Rights (UNCHR), 'Guiding Principles on Internal Displacement' (11 February 1998) UN Doc E/CN.4/1998/53/Add.2, principle 7(3); UNHCR, 'Guidance on Planned Relocation' (n 59); Burson and others (n 105) 400.

[124] See UNHCR, 'Guidance on Planned Relocation' (n 59) paras 15, 21, 28–29.

[125] See 'Toolbox' (n 59).

[126] See The Nansen Initiative, 'Agenda for the Protection of Cross-Border Displaced Persons in the Context of Disasters and Climate Change' vol 1 (2015); ILA (n 41) principles 7, 9.

[127] Fiji Planned Relocation Guidelines (n 50) 9.

willingness to share their land with those resettled from other states.[128] Pacific Islanders also have a pathway to permanent residency in Australia, through the 'Pacific Engagement Visa', with up to 3,000 visas available as of 2024 as part of a broader strategy to strengthen connections in the region. Expanding on other visa programmes for the region, like the Pacific Access Category or the Compact of Free Association, could be considered forms of international assistance, albeit for limited populations. The former permits a certain number of citizens from Tonga, Tuvalu, Kiribati, and Fiji to apply for residence in New Zealand. The latter allows Marshallese to travel freely to the United States. These pathways are not without costs and problems. For example, Marshallese migrants have faced discrimination and difficulty accessing social and health services in the United States.[129] Bilateral agreements are more explicitly engaging with climate change and human mobility. The Australia-Tuvalu Falepili Union treaty recognizes the security challenges arising from climate change, connecting cooperation on climate and security with access to legal migration pathways for citizens of Tuvalu. Under the agreement Australia will provide up to 280 permanent residency visas to Tuvaluans annually, while Tuvalu committed to limiting any future security or defence-related engagement with another state to those mutually agreed upon by Australia.[130]

Interim measures that allow for temporary or circular migration in the region can potentially serve as shorter-term solutions, while planning for more permanent migration continues. These could provide staggered movement and remittances that enable others to stay, which could reduce the need for larger-scale relocation in the future. They reflect decisions that occur at a household level, with some individuals migrating while others remain behind. Existing labour migration programmes also offer a means to adapt—and could be viewed as one form of assistance from developed states like New Zealand and Australia. However, these programmes were not contemplated to address climate change and are likely to subject people to the same risks current migrant workers face: poor work conditions and treatment, restrictions on their movements, and temporary legal status that precludes access to permanent residency by design.[131] Indeed, the requirement of repatriation limits the potential for existing labour mobility programmes to facilitate longer-term adaptive migration. To contribute to building resilience, they would need to more explicitly incorporate the same basic rights for migrants as are provided for other residents.

[128] See 'Statement of United Nations Special Rapporteur David R Boyd on the Conclusion of His Mission to Fiji' (OHCHR 2018) <www.ohchr.org/EN/NewsEvents/Pages/DisplayNews.aspx?NewsID=23964&LangID=E> accessed 31 January 2023. However, the Special Rapporteur characterizes this as a progressive and generous approach, rather than one mandated by law.

[129] See Seiji Yamada, Maxine Burkett, and Gregory G Maskarinec, 'Sea-Level Rise and the Marshallese Diaspora' (2017) 10 Environmental Justice 93.

[130] Australia-Tuvalu Falepili Union (n 93).

[131] See Farbotko and others, 'Transformative Mobilities in the Pacific' (n 48) 397.

More broadly, the Pacific Islands are at the forefront of planning for and collective action to address the risks created by climate change. The Framework for Resilient Development in the Pacific (FRDP) is the first regional integrated strategy for disaster risk management and climate change, and it calls upon states to incorporate human mobility into national policies and measures, including relocation and labour migration policies.[132] The 2018 Boe Declaration on Regional Security reaffirms that climate change poses the 'single greatest threat to the livelihoods, security and well-being of the peoples of the Pacific'. The Declaration expands the conceptualization of security in the region to include environmental, resource, and human security and the protection of the rights, health, and prosperity of Pacific people.[133] The 2050 Strategy for the Blue Pacific Continent also highlights the threats to the region from climate change and disasters, one of its seven interconnected thematic areas. It includes a strategic pathway forward to '[e]nsure the protection and practice of the rights, cultural values and heritage and traditional knowledge of Pacific peoples in global and regional protocols for climate and disaster risk reduction, and mobility including relocation, migration, and displacement'.[134] Efforts to reach a regional agreement on mobility built on these prior strategies and guidance, culminating in the endorsement of a Pacific Regional Framework on Climate Mobility in 2023 by Pacific Island countries. The Framework is the first to focus on climate change and human movement. It takes a collaborative approach to enable the region to 'anticipate, prepare for and respond to' the impacts of climate change, particularly those related to mobility. Adaptation is at the heart of the Framework, with core areas of action that prioritize helping people stay in their homes while also ensuring different forms of rights-based and people-centred movement. The Framework includes commitments to act on planned relocation, migration, displacement, and stranded migrants in the region, and it calls for technical and financial support from development partners.

VI. Conclusion

The Pacific Islands are at the forefront of climate change: they are some of the first countries to face significant impacts, with a long history of advocacy, collective action, and cooperation. There is also a shared connection to land and sea that is deeply engrained in culture and life. Yet there are different strategies for addressing climate impacts in the region, and different perceptions of the role human mobility plays in adapting to climate change. Integrating human rights into adaptation

[132] 'Framework for Resilient Development in the Pacific—An Integrated Approach to Address Climate Change and Disaster Risk Management (FRDP) 2017–2030' (2016) Voluntary Guidelines for the Pacific Islands Region 15.

[133] Pacific Island Forum, 'Boe Declaration on Regional Security' (September 2018).

[134] Pacific Island Forum, '2050 Strategy for the Blue Pacific Continent' (2022).

programmes and taking anticipatory action can enable Pacific Islanders to stay for as long as possible. At the same time, some people will need to move to ensure their continued access to resources and enjoyment of fundamental rights. This chapter argues for adaptation that seeks to balance both, exploring the tensions that arise from planning to stay and to move. It shows the limits to arguments anchored in the climate change regime, which continues to face shortfalls in adequate funding and support despite the legal obligation to do more. It also discusses forms of assistance and cooperation that move beyond finance to enable people to move, particularly across borders, safely and with access to legal rights and protections. The centrality of community-led processes and decision-making feature throughout the analysis, as an exercise of procedural rights, self-determination, and best practice. Indeed, participation will be key to adaptive mobility in the Pacific Islands, to ensure that moving—or not—reflects more choice than compulsion.

9

Drylands: International Mobility in the Sahel

I. Introduction

This chapter explores the relationship between climate change and human mobility in the Sahel. The region is considered one of the most climate-vulnerable in the world. As discussed in the other case studies, this vulnerability stems from multiple factors that include climate impacts and socioeconomic conditions. The chapter first discusses these factors, describing the impacts to drylands in the region, the conditions that give rise to vulnerability, and the risks to people that climate change helps create. It then turns to human mobility in the region, which is a way of life for many in the Sahel. Pastoralists migrate seasonally with shifts in rainfall. Labour migrants follow established networks and, like others, move from areas of drought in search of livelihood opportunities. Often this mobility crosses an international border. The Sahel is also a region with a history of conflict and refugee movement. Although the analysis considers internal mobility, the cross-border nature of much past and predicted movement calls for a focus on international adaptive mobility. Accordingly, the chapter next describes how the integration of human rights duties can shape understandings of adaptation obligations, and whether and how these obligations can lead to the facilitation of international migration. It ends with a discussion of support for such migration and, more generally, for anticipatory adaptive action in the region.

Like the chapters on Bangladesh and the Pacific Islands, this chapter argues for an approach to adaptation that incorporates human rights at all stages, in planning, implementation, and monitoring. Participatory processes are also critical and should be inclusive, iterative, and provide agency for those affected. Together, the case studies establish the need for more anticipatory adaptive action across contexts, the circumstances that require adaptive mobility measures, and the importance of adequate support.

Climate Change, Human Rights, and Adaptive Mobility. Lauren Sakae Nishimura, Oxford University Press.
 DOI: 10.1093/9780198930068.003.0010

II. Climate impacts, vulnerabilities, and risks to people and their rights

The Sahel is a wide swath of land that stretches across Africa from the Atlantic Ocean in the west to the Red Sea in the east. Its semi-arid climate serves as a transition between the drier Saharan desert to the north and the more humid tropical land to the south. It is comprised mainly of varying kinds of drylands. Desert and shrubland in the north give way to grasslands, shifting to savannah woodland towards the south. Further south and along the coast, the climate is wetter and more tropical. In the west near the Atlantic Ocean, much of the population lives less than five metres above sea level.[1] Overall, the Sahel contains diverse ecosystems largely reliant on river systems and interconnected wetlands. These riverine systems nurture the crops and wildlife that many in the region rely on for food and livelihoods, and they act as an important source of water during droughts.[2]

The Sahel includes parts or all of Senegal, Mauritania, Mali, Burkina Faso, Niger, Nigeria, Chad, Sudan, and Eritrea. The Gambia and Guinea-Bissau are also sometimes considered part of the region. Many of the countries in the region are members of the Economic Community of West African States (ECOWAS). The region is often characterized as one of crises and persistent vulnerability.[3] States in the Sahel consistently rank near the bottom of human development measures, based on life expectancy, education, and gross national income per capita. Aside from Nigeria, all countries in the region are classified as Least Developed Countries (LDCs). Poverty is widespread in many of the countries in the Sahel, with over 30 per cent of the population living in extreme poverty, or less than USD 1.90 per day.[4] The region has experienced chronic food shortages since the early 1970s, with a confluence of causes that include political instability, economic crises, desertification, and other impacts of climate change.[5] Access to natural resources in the Sahel is critical, as nearly 80 per cent of residents rely on them for livelihoods and survival. Much of the region is rural. Over two-thirds of the population live in rural areas and depend on agricultural subsistence.[6] Yet urban areas are also growing. Some cities in the region have increased in population by up to 120 per cent in recent decades, and urban populations are projected to be home

[1] Center for International Earth Science and Information Network (CIESIN), Columbia University, 'Low Elevation Coastal Zone (LECZ) Urban-Rural Population and Land Area Estimates, Version 3' (2021) <https://doi.org/10.7927/d1x1-d702> accessed 19 August 2024.

[2] Office for the High Commissioner for Human Rights (OHCHR), 'The Slow Onset Effects of Climate Change and Human Rights Protection for Cross-Border Migrants' (22 March 2018) UN Doc A/HRC/37/CRP.4 (Slow Onset Study) paras 93–94.

[3] See eg UN Office for the Coordination of Humanitarian Affairs (OCHA), '2024 Humanitarian Needs & Requirements Overview: Sahel' (2024).

[4] UNDP, 'Sahel Human Development Report 2023: Executive Summary' (2024) Annex C.

[5] Slow Onset Study (n 2) para 94.

[6] United Nations Environment Programme (UNEP), 'Livelihood Security: Climate Change, Migration and Conflict in the Sahel' (UNEP 2011) 5, 16–17.

to half the population of the western Sahel by 2045.[7] In general, populations in the LDCs are growing 2.5 times faster than rest of the world and are predicted to double by 2050; in some states in the Sahel, populations will nearly triple.[8] People tend to congregate around water sources in the Sahel. The Inner Niger Delta, for example, has nearly seven times the population density as the region overall.[9]

The Sahel is facing a range of climate impacts. Over the past 40 years, temperatures have risen more than in most regions in the world.[10] The Sahel has also experienced an increase in intense precipitation events and extreme storms, which are highly correlated with increased land temperatures.[11] Periods of long-lasting droughts occurred from the 1970s through the 1980s, followed by more recent recovery of annual rainfall amounts. This recovery is accompanied by more extremes: an increase in rainy days and intensity of precipitation, longer and more frequent dry spells, and earlier ends to rainy seasons.[12] Rainfall, or a lack thereof, plays a central role in the composition of the region's ecosystems. Changes in rainfall have led to expansion of grasslands, woodland coverage, and herbaceous vegetation production in some areas, while droughts are linked to the mortality of up to 20 per cent of the trees in the Sahel over a 60-year period.[13] Food production and livelihoods are likewise tied to rainfall, which is key to determining accessibility of food and food prices.[14]

The United Nations Framework Convention on Climate Change (UNFCCC) highlights the African continent as requiring cooperation to protect areas affected by drought, desertification, and floods.[15] Globally, it is estimated 1.6 billion people were affected by droughts over the last several decades, twice as many as by sudden-onset events like storms. Another 2.7 billion were affected by floods.[16]

[7] See World Health Organization (WHO), 'Sahel Region' (2006) Health Action in Crises; Richard Cincotta and Stephen Smith, 'What Future for the Western Sahel? The Region's Demography and Its Implications by 2045' (Atlantic Council 2021) 11.

[8] UN Department of Economic and Social Affairs, Population Division, 'World Population Prospects 2019: Highlights' (2019) ST/ESA/SER.A/423.

[9] F Jane Madgwick and others, 'Wetlands and Human Migration in the Sahel' (Wetlands International 2017) 23.

[10] See Christopher H Trisos and others, 'Africa' in HO Pörtner and others (eds), *Climate Change 2022: Impacts, Adaptation and Vulnerability. Contribution of Working Group II to the Sixth Assessment Report of the Intergovernmental Panel on Climate Change* (Cambridge University Press 2022) 1321, 1325.

[11] Christopher M Taylor and others, 'Frequency of Extreme Sahelian Storms Tripled since 1982 in Satellite Observations' (2017) 544 Nature 475; Trisos and others (n 10) 1325–26.

[12] Rachel Bezner Kerr and others, 'Food, Fibre and Other Ecosystem Products' in *Climate Change 2022: Impacts, Adaptation and Vulnerability* (n 10) 785.

[13] Trisos and others (n 10) 1334; Camille Parmesan and others, 'Terrestrial and Freshwater Ecosystems and Their Services' in *Climate Change 2022: Impacts, Adaptation and Vulnerability* (n 10) 239, 246.

[14] World Food Programme and others, 'Climate Risk and Food Security in Senegal: Analysis of Climate Impacts on Food Security and Livelihoods' (2013) 8.

[15] United Nations Framework Convention on Climate Change (adopted 9 May 1992, entered into force 21 March 1994) 1771 UNTS 107 (UNFCCC) art 4(1)(e).

[16] See International Organization for Migration (IOM), 'Migration, Environment and Climate Change: Assessing the Evidence' (2009); Jiseon Lee and others, 'Water-Related Disasters and Their Health Impacts: A Global Review' (2020) 8 Progress in Disaster Science 100123.

Droughts in the western Sahel are projected to increase in frequency and double in duration if global temperatures surpass the 2°C warming threshold. The length of rainy seasons will be further reduced, while precipitation is predicted to increase in the eastern Sahel.[17] Sea level rise threatens coastal areas, home to cities that are growing in population. Certain areas, including these cities, are considered climate change hotspots where impacts are predicted to contribute to desertification, wetland degradation, coastal erosion, and increased flooding.[18]

The effects of climate change are also contributing to increasing land degradation. Droughts and more erratic rainfall patterns are decreasing agricultural productivity and suitable land for pastoralists and farmers alike. Pastoralists and small-scale farmers are particularly vulnerable to climate change. Around 50 million people in the Sahel rear animals for livelihoods and survival, but available herding land is shrinking. There is competition from farmers who are seeking to expand areas under cultivation in the face of reduced crop yields.[19] An estimated 65 per cent of cultivable land in the western Sahel is already degraded.[20] In Niger, close to 95 per cent of productive land is dryland, and the population has experienced repeated food crises.[21] The stress on land is expected to continue: global warming of 1.5°C will reduce the area suitable for maize production by 40 per cent in the region, and sorghum yields will decline between 2 to 5 per cent in the western Sahel.[22]

The risks people face, however, are not exclusively environmental. Existing socioeconomic conditions interact with climate impacts to affect vulnerability in the Sahel. Many countries in the Sahel, for example, rank at the bottom globally for gender equality.[23] Rural women in the region face obstacles due to a lack of access to land and traditional, patriarchal decision-making processes. These can be exacerbated by intersecting attributes of age, disability, or migrant status that limit choice and control of adaptive options.[24] The burden of domestic labour tends to fall upon women and girls, who are responsible for meal preparation, childcare, and gathering water and fuel and often suffer from nutritional deficits as a result

[17] Trisos and others (n 10) 1326, 1329.

[18] See Isabelle Niang and others, 'Africa' in Vincente R Barros and others (eds), *Climate Change 2014: Impacts, Adaptation, and Vulnerability. Part B: Regional Aspects. Contribution of Working Group II to the Fifth Assessment Report of the Intergovernmental Panel on Climate Change* (Cambridge University Press 2014) 1209, 1211; Carlo Buontempo, 'Sahelian Climate: Past, Current, Projections' (Met Office Hadley Centre 2010).

[19] OHCHR, 'Human Rights, Climate Change, and Migration in the Sahel' (2021).

[20] Koko Warner and others, 'In Search of Shelter: Mapping the Effects of Climate Change on Human Migration and Displacement' (CARE 2009) 9.

[21] UN Human Rights Council (UNHRC), 'Report of the Special Rapporteur on the Right to Food, Jean Ziegler' (10 January 2008) UN Doc A/HRC/7/5, para 51.

[22] J Birkmann and others, 'Poverty, Livelihoods and Sustainable Development', in *Climate Change 2022: Impacts, Adaptation and Vulnerability* (n 10) 1200; Trisos and others (n 10) 1351.

[23] OCHA (n 3) 14.

[24] Trisos and others (n 10) 1318.

of gendered distributions of labour.[25] Differential vulnerability is also mediated by age. Shortages in food and drinking water in the region disproportionately affect children and harm their physical and cognitive growth.[26] Rising temperatures and longer, more intense heatwaves are likely to increase health risks for children and the elderly in a region where there are already widespread malnutrition and health issues.[27]

Socioeconomic conditions further interact with governance issues to reduce adaptive capacity, shaping risks in the Sahel from conflict, land tenure challenges, weak labour markets, and poor development planning and policies.[28] For example, irrigation and dam projects have cut off once fertile areas from water sources, often privileging the needs of urban centres or farmers at the expense of others downstream. Development projects like these have caused significant changes in the region: the loss of seasonal flooding in Mali due to a hydroelectric dam, environmental degradation of Lake Chad from upstream irrigation, and declining water levels in the Inner Niger Delta because of dams on the Niger River. A decline in the diversity and productivity of wetlands accompanied these changes.[29] Political instability also affects governance and adaptive capacity. Since 2020, Sudan and four of the founding ECOWAS countries—Mali, Guinea, Burkina Faso, and Niger—have experienced a military coup, signalling growing unrest and threats to economic progress and cooperation in the region.

Together, climate impacts and the context in which they operate threaten the enjoyment of a multitude of rights. The rights to life and to an adequate standard of living are put at risk when land and food sources are constrained. Water is also crucial for the region. It is an essential resource, access to which is a basic human right. A lack of access to water is an ongoing problem for many, and it will worsen as the climate changes. Reduced water availability will further hamper overall economic conditions. By 2050, the Sahel could lose up to an estimated 11.7 per cent of its GDP from water scarcity alone. Furthermore, if adaptation measures are not undertaken, climate-related shocks could lead to 13.5 million people across much of the Sahel falling into poverty during this same period.[30] As the Special Rapporteur on extreme poverty and human rights warned, without urgent climate action '[m]illions face malnutrition due

[25] CARE, 'Gender Equality and Women's Empowerment in the Context of Food Security and Nutrition' (2020) 17.

[26] OHCHR, 'Analytical Study on the Relationship Between Climate Change and the Full and Effective Enjoyment of the Rights of the Child, Report of OHCHR' (4 May 2017) UN Doc A/HRC/35/13, para 13.

[27] See Trisos and others (n 10) 1318; WHO (n 7).

[28] UNEP 2011 (n 6) 8–9.

[29] See Madgwick and others (n 9); Slow Onset Study (n 2) para 98.

[30] World Bank Group, 'G5 Sahel Country Climate Development Report' (2022); see also World Bank, 'High and Dry: Climate Change, Water, and the Economy' (2016).

to devastating drought, and many more will have to choose between starvation and migration'.[31]

III. Current and predicted mobility

Mobility has long been a means to cope with environmental change in the Sahel. Pastoralists often migrate alongside seasonal shifts in rainfall, those seeking agricultural work follow established networks, and environmental degradation and long-term drought and desertification push people from their homes in search of opportunities elsewhere. This mobility regularly crosses international borders. In Burkina Faso, up to half of adults have moved outside the country for at least some part of the year.[32] Prolonged drought in the late 1960s and early 1970s contributed to the movement of around a million people from the country to other states in the region. During the same period, Malians left their country to move to other areas in the region. Drought led to shifting migration in parts of Senegal, with initial seasonal migration within the country that expanded to nearby states and then to Europe. Those left behind, mainly women and children, faced a heavy economic burden and relied on remittances for income and to finance community social services.[33] Despite varying outcomes, migration continues to be an important form of adaptation in the Sahel that can help build resilience through livelihood diversification. Without planning and resources, however, it can perpetuate further environmental degradation, becoming part of a cycle of worsening conditions, food insecurity, and rights violations that leads to further movement and pressure on resources in places of destination. Such pressure can turn destinations into places migrants leave and can contribute to multiple experiences of displacement. Wetlands in the region, often a place of refuge during dry season and drought, are increasingly sources of out-migration as water scarcity pushes people to seek alternative livelihoods.[34]

The distance people migrate varies and is influenced by the context and duration of environmental change. Those living in border communities may migrate to neighbouring countries to cope with climate impacts.[35] Yet decreases in rainfall and the accompanying reduction of crop yields can limit the ability to move longer distances. Studies of Burkina Faso indicate that initial migration from rural areas facing rainfall deficits tended to be short-term, while more severe deficits

[31] UNHRC 'Climate Change and Poverty: Report of the Special Rapporteur on Extreme Poverty and Human Rights' (17 July 2019) UN Doc A/HRC/41/39, para 4.

[32] Warner and others (n 20) 9.

[33] IOM (n 16) 329–30.

[34] Madgwick and others (n 9) 12.

[35] See OHCHR, 'Human Rights, Climate Change, and Migration in the Sahel' (n 19) 4; Caroline Zickgraf, 'Relational (Im)Mobilities: A Case Study of Senegalese Coastal Fishing Populations' (2022) 48 Journal of Ethnic and Migration Studies 3450.

constrained the ability to move longer distances or internationally.[36] This stands in contrast with reports of wide-scale international movement during long-term droughts. It also underscores the different ways people move in the region and the different responses individuals may have to similar stressors. Simple generalizations therefore belie the complex and diverse experiences of human mobility in the Sahel. More recent research highlights the disruptions that climate impacts can cause to mobility, and the multiplicity of ways people move or stay when confronted with compounding risks. One study found that in Senegal, an urban fishing community facing sea level rise, coastal erosion, and reduced fish stocks responded to the resulting loss of livelihoods and homes by altering patterns of mobility. Sought after for their fishing skills and knowledge, some maritime fishers undertook labour mobility across the border to Mauritania. This international mobility, in turn, enabled more local adaptation: remittances supported family members to stay and to employ smaller-scale movements in response to local land pressures and coastal erosion. The latter included nearby relocation, which allows for regular return to ancestral and family homes.[37]

While movement takes many forms, one clear trend across the region is the growth of urban populations. The second half of the twentieth century featured a rapid increase in internal migration and the significant expansion of cities in the Sahel. For arid and semi-arid areas in Senegal, Mali, Burkina Faso, and Niger, rainfall decreases, land degradation, and violence contributed to such mobility.[38] States in the region are already acknowledging the role climate impacts are playing in continued migration from rural areas to cities. Both Sudan and Mauritania's National Adaptation Programmes of Action (NAPA), for example, highlight the role of drought and climatic stress on movement to urban areas.[39] Drier conditions in sub-Saharan Africa and the shocks to agriculture from these conditions are also linked to population movements and the urban growth of manufacturing centres.[40] As climate impacts worsen, labour and rural to urban migration are expected to increase in drylands, with displacement resulting from increasingly intense and prolonged impacts. Similar patterns will arise in the region's coastal zones, with initial rural to urban migration and more permanent movement away from areas at risk

[36] See Sabine Henry, Bruno Schoumaker, and Cris Beauchemin, 'The Impact of Rainfall on the First Out-Migration: A Multi-Level Event-History Analysis in Burkina Faso' (2004) 25 Population and Environment 423.

[37] Zickgraf (n 35).

[38] See Ton Dietz and Els Veldhuizen, 'Population Dynamics' in AJ Dietz, R Ruben, and A Verhagen (eds), *The Impact of Climate Change on Drylands: With a Focus on West Africa* (Springer 2004); Warner and others (n 20) 9.

[39] Republic of Sudan: Ministry of Environment and Physical Development, 'NAPA' (2007) 36 (those facing droughts have moved into towns from rural areas); Islamic Republic of Mauritania, 'NAPA' (2004) 16 (recognizing 'massive immigration to urban areas of farmers due to climactic stress and impacts on sources of subsistence').

[40] J Vernon Henderson, Adam Storeygard, and Uwe Deichmann, 'Has Climate Change Driven Urbanization in Africa?' (2017) 124 Journal of Development Economics 60.

from sea level rise.[41] Cities such as Dakar, Senegal, are close to sea level or slightly above, exposing millions to coastal climate impacts in the absence of intervention. Like many other urban areas globally, migrants in these cities tend to live in more vulnerable communities. In Dakar, up to 40 per cent of new migrants moved to high-risk flood areas.[42] Indeed, people often move away from places experiencing environmental change to areas that are equally or more environmentally fragile. In the Sahel, many migrating to hazardous urban areas leave behind advancing desert conditions and failing pastoral systems.[43]

Pastoralists' livelihoods are based on livestock keeping and herding, and they often travel great distances through remote, marginal environments to track variable plant growth and grazing land.[44] Nomadic pastoralists generally respond to changes in environment and disruptions by modifying their mobility patterns. As grazing land becomes drier and wetlands and water sources dry up, pastoralists are changing their migration routes. In Sudan, pastoralists are moving southwards as they contend with worsening conditions and loss of herds.[45] Shifts south and across borders may become more permanent, and there is concern that this will replace temporary and seasonal migration as certain climate events—droughts and floods—become more frequent and severe.[46] Fragmentation of nomadic migration due to large-scale agriculture is already disrupting mobility with consequences to well-being and resources.[47] In some cases, policies of enforced sedentarization and fencing that impedes livestock mobility have led to degradation of resources and migration away from traditional areas. These kinds of policies undercut the adaptive capacity pastoralists possess. They undermine the generational knowledge and experience pastoralists have acquired, as does the continued loss of pastures and livestock corridors needed to move from competing land uses.[48]

The Sahel further highlights the risks posed and complex interaction between resource competition, conflict, and mobility. The Lake Chad Basin provides an

[41] See Kanta Kumari Rigaud and others, 'Groundswell: Preparing for Internal Climate Migration' (The World Bank 2018) 30.

[42] W Neil Adger and others, 'Human Security' in Christopher B Field and others (eds), *Climate Change 2014: Impacts, Adaptation, and Vulnerability. Part A: Global and Sectoral Aspects. Contribution of Working Group II to the Fifth Assessment Report of the Intergovernmental Panel on Climate Change* (Cambridge University Press 2014) 768.

[43] UNHRC, 'The Right to Adequate Housing: Report of the Special Rapporteur on Adequate Housing as a Component of the Right to an Adequate Standard of Living, and on the Right to Non-Discrimination in this Context, Raquel Rolnik' (9 August 2009) UN Doc A/64/255, para 29.

[44] Alisher Mirzabaev and others, 'Deserts, Semiarid Areas and Desertification' in *Climate Change 2022: Impacts, Adaptation and Vulnerability* (n 10) 2210.

[45] Sudan's NAPA (n 39) 9, 30.

[46] See UNEP 2011 (n 6) 8.

[47] See Andrew Baldwin and Elisa Fornalé, 'Adaptive Migration: Pluralising the Debate on Climate Change and Migration' (2017) 183 The Geographical Journal 322, 325.

[48] See Alisher Mirzabaev and others, 'Desertification' in PR Shukla and others (eds), *Climate Change and Land: An IPCC Special Report on Climate Change, Desertification, Land Degradation, Sustainable Land Management, Food Security, and Greenhouse Gas Fluxes in Terrestrial Ecosystems* (Cambridge University Press 2019) 276; Mirzabaev and others (n 44) 2210.

example: climate change and water extraction contributed to the shrinking of the lake to a tenth of its former size. This has reduced resources and helped drive conflict.[49] Years of conflict and violence in the Basin area, including attacks by Boko Haram and the Islamic State in West Africa Province, led to the internal displacement of millions and refugee movements to nearby countries.[50] Elsewhere, ongoing conflicts in Mali and Sudan are displacing millions within countries and across borders. Resource constraints and poor management policies can also create conflict between competing land users. Past conflicts between pastoralists and farmers have been linked to competition over natural resources, in expanding agricultural areas where the policies and institutions in place failed to mediate tensions or ensure food security.[51] These conflicts, in turn, can lead to an increase in migration and displacement. Mobility can itself increase vulnerability. Displaced persons and refugees are often highly exposed to environmental hazards. Without permanent legal status or protections, they may remain in precarious protracted situations, residing for prolonged periods in settlements and with statuses meant to be temporary. Such tenuous situations can limit further mobility, creating immobile populations at risk of being overlooked in adaptation planning and in need of proactive measures to reduce economic and social barriers.[52]

More generally, cross-border and longer distance migration is a means to secure livelihoods in the face of changing conditions. For individual migrants from the Sahel, destinations are often north, to Egypt, Tunisia, Morocco, and Libya, with people remaining or transiting beyond to Europe.[53] Slow-onset events can inhibit movement and reduce the distance individuals are able to migrate, if at all. However, while moving over longer distances may be impeded during times of environmental and economic constraint, cross-border migration has been and will continue to be an adaptive strategy.[54] How it is undertaken will depend on the resources, information, and support available. Understanding past movement can also inform where and how future movement might unfold. For example, migrants are likely to prefer moving to places with established migrant populations due to the social, economic, and community networks they offer. In addition, there are instances where mobility is being used by rural households in the drylands, to earn income and mitigate the effects of environmental impacts on crops and livestock

[49] UN Children's Fund (UNICEF), 'No Place to Call Home: Protecting Children's Rights When the Changing Climate Forces Them to Flee' (2017) 18.

[50] United States Agency for International Development (USAID), 'Lake Chad Basin—Complex Emergency' (26 June 2020) Fact Sheet #3 <https://www.ecoi.net/en/file/local/2023391/lake_chad_ce_fs01_01-03-2020.pdf> accessed 12 August 2024.

[51] See Republic of Sudan: Ministry of Environment and Physical Development, 'NAP' (2016) 4; Sudan's NAPA (n 39) 16; Adger and others (n 42) 772–73.

[52] Birkmann and others (n 22) 1182.

[53] See The World Bank, 'Turn Down the Heat: Confronting the New Climate Normal' (2014) 28, 144.

[54] Slow Onset Study (n 2) paras 100–01.

productivity. Yet obstacles to long-term adaptation remain and must be addressed by governmental action and support.[55]

IV. Integration of human rights and adaptive mobility

The current and expected changes to the landscape and mobility in the Sahel are part of the context in which adaptation obligations arise. In the absence of effective, timely adaptation measures, pressure on the land from climate impacts will exacerbate precarious migration, refugee movement, and displacement. Yet known socioeconomic conditions, past and ongoing mobility, and predicted climate impacts can help states to plan for and implement adaptation and to facilitate adaptive mobility where needed. Like sea level rise, drought and desertification are foreseeable and occur over timescales that create both opportunities and obligations to act. They provide the time to take anticipatory action that satisfies obligations and that addresses the causes of precarious mobility or displacement.

Likewise, socioeconomic, political, and governance factors will almost always affect the extent to which an environmental hazard or resource constraint contributes to the risks people face.[56] Social vulnerability can heighten risks, making it more difficult to access resources. Gender or age, for example, contribute to greater likelihood of existing nutritional deficits and health issues, which can compound harm from climate impacts.[57] Urbanization and shifts in demographics can further alter vulnerability to climate impacts, particularly for those migrants already on the margins. Yet there is still time to act, particularly given that some risks are foreseeable. As a region where most people rely on access to natural resources—water and land for subsistence farming and grazing—the depletion or degradation of these resources puts the enjoyment of related rights at risk. These risks make relevant positive obligations to ensure safe, adequate, and equitable access to essential resources.[58] They trigger states' duties, independently as rights obligations, and through adaptive measures when integrated into the interpretation and implementation of adaptation obligations. Thus, in places where these resources can no longer be provided, obligations to act on adaptation combined with the predicted precarious movement that might ensue without such action establish the need to explore adaptive mobility options. These options should be part of longer-term strategies to reduce risk and incorporate more transformational adaptation, which

[55] See Charlotte Wiederkehr, Michael Beckmann, and Kathleen Hermans, 'Environmental Change, Adaptation Strategies and the Relevance of Migration in Sub-Saharan Drylands' (2018) 13 Environmental Research Letters 113003.

[56] See Chapter 1, II.B; see also Matthew Scott, *Climate Change, Disasters, and the Refugee Convention* (Cambridge University Press 2020) ch 2; United Nations Office for Disaster Risk Reduction (UNDRR), 'Sendai Framework for Disaster Risk Reduction 2015–2030' (UNISDR 2015) para 6.

[57] See Section II.

[58] See Chapter 4, III.

are repeated themes of the sixth assessment report of the Intergovernmental Panel on Climate Change (IPCC) on impacts, adaptation, and vulnerability.[59]

As Chapter 2 describes, adaptation planning is a first step in facilitating human rights-based adaptive mobility and accessing financing and assistance. For the climate regime, adaptation planning includes NAPAs and National Adaptation Plans (NAPs). These planning processes allow states to integrate adaptation into domestic planning and include potential adaptation measures that can address vulnerability by, amongst other things, improving livelihood opportunities for herders, access to food and drinking water, or recognising migration as a potential adaptation strategy.[60] Yet only a handful of adaptation planning documents submitted from states in the Sahel propose projects involving planned relocation or migration. Burkina Faso's NAP includes the relocation of populations from flood zones and low-lying areas as an adaptation measure.[61] Sudan's NAP highlights displacement and migration as an impact of climate change, with 'tribal migration route rehabilitation' in part of the country as a form of adaptation.[62] Its Adaptation Communication details adaptation efforts to build resilience in dryland pastoral and farming systems through interventions to improve access to water and rehabilitate communal rangelands.[63] These kinds of measures can enable people to both stay and to move freely in pastoral areas. Nigeria's Adaptation Communication prioritizes 13 sectors for adaptation strategies and policies, one of which is 'Disaster, Migration, and Security'. It calls for the integration of migration and displacement issues in national climate change planning and includes a list of measures to strengthen capacity and anticipation of impacts.[64] Recognitions like these are a step towards adaptation that contemplates mobility. However, more explicit incorporation of adaptive mobility into adaptation plans and programmes is needed to address ongoing risks in the region and to satisfy international obligations.

To date, adaptation projects in the region financed by the Adaptation Fund focus on building resilience to help prevent displacement, rather than considering mobility as a positive means of adaptation. A completed United Nations Development Programme (UNDP) project in Eritrea, for example, recognized migration from extreme conditions as a risk of climate change. The project sought to avoid this risk through adaptation measures that targeted water management and agricultural development, thereby improving food security and providing resilient livelihood opportunities. Likewise, the proposal for a coastal resilience project in Senegal

[59] See IPCC, *Climate Change 2022: Impacts, Adaptation and Vulnerability* (n 10).

[60] See eg Sudan's NAPA (n 39) 43, 45; Eritrea: Ministry of Land, Water and Environment, 'NAPA' (2007) 27.

[61] Ministry of Environment and Fishery Resources, 'Burkina Faso NAP' (2015) 15, 64, 66.

[62] Sudan's NAP (n 51) 28.

[63] Republic of Sudan, 'First Nationally Determined Contribution under the Paris Agreement' (2021) 22.

[64] Federal Republic of Nigeria, 'Adaptation Communication to the United Nations Framework Convention on Climate Change' (2021) 29.

noted that its activities to rehabilitate dikes will protect housing against floods, helping to prevent displacement and relocation. A project in Mali similarly aimed to increase resilience of communities and their water and livelihood systems. The project's adaptation programme targeted particularly vulnerable populations that included small-scale farmers, women, youth, and displaced populations. However, an evaluation of the programme noted that insecurity exacerbated by the coronavirus pandemic could undermine its sustainability and lead to displacement.[65] Mention of human rights in documents submitted to the Adaptation Fund was also sparse, generally referenced in response to the Adaptation Fund's Environmental and Social Policy requirements to assess risks and respect human rights.[66] Human rights issues for these projects were not identified as a risk, with compliance assumed or promised. In the case of the coastal management project in Senegal, the proposal noted that there were 'no relevant human rights issues to consider', although gender equality and labour rights were discussed in greater detail.[67] None of these projects considered mobility in their adaptation responses to address climate-related risks, which will be increasingly necessary in the region even if more robust climate action is undertaken.

Furthermore, effective adaptation and adaptive mobility require the participation of affected people in decision-making. Adaptation is a process that should be informed and led by individuals and communities that will be impacted. As previous chapters explain, at a minimum, people must be provided with information and included in the decision-making that will affect their interests. Such inclusion is required by adaptation obligations that integrate relevant procedural human rights, and it is also essential to transformative adaptation. However, to ensure inclusive and meaningful engagement, it is important to first interrogate who will be affected, what barriers exist to participation, and who is making decisions and carrying out measures.[68] Answers to these questions will vary across contexts but should be considered in all state-facilitated adaptation policies and programmes. Decision-making must therefore be based on participatory processes that incorporate community perspectives into planning and implementation. For mobility

[65] See Kalame Fobissie, Brhan Khiar Saleh, and Eyob Haile, 'Terminal Evaluation Report: Climate Change Adaptation Programme in Water and Agriculture in Anseba Region, Eritrea' (2019) Final Project Report 7 <www.adaptation-fund.org/project/climate-change-adaptation-programme-in-water-and-agriculture-in-anseba-region-eritrea/> accessed 22 August 2024; Youssoufou Congo and Ibrahim Nienta, 'Final Evaluation of the Programme Support for Climate Change Adaptation in the Vulnerable Regions of Mopti and Timbuktu (PACV-MT)' (2020) Final Report 6–7, 11 <www.adaptation-fund.org/project/programme-support-for-climate-change-adaptation-in-the-vulnerable-regions-of-mopti-and-timbuktu-2/> accessed 22 August 2024; Centre de Suivi Écologique, Senegal, 'Reducing Vulnerability and Increasing Resilience of Coastal Communities in the Saloum Islands (Dionewar)' (2017) Project/Programme Proposal to the Adaptation Fund 67 <www.adaptation-fund.org/project/reducing-vulnerability-increasing-resilience-coastal-communities-saloum-islands-dionewar/> accessed 22 August 2024.

[66] See Adaptation Fund Board, 'Environmental and Social Policy' (2016).

[67] Centre de Suivi Écologique (n 65) 65–67.

[68] See Chapters 5, III; 7, IV; 8, III.

in the region, particularly for pastoralists and those they encounter, this will involve processes and cooperation that may cross borders. Those affected should be provided with agency, to help shape the adaptive measures they will be impacted by or undertake. Doing so can help avoid the perception or experience of being forced to act, including by moving. The African Commission for Human and Peoples' Rights (the Commission) has highlighted the connection between human rights, consent, and the prevention of forced movement, its 2009 resolution urging African Union members to 'ensure that human rights standards . . . such as the principle of free, prior and informed consent, be included into any adopted legal text on climate change as preventive measures against forced relocation, unfair dispossession of properties, loss of livelihoods and similar human rights violations'.[69]

Implementation of adaptation and adaptive mobility will also require an increased focus on proactive and pre-emptive solutions. A variety of measures become necessary when considering risks to resources and rights and integrating precaution and positive duties. These include preparing for changes to ecosystems and resources, facilitating mobility for pastoralists and those in vulnerable drylands, and securing land elsewhere. The transboundary movement of pastoralists and their livestock is an enduring feature of the region, one that will require adaptation to address land use competition as climate impacts exacerbate adverse effects to pastoral land. In addition, adequate housing and land tenure, as well as governmental capacity to ensure rights, are crucial to sustainable land use, which in turn fosters access to food and food security and reduces the risk of further climate-related mobility. Conversely, the absence of land tenure security and adequate resources contributes to vulnerability and hinders the ability to adapt, which can contribute to immobility or lead to displacement. It also runs afoul of positive human rights duties to ensure access to essential resources, as well as obligations to adapt that integrate these duties into their interpretation. The latter is, as this book argues, central to ensuring that measures taken by states are truly adaptive and meet broader goals to reduce vulnerability, strengthen resilience, and enhance adaptive capacity.[70]

More specifically, adaptation can enable and build upon the kind of governance that can address the causes of vulnerability. For example, there are programmes in place that incorporate social protection into adaptation. These can support short-term responses to extreme events while helping to build longer-term adaptive capacity. Research in the Sahel found that the use of seasonal forecasting tailored to the local context has the potential to benefit adaptive social protection programmes, but more capacity-building and dialogue between stakeholders are

[69] African Commission for Human and Peoples' Rights (ACHPR), 'Resolution on Climate Change and Human Rights and the Need to Study Its Impact in Africa' (Banjul, The Gambia 25 November 2009) ACHPR/Res.153(XLVI)09.

[70] See Paris Agreement (adopted 12 December 2015, entered into force 4 November 2016) 3156 UNTS 79, art 7(1); Chapter 2, III.

required.[71] Similarly, for mobility, an approach that enables movement in advance of land becoming uninhabitable or rights being violated can contribute more to resilience than waiting until tipping points are reached and people are forced to move. Such an adaptive approach sidesteps problematic categorizations of people on the move and bases responses on context and needs.[72] It can provide rights protection and address risks prior to and throughout adaptive mobility, which in turn can help avoid displacement or involuntary immobility. Such an approach moves beyond duties and guidance on displacement for the Sahel. The Guiding Principles on Internal Displacement and their binding counterpart in the region—the Kampala Convention—include duties to prevent the conditions that lead to displacement, which could apply to adaptation in situ or proactive mobility.[73] While they strengthen arguments to take preventative action, as discussed in Chapter 1, limitations preclude their broad application to the kind of pre-emptive, anticipatory mobility advanced by this book. Other solutions, including free movement or regional mobility agreements, require coordination and cooperation, as discussed next.

V. International assistance and cooperation

International support will be critical to successful adaptation in the Sahel. Financial and technical support are particularly important. They are required as a matter of law within the climate change regime and are needed to implement adaptation measures. Yet adequate support is lacking. The Commission has highlighted the insufficiency of assistance for African countries, expressing concern over developed countries' failure to implement their climate obligations and the absence of 'long-term cooperative action, including a lack of technology transfer and financial assistance for mitigation and adaptation, [which] seriously undermines the capacity of African governments to safeguard human rights in Africa'.[74] For over a decade, the Commission has also repeatedly linked climate change and human rights in resolutions, encouraged measures to protect vulnerable groups, requested a study on climate impacts, and called for the integration of human rights and climate

[71] Joseph Daron and others, 'Integrating Seasonal Climate Forecasts into Adaptive Social Protection in the Sahel' (2021) 13 Climate and Development 543; Bezner Kerr and others (n 12) 801.

[72] Chapter 1, IV describes the limitations of approaches that provide protection based on a category of migrant.

[73] United Nations Commission on Human Rights (UNCHR), 'Guiding Principles on Internal Displacement' (11 February 1998) UN Doc E/CN.4/1998/53/Add.2, principles 5, 6; African Union Convention for the Protection and Assistance of Internally Displaced Persons in Africa (adopted 23 October 2009, entered into force 6 December 2012) (Kampala Convention) arts 3(1)(a), 4.

[74] ACHPR, 'Resolution on Climate Change and Human Rights in Africa' (Banjul, The Gambia 20 April 2016) ACHPR/Res.342(LVIII)2016.

change into development plans.[75] There has yet to be a climate-specific case before a regional human rights body in Africa. But it remains an issue of priority. In 2023, the Commission conducted a draft study on climate change and human rights in Africa, a study initially called for in a 2009 resolution. The Commission recognized that the adverse effects of climate change are exacerbated by indiscriminate resource extraction and exploitation. The resulting impacts threaten the lives, integrity, and livelihoods of people, with disproportionate effect on those most vulnerable.[76] Addressing these threats will require both regional and domestic efforts, as well as the international support lacking from developed countries.

Cooperation and assistance are needed, in part, because of the capacity and status of LDCs in the Sahel. Even with foresight and planning, implementation of adaptation programmes in the region will be difficult without adequate support. Instability in governance, social structures, and economic conditions in the Sahel reduce adaptive capacity, as do issues with land tenure and development policies. The climate change regime singles out LDCs and creates an obligation of all parties to take into account their 'specific needs and special situations' in funding and technology transfer, both of which are required forms of adaptation assistance.[77] These needs and situations help justify and are reinforced by differentiation in the regime and integration of the common but differentiated responsibilities and respective capabilities (CBDR-RC) principle into obligations of assistance for LDCs.[78] Cooperation is also a key element of complementary initiatives arising out of other regimes. The Great Green Wall, for example, is a transboundary project launched by the African Union in 2007 to restore degraded landscapes across the Sahel and beyond. The project is being implemented in 22 countries in cooperation with the United Nations Convention to Combat Desertification (UNCCD), the World Bank, and the Global Environment Facility (GEF). More than USD 14 billion has been mobilized, although not even 20 per cent of the target for restoration had been reached as of 2023.[79] Nonetheless, the project is lauded by the IPCC

[75] ACHPR, 'Resolution on Climate Change and Human Rights and the Need to Study Its Impact in Africa' (n 69); ACHPR, 'Resolution on Climate Change in Africa' (Luanda, Angola 12 May 2014) ACHPR/Res.271(LV)2014; ACHPR, 'Resolution on Climate Change and Human Rights in Africa' (Banjul, The Gambia) (n 74); ACHPR, 'Resolution on the Human Rights Impacts of Extreme Weather in Eastern and Southern Africa Due to Climate Change' (Sharm el Sheikh, Arab Republic of Egypt 14 May 2019) ACHPR/Res.417(LXIV)2019.

[76] See 'Call for Comments to the "Study on the Impact of Climate Change on Human and Peoples" Rights in Africa" (African Commission on Human and Peoples' Rights, 23 October 2023) <https://achpr.au.int/en/news/press-releases/2023-10-28/call-comments-study-impact-climate-change-human-and-peo> accessed 19 August 2024.

[77] UNFCCC, art 4(9); Paris Agreement, art 9(4). Discussed further in Chapter 2, IV.B.

[78] See Chapter 6, III.A.

[79] See Spoorthy Raman, 'Progress Is Slow on Africa's Great Green Wall, but Some Bright Spots Bloom' *Mongabay Environmental News* (3 August 2023) <https://news.mongabay.com/2023/08/progress-is-slow-on-africas-great-green-wall-but-some-bright-spots-bloom/> accessed 20 August 2024; UNCCD, 'The Great Green Wall Implementation Status and Way Ahead to 2030' (2020) <https://www.unccd.int/sites/default/files/2024-08/1551_GGW_Report_ENG_Final_040920.pdf> accessed 20 August 2024.

as demonstrative of a transboundary approach to managing and adapting to risks that are of 'common concern', which will require continued cooperative actions.[80]

Cooperation can help states in their efforts to coordinate action, balance competing interests, and avoid maladaptation and violations of human rights. The need to ensure minimum essential resources creates clear mandates to prioritize certain rights: survival and subsistence rights should come before protection of property rights. Other circumstances may require a different balance, which means making difficult decisions about which adaptation initiative gets support and who benefits or might be harmed. For example, a state may need to decide whether small-scale irrigation measures that divert water upstream should go forward, even if they will reduce access to water for multiple users downstream.[81] Coordination and cooperation between states will be needed where water sources cross borders. Adaptation is unlikely to provide equal benefits or resources for all, which makes an approach focused on fundamental rights and participation essential. Such an approach must incorporate non-discrimination and equality, core principles of human rights law. These principles underlie states' obligations and individuals' exercise of rights, and they require an accounting for existing marginalization and vulnerabilities in measures to address climate change. At a minimum, these principles create obligations for states not to discriminate in their adaptation plans, policies, and programmes. They further require that those most vulnerable be prioritized in adaptation. For adaptive mobility, ensuring rights and addressing those most vulnerable translates into measures that consider risks to migrants and their families, as well as those who might be unable to move.[82] In the Sahel, effective adaptive mobility measures will involve prioritizing movement that allows for continued access to basic rights and resources, for pastoralists and those seeking livelihood opportunities. These measures are likely to require coordination that crosses borders and support beyond what local and state governments can provide on their own.

The need for adequate resources raises two further considerations. The first concerns funding reaching local communities and the second that these communities drive the processes of adaptation. Beyond a general shortfall in financing, there are barriers to adaptation funding in the Sahel, and to its use for the benefits of local communities. Global funds aim to support vulnerable developing countries, in line with the climate regime's obligations and the CBDR-RC principle. However, accessing funding that is available can prove difficult, particularly when bureaucratic hurdles create onerous processes for proposals to go forward. The

[80] Trisos and others (n 10) 1312.

[81] See AK Magnan and others, 'Addressing the Risk of Maladaptation to Climate Change' (2016) 7 Wiley Interdisciplinary Reviews: Climate Change 646, 650.

[82] See Chapter 5, II.B; OHCHR, 'Key Messages on Human Rights, Climate Change, and Migration' <www.ohchr.org/Documents/Issues/ClimateChange/Key_Messages_HR_CC_Migration.pdf> accessed 17 May 2017.

Adaptation Fund, for example, introduced a direct access modality to better enable developing countries to obtain and manage funding. It was the first of its kind for climate funds, and it aims to reduce transaction costs and enhance national agency and ownership of adaptation. Projects arising from this modality tend to be more focused on communities, which improves the likelihood of local benefits. Yet countries in Africa are struggling to get accreditation for direct access due to a long, cumbersome accreditation process. Most developing countries are not accredited.[83]

As of June 2020, there were 32 national implementing entities approved for direct access funding. Among these, only Senegal and Niger have approval in the Sahel.[84] Analysis of the Green Climate Fund (GCF) similarly found that adaptation financing is not always reaching highly vulnerable countries, particularly LDCs in Africa. Direct access to funds remains elusive for some of these countries, with the lack of institutional capacity playing a role in both vulnerability and access.[85] For communities in need of adaptation support, this can mean that funding is either not available or when received, not disbursed. There must also be a shift from readying countries to receive funds to concrete action and implementation of adaptation measures, which should better incorporate local and subnational actors. These actors are key to effective adaptation, and their engagement and empowerment are essential to finance reaching its intended targets.[86] Engagement raises the second concern, that of community involvement. As Chapter 5 and the prior section explain, adaptation is a process that should be driven by affected individuals and communities. Participation is required by international legal obligations, and it should be provided through inclusive, meaningful engagement that provides agency for the people who will be implementing and impacted by adaptation.

In addition, regional cooperation will be necessary to better implement or expand on existing policies and mechanisms that affect migration. A general disaster risk reduction (DRR) framework and a programme of action exist for the African continent, which aim to improve coherence and integration between adaptation, DRR, addressing conflict and fragility, and other development priorities. They also seek to promote inter-country cooperation.[87] Each of these areas of action affects mobility, and cooperative efforts to implement them can help build resilience and

[83] Trisos and others (n 10) 1308.

[84] Adaptation Fund, 'Climate Adaptation Finance: Direct Access' (2020).

[85] See Matthias Garschagen and Deepal Doshi, 'Does Funds-Based Adaptation Finance Reach the Most Vulnerable Countries?' (2022) 73 Global Environmental Change 102450.

[86] Kulthoum Omar-Matsumi, Mandy Barnett, and Liane Schalatek, 'Broken Connections and Systemic Barriers: Overcoming the Challenge of the "Missing Middle" in Adaptation Finance' Global Commission on Adaptation Background Paper.

[87] African Union, 'Africa Regional Strategy for Disaster Risk Reduction' (ARSDRR) (July 2004); African Union, 'Programme of Action for the Implementation of the Sendai Framework for Disaster Risk Reduction 2015–2030 in Africa: In Line with the ARSDRR' (November 2016).

reduce precarious migration. Regional adaptation projects are also underway that require cooperation on shared ecosystems. A project in Burkina Faso, Niger, and Benin that covers the W-Arly-Pendjari (WAP) transboundary complex aims to strengthen the resilience of ecosystems and improve people's livelihoods through concrete adaptation measures. The WAP complex is a network of protected areas that spans approximately 50,000 square kilometres across the three countries. Its biological diversity contributes to the well-being of the area and its one million inhabitants, put at risk from multiple stressors including overgrazing, climate impacts and bushfires, water pollution, transhumance, and unsustainable fishery and timber practices. The project conceptualization considers 'climate migration' as a threat to resources and ecosystems, and transboundary transhumance—the seasonal movement of livestock to new grazing areas—as a source of degradation, destruction of flora, and economic and social conflicts. It focuses, in part, on addressing the droughts that lead 'to uncontrolled movement of pastoralists', flooding, and expansion of agricultural lands through 'urgent adaptation' actions.[88] Like other country-specific projects in the region, however, it does not specifically contemplate migration as a means to build resilience, instead focusing on measures that enable local communities to remain. This is a necessary first step for most adaptation programmes. Yet the recognition of 'climate migration' and increasingly unpredictable movement of pastoralists across borders underscores the need for further steps that plan for mobility and that enable migration that continues to ensure human rights, as projects funded by the Adaptation Fund promise to do. Cooperation and support will be necessary to achieve rights-protective mobility and to facilitate continued migration pathways for pastoralists throughout the region, while also balancing agricultural needs and interests.

Furthermore, international adaptive mobility requires legal pathways for migrants and cooperation to provide concrete solutions. The expansion and liberalization of regional mobility or free movement agreements is one means suggested to regularize migration between states and promote the integration of migrants.[89] Many of the countries in the Sahel were founding members of ECOWAS, which has adopted several free movement protocols.[90] Yet increasing political instability

[88] Sahara and Sahel Observatory, 'Integration of Climate Change Adaptation Measures in the Concerted Management of the WAP Transboundary Complex: ADAPT-WAP' (2019) Regional Project Proposal 7, 24, 28, 58 <https://www.adaptation-fund.org/project/integration-climate-change-adaptation-measures-concerted-management-wap-transboundary-complex-adapt-wap-benin-burkina-faso-niger/> accessed 21 August 2024.

[89] See Tamara Wood, 'The Role of Free Movement of Persons Agreements in Addressing Disaster Displacement: A Study of Africa' (Platform on Disaster Displacement 2019); Ama Francis, 'Free Movement Agreements & Climate-Induced Migration: A Caribbean Case Study' (Sabin Center for Climate Change Law, Columbia Law School 2019).

[90] Protocol Relating to Free Movement of Persons, Residence and Establishment (adopted 29 May 1979, entered into force 8 April 1980) A/P.1/5/79 (ECOWAS Protocol). The Community of Sahel-Saharan States also sought to establish a free movement agreement, which was drafted but never adopted.

threatens cooperation in the region. The withdrawal of Burkina Faso, Mali, and Niger from ECOWAS in January 2024, each of which had experienced recent military coups, puts in question how these countries will navigate regional cooperation going forward. These countries, controlled by juntas, created a defence pact and an Alliance of Sahel States centred on security issues. Such a shift evinces efforts to move away from external influence and support from Western countries, and it leaves open issues of cross-border and regional migration.[91]

Nearby, the Intergovernmental Authority on Development (IGAD) has made steady progress on the implementation of a free movement protocol between its member states. IGAD is comprised of eight countries, which include Eritrea and Sudan. With funding from the European Union, it began developing a protocol in 2017 that was finalized in 2020. The resulting Protocol on Free Movement of Persons in the IGAD Region is the first to explicitly incorporate climate change and the movement of persons affected by disasters. It allows for the movement of people 'in anticipation of, during or in the aftermath of disaster', as well as the stay for those who cannot return due to disaster. It also encourages coordination and cooperation to ensure facilitation of free, safe, and orderly movement.[92] An additional IGAD Protocol on Transhumance was adopted in 2021, which allows for the seasonal movement of livestock and herders seeking water and pasture across borders. This Transhumance Protocol recognizes such mobility as 'an adaption and survival mechanism'.[93] At the time of writing, seven of the eight IGAD members had officially signed at least one of these protocols: Sudan, South Sudan, and Ethiopia signed both; Kenya and Djibouti the Transhumance Protocol; and Somalia and Uganda the Free Movement Protocol.

In principle, protocols and free movement agreements like those in IGAD and ECOWAS enable citizens of member states to enter and reside in other member states. However, the extent to which free movement agreements can help climate-vulnerable populations depends on how well they are respected by national governments. These agreements offer a means to assist and to address international mobility through an adaptive lens. They include benefits that fill gaps in protection and international law: they can offer access to international territory; permit entry, stay, and work; and allow for permanent or regularized status. These features are not automatic: if there are too many bureaucratic hurdles and human rights protections are not explicit, then the protections they could afford will be limited.[94]

[91] See 'Niger, Mali and Burkina Faso Military Leaders Sign New Pact, Rebuff ECOWAS' *Al Jazeera* (6 July 2024) <www.aljazeera.com/news/2024/7/6/niger-mali-and-burkina-faso-military-leaders-sign-new-pact-rebuff-ecowas> accessed 19 August 2024.

[92] Protocol on Free Movement of Persons in the IGAD Region (endorsed 26 February 2020, adopted by Council of Ministers of Member States 24 June 2021) arts 3, 16, 30.

[93] Intergovernmental Authority on Development (IGAD) Protocol on Transhumance (endorsed 27 February 2020, adopted by Council of Ministers of Member States 24 June 2021) art 3.

[94] See Wood (n 89) 7–8.

Similarly, and as discussed for the Pacific Islands, a system that facilitates visas or allows entry is a way for states to fulfil obligations to cooperate and assist in adaptation. The admittance of international migrants can be tailored to the specific risks that adaptive mobility measures in a state are addressing, thereby alleviating some concerns that such a system would create a duty to admit all migrants. Instead, the focus should be on those most vulnerable to climate change and the foreseeable risks to the enjoyment of their rights. Given this focus, states could consider which populations face greater risks from climate change, using a rights-based approach to strike the balance in establishing priorities. Scaled up, such a system could operationalize the differentiation within the climate regime—to share the burden of adaptation while distributing its costs depending on the status and capacity of states. This would result in different obligations for states, determined by whether they are taking on international migrants or supporting in such efforts.

VI. Conclusion

This chapter explores the impacts of climate change on the Sahel region, and the influence these impacts will have on human mobility. It is a region where most rely on the land for survival and subsistence, and where constraints imposed on shared resources from water scarcity and drought are affecting agriculture and pastoral livelihoods. These constraints are contributing to changes in mobility, altering nomadic lifestyles and increasing migration in the face of environmental degradation. To address predicted impacts, concerted efforts will be needed in countries where capacity and governance will be tested by climate change. Adaptive action can and will include measures to build resilience for communities, to enable continued mobility for pastoralists, and to provide pathways for migration within countries and across borders. Cooperation and support are needed, both regionally and internationally. International support is required within the climate regime, at a minimum to provide financial and technical assistance in adaptation. Regional cooperation and coordination are also needed, to ensure participation of affected people and that support reaches those it is intended for.

Together with the other case studies, this chapter makes the case for the integration of human rights throughout the process of adaptation. It advances a systemic understanding of adaptation obligations, which can lead to anticipatory adaptive action that addresses foreseeable risks to people and communities. It also recognizes the need for more direct access to support, so that states in the Sahel can plan for and undertake adaptation measures that adequately respond to these risks. States should determine what measures are appropriate, but such discretion is not unfettered: as the book and this chapter argue, in some circumstances facilitating migration or planned relocation will be necessary to satisfy adaptation obligations. Doing so, through processes that are inclusive and participatory, better ensures the rights of some of those most vulnerable to climate impacts in the region. It can help avoid displacement while enabling people to move with greater agency and access to rights and resources.

Conclusion

This book argues that adaptation obligations in the climate change regime can play a role in shaping and enabling proactive, anticipatory measures to plan for and implement adaptive mobility. In doing so, it helps fill the gaps in legal protection that Chapter 1 identifies, gaps created, in part, by looking first to frameworks developed for human mobility and then applying them to the context of climate change. The book reverses this approach and looks first to the climate change regime. As Chapter 2 describes, this affords a reliance on the dynamic nature of adaptation and obligations on adaptation. Adaptation is not contemplated as a one-time measure in climate treaties or the climate regime. It is, instead, a process, one that preceding chapters established should address a confluence of risks and conditions that contribute to vulnerability. Adaptation must also necessarily shift over time, to anticipate or respond to changing conditions. In certain circumstances, it may require enabling mobility as part of the process of adaptation.

The tools of treaty interpretation provide the methodological means to reach this conclusion. As Chapter 3 explains, interpreting adaptation obligations requires both incorporating principles of precaution and differentiation and integrating relevant international human rights law. This interpretive process can help concretize adaptation obligations, which are broad and offer states discretion in determining appropriate adaptation measures. Chapter 4 focuses on the integration of key substantive rights and their positive duties—the rights to life and to an adequate standard of living—that are critical to survival, subsistence, and the ability and choice to stay or move. It establishes the general relevance of these rights and duties, as well as how they influence the interpretation of obligations to act on adaptation. When climate impacts put the enjoyment of these rights at risk, states must take adaptive action that includes measures to enable mobility—adaptive mobility. Chapter 5 turns to the integration of procedural rights. Rights to information and participation provide the foundation for participatory processes that include individuals and communities who are impacted by, and must often implement, adaptation measures. Moreover, as the chapter argues, these processes should be inclusive, offer options, and strive for consent. Like the adaptation that will result, participatory processes must also be precautionary, incorporate principles of non-discrimination and equality, and take into account the differential ways that climate change affects individuals and communities. This aligns with the book's approach and its advocacy for decisions that are built upon the self-determination and agency of those affected.

Climate Change, Human Rights, and Adaptive Mobility. Lauren Sakae Nishimura, Oxford University Press.
 DOI: 10.1093/9780198930068.003.0011

Implementation of such an approach, and the processes and adaptive actions it leads to, will require cooperation and assistance. Chapter 6 shows how grounding adaptive mobility in the climate regime provides a legal basis to argue for such support. Differentiated obligations bolster claims that developed states, particularly those wealthy countries that have contributed to and benefited most from the causes of climate change, must provide material and meaningful support for developing countries in their adaptation efforts. Support cannot be incidental or one-off but must be continuous and enhanced, as the Paris Agreement reminds parties. Furthermore, the cooperation required by climate treaties can help foster the collective action needed to share and distribute the costs involved in efforts to urgently address the impacts of climate change.

The case studies of Bangladesh, the Pacific Islands, and the Sahel region in Chapters 7 to 9 help illustrate the book's legal arguments. They establish the stakes for inaction and the kinds of impacts that can and will continue to occur absent concerted climate action. They also show the importance of contextual factors and community preferences, which affect when and how adaptation obligations are relevant. Indeed, the process of interpretation is not fixed, just as the adaptation measures that satisfy obligations will necessarily evolve over time. As each case study discusses, there will be circumstances where foreseeable risks require positive action to ensure the rights of those vulnerable to climate change, and this will include action on adaptive mobility. Such action can take different forms, from including mobility in planning, while taking steps to adapt in place, to the facilitation of planned relocation or migration, within or across borders.

Overall, the kind of adaptive mobility advanced in the book requires planning for and implementing measures *in advance* of serious rights violations, while ensuring the meaningful participation of those affected. It offers a novel approach to human mobility, one that departs from current approaches under international law, particularly their focus on movement that is forced and on protections that are triggered only after people move or cross borders. This timing is critical: the book's approach aims to compel more anticipatory action by states, which can lead to better outcomes for vulnerable people and communities. In the process, it can help to mitigate future scenarios of wide-scale displacement and migration crises, while better reflecting the reality of how human mobility often arises, with multiple factors and influences contributing to decisions about how and whether to move.

I. Benefits of the book's approach

There are a range of benefits to the approach developed in this book, which arise from the conceptual and legal clarity its arguments can provide to law- and policymakers, and the claims it makes available for individuals and communities. Foremost, and in general, the integration of human rights law into adaptation obligations contributes to a more concrete and contextual understanding of adaptation, which can help shape adaptive mobility. This integration directly connects human rights law with adaptation obligations to help guide and compel certain adaptation measures. Doing so puts people and their rights at the centre of solutions; it incorporates concern for harm to individuals and allows for the introduction of duties owed to people. As a result, action taken under the climate regime must not only target state governance and relations, but also measures that affect people and their rights. More specifically, human rights law has the potential to be a powerful source of protection for those who move in the context of climate change. This book focuses on how rights can inform adaptation obligations and prevent harm through states' adaptation measures: to plan and prepare for climate impacts that put the enjoyment of rights at risk, to better prevent displacement through the protection of rights, and to facilitate migration or planned relocation before risks become harm. The legal basis for these anticipatory actions comes from the integration of positive rights duties, which prompts state action to address foreseeable risks to certain essential resources.

Positive duties, and the need to address known and foreseeable risks, further require that adaptation target those societal conditions that contribute to these risks. Conditions include those factors that render a person vulnerable to climate impacts—environmental degradation and a lack of access to 'essential goods and services' such as food, water, healthcare, and housing.[1] An understanding of vulnerability that accounts for these known conditions can help create priorities in adaptation. Likewise, the interaction of these conditions with climate impacts contributes to decisions on mobility. Acknowledging the multi-causality of mobility affords a focus on the confluence of risks created by climate change. It calls for a shift beyond protection based on forced movement or a category of migrant and towards the recognition of the rights, needs, and agency of individuals throughout the process of migration.

An approach grounded in human rights also takes advantage of the broad applicability of rights, which are not tied to migration status. Such breadth allows for measures to ensure the rights of all, while prioritizing those marginalized or most vulnerable. Measures should therefore include action to address existing inequalities and to empower affected people in decision-making. The latter is a key addition to adaptation obligations; procedural rights require consideration of individuals who might otherwise be ignored in adaptation decision-making and policies. Thus, an approach based in human rights provides a degree of certainty to people: once risks to rights are identified, then a means to ensure these rights can be explored, and actions taken can and should include input from those affected.[2]

[1] Human Rights Committee (HRC), 'General Comment No 36: Article 6 (Right to Life)' (2019) UN Doc CCPR/C/GC/36, para 26; Chapter 4, II.

[2] See Chapter 5, II.

More generally, the integration of human rights law can help transform the climate change regime to provide more holistic responses. As others have argued, human rights can provide 'the ethical moorings' for meaningful climate action.[3] They can also provide a legal basis for compelling and shaping this action.

The second benefit of anchoring arguments for adaptive mobility within the climate change regime draws on the principle of precaution. As an operative principle, precaution guides the interpretation and application of obligations. For adaptation obligations, it necessitates measures to address risks and take anticipatory action in the face of uncertainty. It strengthens arguments to shift adaptive actions earlier, and to a time that is sooner than might be required by human rights law alone. Thus, measures must be taken to plan for, implement, and monitor adaptive mobility measures when risks to the enjoyment of rights, or their determinants (access to adequate food, water, housing), are foreseeable. Yet, as the book clarifies, such foreseeability need not be imminent or certain. Flexibility in timing and pre-emptive action accords with knowledge about the future impacts of climate change that is increasing in certainty but necessarily predictive. Such knowledge offers the opportunity to act now and plan for mobility, which will be crucial to facilitating migration or planned relocations. More proactive, timely measures on mobility will also help mitigate or avoid displacement.

Together, human rights law and the precautionary principle provide clearer thresholds for when action must be taken: states must act to ensure certain core and non-derogable rights, and they should act when the enjoyment of any number of rights is at risk. Action informed by human rights and precaution does not, however, dictate specific measures across contexts. Instead, it shapes when and how adaptive mobility should occur, to proactively protect rights before, during, and after movement. The role of human rights law is not static, and their integration into adaptation underscores that adjusting to climate impacts is an ongoing, iterative process. Precaution and human rights can help justify legal protection before people experience significant harm and are forced to move, in contrast with current approaches to human mobility under international law. Furthermore, when action is taken to provide anticipatory pathways for mobility rather than individual status-based protection after movement, the need to establish climate change as the cause of movement can be avoided. That climate impacts will be a direct or indirect driver of mobility is widely acknowledged,[4] but, as the case studies show, it is impacts in concert with specific conditions and contexts that ultimately lead to the need to move. Waiting for forced movement and failing to consider the risks to those in areas where, for example, drinking water is no longer safe or it is difficult to pursue one's livelihood is a remarkably insufficient—if not

[3] See Sheila R Foster and Paolo Galizzi, 'Human Rights and Climate Change: Building Synergies for a Common Future' in Daniel A Farber and Marjan Peeters (eds), *Climate Change Law*, vol I (Edward Elgar 2016) 45.

[4] See Chapter 1, III.

disingenuous—approach to addressing climate impacts and related mobility. Rather than ensuring rights protection for those who *must* move or who have already moved, the book's approach argues for enabling movement in advance of violations of human rights and before land is no longer able to sustain livelihoods or access to essential resources.

The third benefit of the book's approach is that it embeds adaptive mobility within a collective framework. One limitation of current approaches to climate change and human mobility is their recourse to international protection frameworks that individualize what should be a collective problem. They make legal status and protection contingent on meeting specific criteria or categories, placing the burden on the individual to prove they meet these requirements. Generally, this means that the migrant must navigate bureaucratic systems for themselves, rather than treating the role of climate change in mobility as an issue demanding collective solutions.[5] As the case studies illustrate, when large areas and whole communities are at risk and may need to move, solutions must go beyond requiring people to act on their own and in search of protection that often comes too late. Collective action is critical and, within international law, can be found through an approach that starts with a focus on climate change and adaptation obligations. Adaptation obligations include those to cooperate and assist, in treaties that create state duties and entitlements. Differentiation in these treaties makes clear that developed countries owe support for adaptation to developing countries. Adaptation is a cooperative effort, one that the Paris Agreement recognizes spans multiple levels of governance and actors.[6] Exploring mobility through the climate regime also incorporates a crucial element of justice and responsibility that is often overlooked in discussions of migration: developed countries that contributed most to the causes of climate change are obliged to provide support. Human rights law bolsters arguments for developed country responsibility, requiring assistance to developing countries in their efforts to adapt and protect people and their rights.

In addition, the climate regime provides the institutional mechanisms necessary to deploy certain forms of support—namely financial and technical—which can make burden sharing more than a mere proposition. However, progress must be made to develop these mechanisms, and to do so in a way that responds to shortfalls in and barriers to adaptation support. The regime's mechanisms and other institutions could also provide a potential forum to work out practical issues on adaptive mobility. They could, for example, be used as a space to negotiate agreements on safe and legal pathways across borders, one of the means proposed to meet adaptation obligations in the case studies.[7]

[5] See Introduction Chapter.

[6] See Paris Agreement (adopted 12 December 2015, entered into force 4 November 2016) 3156 UNTS 79, art 7(2).

[7] See Chapters 8, 9.

Finally, the book seeks to contribute to the broader discourse on climate change and human mobility. It frames mobility as adaptive and potentially positive, and it highlights steps that can be taken to improve outcomes for migrants. In doing so, it shifts the analytical focus away from the threat that mobility poses to national security, and it instead explores the ways that adaptive mobility measures can reduce risks to people and help avoid migration crises. Much of the work to change perceptions of mobility and state interests will be political and diplomatic.[8] Yet like the central concepts within this book, perceptions and interests can evolve, and legal arguments can contribute to altering the discourse. Debates about international law are continually analysing and informing cooperation and coordination between states, treaty regimes, and the rules they create. International law is, itself, a reflection of a pluralistic political reality that can create clashes and conflicts. However, as the arguments for treaty interpretation emphasize, the system of international law can strive for coherence and achieve points of convergence. For adaptation obligations, this can be accomplished through the integration of human rights. This seems like a timely progression. The impacts of climate change are becoming more frequent and severe. Urgent climate action is required to address these impacts. Adaptation, and mobility as an adaptive measure, are increasingly necessary, and human rights law can provide guidance that informs the manner and means by which such mobility occurs.

II. Open questions and ways forward

While the book makes the conceptual and legal case for an anticipatory, rights-based approach to mobility and climate change, it does not presume that setting out such an approach automatically leads to its implementation. Clarifying legal obligations is an important step, though there remain open questions about how to mobilize, much less compel, states to implement their international obligations. These are substantial questions that the book does not answer, which are likely to necessitate inquiries into state practices and influences on these practices. But while translating commitments into action involves political processes, this does not determine the soundness of legal arguments. Nor should the horizons towards which we strive be determined by assessments of political possibility. New and nuanced legal arguments of the kind in this book can also contribute to shifting possibilities. They can open up the range of potential solutions, reframing inquiries

[8] See Benoît Mayer, *The Concept of Climate Migration: Advocacy and Its Prospects* (Edward Elgar 2016) 268; François Crépeau, 'Towards a Mobile and Diverse World: "Facilitating Mobility" as a Central Objective of the Global Compact on Migration' (2018) 30 International Journal of Refugee Law 650, 651 (migration crises will not be addressed until politicians set out a long-term, rights-based vision of mobility).

for scholars and remaining available for use by the people most affected by climate change and others working with them.

There are also accountability and enforcement issues that follow on from the integration of human rights into adaptation obligations. Human rights law entitles individuals whose rights have been violated access to an effective remedy. Enforcement processes could take this further, to allow individuals to seek action or protection from states through adaptation, rather than relegating them to being passive targets in state adaptation plans, policies, and measures. This would require some mechanism for making requests or claims, monitoring impacts, holding states accountable, and offering a remedy or restitution for any losses. The institutions within the climate change regime currently do not offer such a mechanism. Thus, there is a need for its development or for greater reliance on current human rights bodies and institutions. An individual's enforcement of state climate obligations is not currently contemplated within the climate change regime, which provides for dispute settlement between state parties. Climate treaties created a multilateral consultative process (UNFCCC) and a committee (Paris Agreement) to resolve questions on and promote implementation and compliance.[9] These dispute settlement mechanisms accord with most environmental regimes, which focus on facilitating compliance rather than enforcement. However, stronger and clearer commitments may necessitate more robust enforcement options and a review of individual state compliance.[10] Other suggestions for accountability and coordination for climate-related mobility include the use of the Task Force on Displacement or an external quasi-judicial body, but significant development of either would be required.[11]

Accountability and compliance raise additional questions about the role of litigation and secondary or remedial obligations, use of which is an additional step beyond the analysis of primary adaptation obligations that the book undertakes. If responsibility for harm from climate change can be shown, which would require establishing breach of inter alia a climate obligation, human rights law, or the no-harm principle,[12] then adaptation, adaptive mobility, and support for adaptation measures could become a remedy. This would take multiple steps to achieve, and it introduces obstacles in establishing causation that are avoided by arguments in the preceding chapters. To date, most human rights and climate change cases

[9] The UNFCCC consultative process was never developed. However, the first formal meeting of the Paris Agreement Implementation and Compliance Committee (PAICC) was held in June 2020, with two to three meetings held each year since that time.

[10] See generally Daniel Bodansky and Harro van Asselt, 'International Carrots and Sticks' in Daniel Bodansky and Harro van Asselt (eds), *The Art and Craft of International Environmental Law* (Oxford University Press 2024).

[11] See Maxine Burkett, 'Reading Between the Red Lines: Loss and Damage and the Paris Outcome' [2016] Climate Law 118, 125–26; Mayer (n 8) 134.

[12] For more on the elements that establish responsibility, see International Law Commission (ILC), 'Draft Articles on Responsibility of States for Internationally Wrongful Acts' (November 2001) Supplement No 10 UN Doc A/56/10, chp.IV.E.1.

target mitigation policies and projects, rather than adaptation—or lack thereof.[13] Nonetheless, litigation on climate and human rights issues is becoming more prevalent, as are calls for more timely adaptive action.[14]

Reparations are a further tool that is being explored to respond to the harm caused by climate change. There are general arguments to address the intersecting causes and consequences of climate change and debt crises, for example, grounded in the environmental and ecological debt owed to vulnerable developing countries from developed countries for their past and continued extraction and exploitation.[15] There are also specific arguments for migration as reparations for climate change, which could lead to similar ends as the arguments laid out in the book, through community-led processes and choice about where and whether to move.[16] These approaches draw on other forms of obligations—moral and ethical—that connect the changing climate with colonial legacies and racial justice. Strong legal claims can bolster these obligations, and the book's arguments for assistance obligations based in international law can therefore play a complementary role. If met, these adaptation obligations are a step towards more just outcomes, with scaled-up support often one demand in a broader climate reparations project. The incorporation of principles of differentiated responsibility and special circumstances also provides another basis to highlight the historical underpinnings for the harm caused by climate change.[17]

Additionally, and unlike remedial processes that seek to address damages, this book uses legal arguments to intervene at a different temporal moment. It focuses on how law can be understood to shape action that is anticipatory, while building on the recognition that harm from climate impacts is already occurring and will continue to occur. This enables a deliberate reframing of the problem of climate mobility, one that both draws on international law as part of the wider discussion

[13] See César Rodríguez-Garavito, 'Litigating the Climate Emergency: The Global Rise of Human Rights-Based Litigation for Climate Action' in César Rodríguez-Garavito (ed), *Litigating the Climate Emergency: How Human Rights, Courts, and Legal Mobilization Can Bolster Climate Action* (Cambridge University Press 2022) 34–35.

[14] See eg Chapter 4, II; HRC, 'Views Adopted by the Committee under Article 5(4) of the Optional Protocol, Concerning Communication No. 3624/2019 (*Daniel Billy et al v Australia*)' (2022) CCPR/C/135/D/3624/2019. Jacqueline Peel and Hari M Osofsky, 'A Rights Turn in Climate Change Litigation?' (2018) 7 Transnational Environmental Law 37; Jolene Lin and Jacqueline Peel, *Litigating Climate Change in the Global South* (Oxford University Press 2024) 63–67, 79–81.

[15] See eg Olúfẹ́mi O Táíwò and Patrick Bigger, 'Debt Justice for Climate Reparations' (Climate and Community Project 2022) <www.climateandcommunity.org/debt-justice-for-climate-reparations> accessed 9 September 2024; Ketaki Zodgekar and others, 'A Dangerous Debt-Climate Nexus: In a Warming World, Technocratic Fixes Are Inadequate Responses to Vulnerability. Achieving Climate and Economic Justice in the Caribbean Calls for Reparative Solutions' (2023) 55 NACLA Report on the Americas 319; see also Olúfẹ́mi O Táíwò, *Reconsidering Reparations* (Oxford University Press 2022); Maxine Burkett, 'Climate Reparations' (2009) 10 Melbourne Journal of International Law 509.

[16] See eg Carmen G Gonzalez, 'Migration as Reparation: Climate Change and the Disruption of Borders' (2020) 66 Loyola Law Review 401; see also E Tendayi Achiume, 'Migration as Decolonization' (2019) 71 Stanford Law Review 1509 (arguing that former colonial powers are obligated to open their borders to former colonial subjects).

[17] Chapter 6, V.

of solutions to address climate change and that demands cooperative action before people are forced to move. Such a demand is aligned with the more general need for anticipatory action that extends beyond mobility, within the law and outside it.

Further research would also complement and extend the reach of the arguments in the book. Its interpretive process could be applied to other rights, including those that are more contested. The rights of Indigenous peoples and the right to self-determination, amongst others, are likely to play an important role in mobility, which requires further elaboration than what the book provides about their content and applicability. The use of other operative principles or goals of the regime may also provide indirect support for or affect the scope of adaptive mobility. Adaptation is linked to broader goals of poverty alleviation and sustainable development. The inclusion of these goals in climate treaties establishes that other international norms or laws govern action on adaptation.[18] Measures on adaptive mobility will therefore need to contend with them, which can be both complementary or contradictory depending on the context and development agenda. And how these broader goals interact with obligations could affect their interpretation, much as human rights law shaped arguments for adaptive mobility. As a result, they could raise additional areas of international regime interaction beyond human rights law that affect climate action and governance.

Climate change poses a tremendous—if not the biggest—problem facing people, ecosystems, and the planet. These challenges are foreseeable, and solutions will require concerted action of an unprecedented scope. Such action should be cooperative, but it will often involve social and political struggles and a myriad of strategies, including recourse to the law and the development of innovative legal arguments. This book offers one such innovation, circumscribed by international law but underscoring its potential. It uses human rights law to shape adaptation obligations, providing a direct means to integrate mobility as adaptation into the climate regime. This approach focuses attention on the rights and needs of vulnerable people and communities, responds to risks from climate impacts, and prompts anticipatory and proactive measures. At the same time, adaptation obligations of assistance and cooperation provide a means to share the burden of adaptation and marshal concrete forms of support for countries and communities facing the brunt of climate impacts. That such support is required, and an entitlement, is further underpinned by collective commitments made in climate treaties. Wealthy countries must provide more funding and assistance, without conditions or expectation of repayment. Doing so can help enable more

[18] See United Nations Framework Convention on Climate Change (adopted 9 May 1992, entered into force 21 March 1994) 1771 UNTS 107 (UNFCCC) preamble, arts 3(4), (5), 4(7); Paris Agreement preamble, arts 2(1), 7(1).

proactive, anticipatory adaptation that allows people to stay or move with greater agency and access to resources. Measures to facilitate adaptive mobility, with their longer-term planning and inclusion of human rights, can better account for the differential impacts of climate change and improve conditions before, during, and after movement.

Bibliography

Books, book chapters, and journal articles

Achiume ET, 'Migration as Decolonization' (2019) 71 Stanford Law Review 1509

Ahmed I and others, 'Climate Change, Environmental Stress and Loss of Livelihoods Can Push People towards Illegal Activities: A Case Study from Coastal Bangladesh' (2019) 11 Climate and Development 907

Ahuja N and others, 'Advancing Human Rights through Environmental Rule of Law' in James R May and Erin Daly (eds), *Human Rights and the Environment: Legality, Indivisibility, Dignity and Geography*, vol VII (Edward Elgar 2019)

Aktypis S, Decaux E, and Leroy B, 'Systemic Integration between Climate Change and Human Rights at the United Nations?' in Ottavio Quirico and Mouloud Boumghar (eds), *Climate Change and Human Rights: An International and Comparative Law Perspective* (Routledge 2015)

Alston P, 'International Law and the Human Right to Food' in Philip Alston and Katarina Tomaševski (eds), *The Right to Food* (Martinus Nijhoff 1984)

Anderson A and others, 'Imminence in Refugee and Human Rights Law: A Misplaced Notion for International Protection' (2019) 68 International and Comparative Law Quarterly 111

Andrew NL and others, 'Coastal Proximity of Populations in 22 Pacific Island Countries and Territories' (2019) 14 PLOS ONE e0223249

Arevalo L, 'The Work of the International Law Commission in the Field of International Environmental Law' (2005) 32 Boston College Environmental Affairs Law Review 493

Arnold CA, 'Adaptive Law' in Rosemary Lyster and Robert RM Verchick (eds), *Research Handbook on Climate Disaster Law: Barriers and Opportunities* (Edward Elgar 2018)

Asafo D, 'The Racism in Climate Change Law: Critiquing the Law on Climate Change-Related Displacement with Critical Race Theory' (2022) 39 Berkeley Journal of International Law 249

Atapattu S, 'Climate Change, Human Rights, and Forced Migration: Implications for International Law' (2009) 27 Wisconsin International Law Journal 607

Ayeb-Karlsson S, Baldwin AW, and Kniveton D, 'Who Is the Climate-Induced Trapped Figure?' (2022) 13 WIREs Climate Change e803

—— Kniveton D, and Cannon T, 'Trapped in the Prison of the Mind: Notions of Climate-Induced (Im)Mobility Decision-Making and Wellbeing from an Urban Informal Settlement in Bangladesh' (2020) 6 Palgrave Communications 1

Baldwin A and Fornalé E, 'Adaptive Migration: Pluralising the Debate on Climate Change and Migration' (2017) 183 The Geographical Journal 322

Barnett J and McMichael C, 'The Effects of Climate Change on the Geography and Timing of Human Mobility' (2018) 39 Population and Environment 339

Bettini G and Gioli G, 'Waltz with Development: Insights on the Developmentalization of Climate-Induced Migration' (2016) 5 Migration and Development 171

Betts A, *Survival Migration: Failed Governance and the Crisis of Displacement* (Cornell University Press 2013)

Biermann F and Boas I, 'Towards a Global Governance System to Protect Climate Migrants: Taking Stock' in Benoît Mayer and François Crépeau (eds), *Research Handbook on Climate Change, Migration and the Law* (Edward Elgar 2017)

Bjorge E, *The Evolutionary Interpretation of Treaties* (Oxford University Press 2014)

Boas I, 'Climate Migration Myths' (2019) 9 Nature Climate Change 901

—— '"Climate Mobility" Is a Proper Subject of Research and Governance' in Alexander Zahar and Benoît Mayer (eds), *Debating Climate Law* (Cambridge University Press 2021)

Bodansky D, 'The United Nations Framework Convention on Climate Change: A Commentary' (1993) 18 Yale Journal of International Law 451

—— 'Climate Change and Human Rights: Unpacking the Issues' (2010) 38 Georgia Journal of International & Comparative Law 511

—— and van Asselt H, 'International Carrots and Sticks' in Daniel Bodansky and Harro van Asselt (eds), *The Art and Craft of International Environmental Law* (Oxford University Press 2024)

—— Brunnée J, and Rajamani L, *International Climate Change Law* (Oxford University Press 2017)

Boisson de Chazournes L, 'Climate Change and Financial Assistance: A Fragmented, Unified or Coordinated Approach?' in *International Law and Developing Countries* (Brill | Nijhoff 2014)

—— 'Financial Assistance' in Lavanya Rajamani and Jacqueline Peel (eds), *The Oxford Handbook of International Environmental Law* (Oxford University Press 2021)

Bower E, Harrington-Abrams R, and Priem B, 'Complicating "Community" Engagement: Reckoning with an Elusive Concept in Climate-Related Planned Relocation' (2024) 88 Global Environmental Change 102913

Bower ER and others, 'Enabling Pathways for Sustainable Livelihoods in Planned Relocation' (2023) 13 Nature Climate Change 919

Boyle A, 'Human Rights and the Environment: Where Next?' in Ben Boer (ed), *Environmental Law Dimensions of Human Rights* (Oxford University Press 2015)

—— 'Climate Change, the Paris Agreement and Human Rights' (2018) 67 International and Comparative Law Quarterly 759

—— 'Soft Law in International Law-Making' in Malcolm D Evans (ed), *International Law* (5th edn, Oxford University Press 2018)

—— and Redgwell C, *Birnie, Boyle, & Redgwell's International Law and the Environment* (4th edn, Oxford University Press 2021)

Bratspies R and Lamdan S, 'The Human Right to Environmental Information' in James R May and Erin Daly (eds), *Human Rights and the Environment: Legality, Indivisibility, Dignity and Geography*, vol VII (Edward Elgar 2019)

Bronen R, 'Climate-Induced Community Relocations: Creating An Adaptive Governance Framework Based In Human Rights Doctrine' (2011) 35 NYU Review of Law and Social Change 357

Brunnée J, 'COPing with Consent: Law-Making Under Multilateral Environmental Agreements' (2002) 15 Leiden Journal of International Law 1

Burkett M, 'Climate Reparations' (2009) 10 Melbourne Journal of International Law 509

—— 'The Nation Ex-Situ: On Climate Change, Deterritorialized Nationhood and the Post-Climate Era' (2011) 2 Climate Law 345

—— 'Lessons from Contemporary Resettlement in the South Pacific' (2015) 68 Journal of International Affairs 75

—— 'Reading Between the Red Lines: Loss and Damage and the Paris Outcome' [2016] Climate Law 118

—— 'Behind the Veil: Climate Migration, Regime Shift, and a New Theory of Justice' (2018) 53 Harvard Civil Rights-Civil Liberties Law Review 445

Burson B and others, 'The Duty to Move People Out of Harm's Way in the Context of Climate Change and Disasters' (2018) 37 Refugee Survey Quarterly 379

Campbell J, 'Climate-Induced Community Relocation in the Pacific: The Meaning and Importance of Land' in Jane McAdam (ed), *Climate Change and Displacement: Multidisciplinary Perspectives* (Hart 2010)

Cançado Trindade AA, 'Principle 15' in Jorge E Viñuales (ed), *The Rio Declaration on Environment and Development: A Commentary* (Oxford University Press 2015)

Cantor DJ, 'Law, Policy and Practice Concerning the Humanitarian Protection of Aliens on a Temporary Basis in the Context of Disasters' (The Nansen Initiative 2015) Background Paper

Carazo MP, 'Contextual Provisions (Preamble and Article 1)' in Daniel Klein and others (eds), *The Paris Agreement on Climate Change: Analysis and Commentary* (Oxford University Press 2017)

—— 'Conceptualising "Relocation" Across Displacement Contexts' (2024) 15 Journal of International Humanitarian Legal Studies 23, 36

Carlarne CP and Colavecchio JD, 'Balancing Equity and Effectiveness: The Paris Agreement & the Future of International Climate Change Law' [2019] New York University Environmental Law Journal 107

Chan N, '"Special Circumstances" and the Politics of Climate Vulnerability: African Agency in the UN Climate Change Negotiations' (2021) 56 Africa Spectrum 314

Cheng L and others, 'Another Year of Record Heat for the Oceans' (2023) 40 Advances in Atmospheric Sciences 963

Connell J, 'Last Days in the Carteret Islands? Climate Change, Livelihoods and Migration on Coral Atolls' (2016) 57 Asia Pacific Viewpoint 3

Costello C and Foster M, 'Non-Refoulement as Custom and Jus Cogens—Putting the Prohibition to the Test' in *Netherlands Yearbook of International Law 2015* (TMC Asser Press 2016)

Craig RK, '"Stationarity Is Dead"—Long Live Transformation: Five Principles for Climate Change Adaptation Law' (2010) 34 Harvard Environmental Law Review 10

Craven M, 'The Violence of Dispossession: Extra-Territoriality and Economic, Social, and Cultural Rights' in Mashood Baderin and Robert McCorquodale (eds), *Economic, Social, and Cultural Rights in Action* (Oxford University Press 2007)

Crawford J, *The Creation of States in International Law* (2nd edn, Clarendon Press 2006)

Crépeau F, 'Towards a Mobile and Diverse World: "Facilitating Mobility" as a Central Objective of the Global Compact on Migration' (2018) 30 International Journal of Refugee Law 650

Cross KW and Kingi P, '"*Fonua*" Cultural Statelessness in the Pacific and the Effects of Climate Change' in Tendayi Bloom and Lindsey N Kingston (eds), *Statelessness, Governance, and the Problem of Citizenship* (Manchester University Press 2021)

Cullet P, 'Principle 7' in Jorge E Viñuales (ed), *The Rio Declaration on Environment and Development: A Commentary* (Oxford University Press 2015)

—— 'Common but Differentiated Responsibilities' in Malgosia Fitzmaurice, Marcellinus Brus, and Panagiotis Merkouris (eds), *Research Handbook on International Environmental Law* (Edward Elgar 2021)

Cundill G and others, 'Toward a Climate Mobilities Research Agenda: Intersectionality, Immobility, and Policy Responses' (2021) 69 Global Environmental Change 102315

Daly E and May JR, 'Indivisibility of Human and Environmental Rights' in James R May and Erin Daly (eds), *Human Rights and the Environment: Legality, Indivisibility, Dignity and Geography*, vol VII (Edward Elgar 2019)

Daron J and others, 'Integrating Seasonal Climate Forecasts into Adaptive Social Protection in the Sahel' (2021) 13 Climate and Development 543

de Coninck H and Sagar A, 'Technology Development and Transfer (Article 10)' in Daniel Klein and others (eds), *The Paris Agreement on Climate Change: Analysis and Commentary* (Oxford University Press 2017)

de Sadeleer N, 'The Precautionary Principle and Climate Change' in Daniel A Farber and Marjan Peeters (eds), *Climate Change Law*, vol I (Edward Elgar 2016)

De Vido S, 'The Right to Water as an International Custom: The Implications in Climate Change Adaptation Measures' (2012) 6 Carbon & Climate Law Review 221

Dietz T and Veldhuizen E, 'Population Dynamics' in AJ Dietz, R Ruben, and A Verhagen (eds), *The Impact of Climate Change on Drylands: With a Focus on West Africa* (Springer 2004)

Docherty B and Giannini T, 'Confronting a Rising Tide: A Proposal for A Convention on Climate Change Refugees' (2009) 33 Harvard Environmental Law Review 349

Durand-Delacre D and others, 'Climate Migration Is About People, Not Numbers' in Steffen Böhm and Sian Sullivan (eds), *Negotiating Climate Change in Crisis* (Open Book 2021)

Duyck Sébastien, 'Delivering on the Paris Promises? Review of the Paris Agreement's Implementing Guidelines from a Human Rights Perspective' [2019] Climate Law 202

Ebbesson J, 'Principle 10' in Jorge E Viñuales (ed), *The Rio Declaration on Environment and Development: A Commentary* (Oxford University Press 2015)

Eckersley R, 'The Common but Differentiated Responsibilities of States to Assist and Receive "Climate Refugees"' [2015] European Journal of Political Theory 481

Eide A, 'Adequate Standard of Living' in Daniel Moeckli and others (eds), *International Human Rights Law* (3rd edn, Oxford University Press 2017)

Farbotko C and Campbell J, 'Who Defines Atoll "Uninhabitability"?' (2022) 138 Environmental Science & Policy 182

—— and McMichael C, 'Voluntary Immobility and Existential Security in a Changing Climate in the Pacific' (2019) 60 Asia Pacific Viewpoint 148

—— and others, 'Transformative Mobilities in the Pacific: Promoting Adaptation and Development in a Changing Climate' (2018) 5 Asia & The Pacific Policy Studies 393

—— and others, 'Relocation Planning Must Address Voluntary Immobility' (2020) 10 Nature Climate Change 702

Ferris E, 'The Relevance of the Guiding Principles on Internal Displacement for the Climate-Migration Nexus' in Benoît Mayer and François Crépeau (eds), *Research Handbook on Climate Change, Migration and the Law* (Edward Elgar 2017)

Foster M, *International Refugee Law and Socio-Economic Rights: Refuge from Deprivation* (Cambridge University Press 2009)

—— and McAdam J, 'Analysis of "Imminence" in International Protection Claims: Teitiota v New Zealand and Beyond' (2022) 71 International and Comparative Law Quarterly 975

Foster SR and Galizzi P, 'Human Rights and Climate Change: Building Synergies for a Common Future' in Daniel A Farber and Marjan Peeters (eds), *Climate Change Law*, vol I (Edward Elgar 2016)

Fredman S, 'Redistribution and Recognition: Reconciling Inequalities' (2007) 23 South African Journal on Human Rights 214

—— 'Substantive Equality Revisited' (2016) 14 International Journal of Constitutional Law 712

Gardiner RK, *Treaty Interpretation* (2nd edn, Oxford University Press 2015)

Garschagen M and Doshi D, 'Does Funds-Based Adaptation Finance Reach the Most Vulnerable Countries?' (2022) 73 Global Environmental Change 102450

Gastelumendi J and Gnittke I, 'Climate Finance (Article 9)' in Daniel Klein and others (eds), *The Paris Agreement on Climate Change: Analysis and Commentary* (Oxford University Press 2017)

Gemenne F, 'Why the Numbers Don't Add up: A Review of Estimates and Predictions of People Displaced by Environmental Changes' (2011) 21 Global Environmental Change. Part A, Human and Policy Dimensions S41

—— 'The Refugees of the Anthropocene' in Benoît Mayer and François Crépeau (eds), *Research Handbook on Climate Change, Migration and the Law* (Edward Elgar 2017)

Gharbaoui D and Blocher J, 'The Reason Land Matters: Relocation as Adaptation to Climate Change in Fiji Islands' in Andrea Milan and others (eds), *Migration, Risk Management and Climate Change: Evidence and Policy Responses*, vol 6 (Springer 2016)

Gonzalez CG, 'Climate Change, Race, and Migration' (2020) 1 Journal of Law and Political Economy 109

—— 'Migration as Reparation: Climate Change and the Disruption of Borders' (2020) 66 Loyola Law Review 401

Goodwin-Gill GS, 'Non-Refoulement, Temporary Refuge, and the "New" Asylum Seekers' in David James Cantor and Jean-François Durieux (eds), *Refuge from Inhumanity? War Refugees and International Humanitarian Law* (Brill 2014)

Gormley WP, 'The Right to Life and the Rule of Non-Derogability: Peremptory Norms of Jus Cogens' in BG Ramcharan (ed), *The Right to Life in International Law* (Martinus Nijhoff 1985)

Grasso M, *Justice in Funding Adaptation under the International Climate Change Regime* (Springer Science & Business Media 2009)

Gromilova M, 'Revisiting Planned Relocation as a Climate Change Adaptation Strategy: The Added Value of a Human Rights-Based Approach' (2014) 10 Utrecht Law Review 76

—— 'Finding Opportunities to Combat the Climate Change Migration Crisis: The Potential of the "Adaptation Approach"' (2016) 33 Pace Environmental Law Review 105

Gros C and others, 'Household-Level Effects of Providing Forecast-Based Cash in Anticipation of Extreme Weather Events: Quasi-Experimental Evidence from Humanitarian Interventions in the 2017 Floods in Bangladesh' (2019) 41 International Journal of Disaster Risk Reduction 101275

Hall MJ and Weiss DC, 'Avoiding Adaptation Apartheid: Climate Change Adaptation and Human Rights Law' (2012) 37 The Yale Journal of International Law 309

Hannum H, 'The Status of the Universal Declaration of Human Rights in National and International Law' (2014) 25 Georgia Journal of International & Comparative Law 287

Hansen J and others, 'Ice Melt, Sea Level Rise and Superstorms: Evidence from Paleoclimate Data, Climate Modeling, and Modern Observations That 2°C Global Warming Could Be Dangerous' (2016) 16 Atmospheric Chemistry and Physics 3761

Haritz M, 'Liability with and Liability from the Precautionary Principle in Climate Change Cases' in Michael Faure and Marjan Peeters (eds), *Climate Change Liability* (Edward Elgar 2011)

Hartmann B, 'Rethinking Climate Refugees and Climate Conflict: Rhetoric, Reality and the Politics of Policy Discourse' (2010) 22 Journal of International Development 233

Harvey C and Barnidge RP Jr, 'Human Rights, Free Movement, and the Right to Leave in International Law' (2007) 19 International Journal of Refugee Law 1

Hathaway JC, *The Rights of Refugees under International Law* (Cambridge University Press 2005)

—— 'Leveraging Asylum' (2010) 45 Texas International Law Journal 503

—— 'Afterword' in Benoît Mayer and François Crépeau (eds), *Research Handbook on Climate Change, Migration and the Law* (Edward Elgar 2017)

—— and Foster M, *The Law of Refugee Status* (Cambridge University Press 2014)

Hedahl M, 'Directional Climate Justice: The Normative Relationship between Moral Claim Rights and Directed Obligations' (2014) 5 Journal of Human Rights and the Environment 35

Henderson JV, Storeygard A, and Deichmann U, 'Has Climate Change Driven Urbanization in Africa?' (2017) 124 Journal of Development Economics 60

Henry S, Schoumaker B, and Beauchemin C, 'The Impact of Rainfall on the First Out-Migration: A Multi-Level Event-History Analysis in Burkina Faso' (2004) 25 Population and Environment 423

Herman-Mercer N and others, 'Changing Times, Changing Stories: Generational Differences in Climate Change Perspectives from Four Remote Indigenous Communities in Subarctic Alaska' (2016) 21 Ecology and Society 28

Hermann E and Kempf W, 'Climate Change and the Imagining of Migration: Emerging Discourses on Kiribati's Land Purchase in Fiji' (2017) 29 The Contemporary Pacific 231

Herrmann SM and Hutchinson CF, 'The Scientific Basis: Links between Land Degradation, Drought and Desertification' in Pierre Marc Johnson, Karel Mayrand, and Marc Paquin (eds), *Governing Global Desertification: Linking Environmental Degradation, Poverty and Participation* (Routledge 2006)

Hodgkinson D and others, '"The Hour When the Ship Comes In": A Convention for Persons Displaced by Climate Change' (2010) 36 Monash University Law Review 69

Hossain MAR and others, 'Impacts and Responses to Environmental Change in Coastal Livelihoods of South-West Bangladesh' (2018) 637–638 Science of The Total Environment 954

Hugo G, 'Environmental Concerns and International Migration' (1996) 30 The International Migration Review 105

—— 'Lessons from Past Forced Resettlement for Climate Change Migration' in Etienne Piguet, Antoine Pécoud, and Paul de Guchteneire (eds), *Migration and Climate Change* (Cambridge University Press 2011)

Islam MA, Islam MS, and Wahab MA, 'Impacts of Climate Change on Shrimp Farming in the South-West Coastal Region of Bangladesh' (2016) 3 Research in Agriculture Livestock and Fisheries 227

Iyer M and Schewel K, 'Articulating and Claiming the Right to Stay in the Context of Climate Change' (2024) 38 Georgetown Immigration Law Journal 207

Kälin W, 'The Human Rights Dimension of Natural or Human-Made Disasters Focus: Disaster Preparedness and Response' (2012) 55 German Yearbook of International Law 119

—— and Künzli J, *The Law of International Human Rights Protection* (2nd edn, Oxford University Press 2019)

Kaswan A, 'Adaptation Justice' in Daniel A Farber and Marjan Peeters (eds), *Climate Change Law*, vol I (Edward Elgar 2016)

Kelman I, 'No Change from Climate Change: Vulnerability and Small Island Developing States' (2014) 180 The Geographical Journal 120

—— 'Imaginary Numbers of Climate Change Migrants?' (2019) 8 Social Sciences 1

Kent A and Behrman S, *Facilitating the Resettlement and Rights of Climate Refugees: An Argument for Developing Existing Principles and Practices* (Routledge 2018)

Kirezci E and others, 'Projections of Global-Scale Extreme Sea Levels and Resulting Episodic Coastal Flooding over the 21st Century' (2020) 10 Scientific Reports 11629

Kitara T, Suliman S, and Farbotko C, '*Fale Pili*: A Tuvaluan Perspective on Mobility Justice' (2024) 9 Applied Mobilities 233

Klepp S, 'Framing Climate Change Adaptation from a Pacific Island Perspective—The Anthropology of Emerging Legal Orders' (2018) 68 Sociologus 149

—— and Chavez-Rodriguez L, 'Governing Climate Change: The Power of Adaptation Discourses, Policies, and Practices' in Silja Klepp and Libertad Chavez-Rodriguez (eds), *A Critical Approach to Climate Change Adaptation: Discourses, Policies, and Practices* (Routledge 2018)

Knox JH, 'Human Rights Principles and Climate Change' in Cinnamon P Carlarne, Kevin R Gray, and Richard Tarasofsky (eds), *The Oxford Handbook of International Climate Change Law* (Oxford University Press 2016)

—— 'Bringing Human Rights to Bear on Climate Change' (2019) 9 Climate Law 165

Koskenniemi M, *From Apology to Utopia: The Structure of International Legal Argument* (Cambridge University Press 2006)

Kumar L and Taylor S, 'Exposure of Coastal Built Assets in the South Pacific to Climate Risks' (2015) 5 Nature Climate Change 992

Lambek N, 'Respecting and Protecting the Right to Food: When States Must Get Out of the Kitchen' in Nadia Lambek and others (eds), *Rethinking Food Systems: Structural Challenges, New Strategies and the Law* (Springer 2014)

Lane HM and others, 'Historical Redlining Is Associated with Present-Day Air Pollution Disparities in U.S. Cities' (2022) 9 Environmental Science & Technology Letters 345

Lauria V and others, 'Importance of Fisheries for Food Security across Three Climate Change Vulnerable Deltas' (2018) 640–641 Science of The Total Environment 1566

Lauta KC and Rytter JE, 'A Landslide on a Mudslide? Natural Hazards and the Right to Life under the European Convention on Human Rights' (2016) 7 Journal of Human Rights and the Environment 111

Le Moli G, 'The Human Rights Committee, Environmental Protection and the Right to Life' (2020) 69 International and Comparative Law Quarterly 735

Lee J and others, 'Water-Related Disasters and Their Health Impacts: A Global Review' (2020) 8 Progress in Disaster Science 100123

Lesnikowski A and others, 'What Does the Paris Agreement Mean for Adaptation?' (2017) 17 Climate Policy 825

Lewis B, 'Balancing Human Rights in Climate Policies' in Ottavio Quirico and Mouloud Boumghar (eds), *Climate Change and Human Rights: An International and Comparative Law Perspective* (Routledge 2015)

Lyster R, 'Climate Justice, Adaptation and the Paris Agreement: A Recipe for Disasters?' (2017) 26 Environmental Politics 438

—— and Burkett M, 'Climate-Induced Displacement and Climate Disaster Law: Barriers and Opportunities' in Rosemary Lyster and Robert RM Verchick (eds), *Research Handbook on Climate Disaster Law: Barriers and Opportunities* (Edward Elgar 2018)

Mace MJ, 'Adaptation under the UN Framework Convention on Climate Change: The International Legal Framework' in W Neil Adger and others (eds), *Fairness in Adaptation to Climate Change* (The MIT Press 2006)

Magnan AK and others, 'Addressing the Risk of Maladaptation to Climate Change' (2016) 7 Wiley Interdisciplinary Reviews: Climate Change 646

Maharjan A and others, 'Migration and Household Adaptation in Climate-Sensitive Hotspots in South Asia' (2020) 6 Current Climate Change Reports 1

Maldonado JK and others, 'The Impact of Climate Change on Tribal Communities in the US: Displacement, Relocation, and Human Rights' [2013] Climate Change 601

Martin PCM and others, 'Responding to Multiple Climate-Linked Stressors in a Remote Island Context: The Example of Yadua Island, Fiji' (2018) 21 Climate Risk Management 7

Martin SF, 'Toward an Extension of Complementary Protection?' in Benoît Mayer and François Crépeau (eds), *Research Handbook on Climate Change, Migration and the Law* (Edward Elgar 2017)

—— Weerasinghe S, and Taylor A (eds), *Humanitarian Crises and Migration: Causes, Consequences and Responses* (Routledge 2014)

Mayer B, *The Concept of Climate Migration: Advocacy and Its Prospects* (Edward Elgar 2016)

—— 'Who Are "Climate Refugees"?' in Avidan Kent and Simon Behrman (eds), *Climate Refugees: Beyond the Legal Impasse?* (Routledge 2018)

—— 'Climate Change Adaptation and the Law' (2021) 39 Virginia Environmental Law Journal 141

—— 'Climate Change Adaptation Law: Is There Such a Thing?' in Alexander Zahar and Benoît Mayer (eds), *Debating Climate Law* (Cambridge University Press 2021)

—— and Cournil C, 'Climate Change, Migration and Human Rights: Towards Group-Specific Protection?' in Ottavio Quirico and Mouloud Boumghar (eds), *Climate Change and Human Rights: An International and Comparative Law Perspective* (Routledge 2015)

McAdam J, 'Swimming against the Tide: Why a Climate Change Displacement Treaty Is Not the Answer' (2011) 23 International Journal of Refugee Law 2

—— *Climate Change, Forced Migration, and International Law* (Oxford University Press 2012)

—— 'Historical Cross-Border Relocations in the Pacific: Lessons for Planned Relocations in the Context of Climate Change' (2014) 49 The Journal of Pacific History 301

—— 'Relocation and Resettlement from Colonisation to Climate Change: The Perennial Solution to "Danger Zones"' (2015) 3 London Review of International Law 93

—— and Ferris E, 'Planned Relocations in the Context of Climate Change: Unpacking the Legal and Conceptual Issues' (2015) 4 Cambridge Journal of International and Comparative Law 137

McAnaney SC, 'Sinking Islands? Formulating a Realistic Solution to Climate Change Displacement' (2012) 87 NYU Law Review 1172

McGranahan G, Balk D, and Anderson B, 'The Rising Tide: Assessing the Risks of Climate Change and Human Settlements in Low Elevation Coastal Zones' (2007) 19 Environment and Urbanization 17

McLachlan C, 'The Principle of Systemic Integration and Article 31(3)(C) of the Vienna Convention' (2005) 54 International and Comparative Law Quarterly 279

McMichael C, Katonivualiku M, and Powell T, 'Planned Relocation and Everyday Agency in Low-Lying Coastal Villages in Fiji' (2019) 185 The Geographical Journal 325

McNamara KE and Combes HJD, 'Planning for Community Relocations Due to Climate Change in Fiji' (2015) 6 International Journal of Disaster Risk Science 315

—— Westoby R, and Chandra A, 'Exploring Climate-Driven Non-Economic Loss and Damage in the Pacific Islands' (2021) 50 Current Opinion in Environmental Sustainability 1

Merkouris P, *Article 31(3)(c) VCLT and the Principle of Systemic Integration: Normative Shadows in Plato's Cave* (Brill 2015)

Meyers N, 'Environmental Refugees' (1997) 19 Population and Environment 167

Millar I and Wilson K, 'Towards a Climate Change Displacement Facility' in Benoît Mayer and François Crépeau (eds), *Research Handbook on Climate Change, Migration and the Law* (Edward Elgar 2017)

Misiedjan D and McKenzie SO, 'The Human Right to Water' in James R May and Erin Daly (eds), *Human Rights and the Environment: Legality, Indivisibility, Dignity and Geography*, vol VII (Edward Elgar 2019)

Molinari C, 'Principle 3: From a Right to Development to Intergenerational Equity' in Jorge E Viñuales (ed), *The Rio Declaration on Environment and Development: A Commentary* (Oxford University Press 2015)

Monson R and Fitzpatrick D, 'Negotiating Relocation in a Weak State: Land Tenure and Adaptation to Sea-Level Rise in Solomon Islands' in Susanna Price and Jane Singer (eds), *Global Implications of Development, Disasters and Climate Change* (Routledge 2015)

—— and others, 'Planning for Climate Mobility: Developing a Relational Approach to Planned Community Relocation in Solomon Islands' in Andreas Neef, Natasha Pauli, and Bukola Salami (eds), *De Gruyter Handbook of Climate Migration and Climate Mobility Justice* (De Gruyter 2024)

Musto C and Redgwell C, 'US — Import Prohibition of Certain Shrimp and Shrimp Products (1998)' in Eirik Bjorge and Cameron Miles (eds), *Landmark Cases in Public International Law* (Hart 2017)

Ndeunyema N, 'Unmuddying the Waters: Evaluating the Legal Basis of the Human Right to Water Under Treaty Law, Customary International Law, and the General Principles of Law' (2020) 41 Michigan Journal of International Law 455

Nevitt M, 'Climate Change and the Specter of Statelessness' (2023) 35 The Georgetown Environmental Law Review 331

Nicholson CT, '"Climate Mobility" Is Not a Proper Subject of Research and Governance' in Alexander Zahar and Benoît Mayer (eds), *Debating Climate Law* (Cambridge University Press 2021)

—— and Mayer B, 'Introduction' in Calum TM Nicholson and Benoît Mayer (eds), *Climate Migration: Critical Perspectives for Law, Policy, and Research* (Hart 2023)

Nishimura L, '"Climate Change Migrants": Impediments to a Protection Framework and the Need to Incorporate Migration into Climate Change Adaptation Strategies' (2015) 27 International Journal of Refugee Law 107

—— 'Adaptation and Anticipatory Action: Integrating Human Rights Duties into the Climate Change Regime' (2022) 12 Climate Law 99

—— 'Adaptation Obligations and Adaptive Mobility' (2022) 69 Forced Migration Review 39

—— 'The Paris Agreement's Contribution to the Development of the Law on Adaptation' in Alexander Zahar (ed), *Research Handbook on the Law of the Paris Agreement* (Edward Elgar 2024)

Oakes R, 'Culture, Climate Change and Mobility Decisions in Pacific Small Island Developing States' (2019) 40 Population and Environment 480

Orakhelashvili A, 'Restrictive Interpretation of Human Rights Treaties in the Recent Jurisprudence of the European Court of Human Rights' (2003) 14 European Journal of International Law 529

Panda A, 'Chapter 11: Climate-Induced Migration and Interdependent Vulnerabilities between Bangladesh and India' in Bishawjit Mallick and Benjamin Etzold (eds), *Environment Migration and Adaptation: Evidence and Politics of Climate Change in Bangladesh* (A H Development Publishing House 2015)

Pauwelyn J, *Conflict of Norms in Public International Law: How WTO Law Relates to Other Rules of International Law* (Cambridge University Press 2009)

Peel J, *Science and Risk Regulation in International Law* (Cambridge University Press 2013)

—— 'International Environmental Law and Climate Disasters' in Rosemary Lyster and Robert RM Verchick (eds), *Research Handbook on Climate Disaster Law: Barriers and Opportunities* (Edward Elgar 2018)

—— and Lin J, *Litigating Climate Change in the Global South* (Oxford University Press 2024)

—— and Osofsky HM, 'A Rights Turn in Climate Change Litigation?' (2018) 7 Transnational Environmental Law 37

Pérez IS and Kallhauge AC, 'Adaptation (Article 7)' in Daniel Klein and others (eds), *The Paris Agreement on Climate Change: Analysis and Commentary* (Oxford University Press 2017)

Piggott-McKellar A, McMichael C, and Powell T, 'Generational Retreat: Locally Driven Adaption to Coastal Hazard Risk in Two Indigenous Communities in Fiji' (2021) 21 Regional Environmental Change 1

—— and others, 'Moving People in a Changing Climate: Lessons from Two Case Studies in Fiji' (2019) 8 Social Sciences 133

Piguet E, 'Climatic Statelessness: Risk Assessment and Policy Options' (2019) 45 Population and Development Review 865

Pobjoy JM, *The Child in International Refugee Law* (Cambridge University Press 2017)

Rahman A and others, 'A Review and Analysis of Water Research, Development, and Management in Bangladesh' (2022) 14 Water 1834

Rahman HMT and Hickey GM, 'What Does Autonomous Adaptation to Climate Change Have to Teach Public Policy and Planning About Avoiding the Risks of Maladaptation in Bangladesh?' (2019) 7 Frontiers in Environmental Science 2

Rajamani L, 'The Increasing Currency and Relevance of Rights-Based Perspectives in the International Negotiations on Climate Change' (2010) 22 Journal of Environmental Law 391

—— 'Ambition and Differentiation in the 2015 Paris Agreement: Interpretative Possibilities and Underlying Politics' (2016) 65 International and Comparative Law Quarterly 493

—— 'The 2015 Paris Agreement: Interplay Between Hard, Soft and Non-Obligations' (2016) 28 Journal of Environmental Law 337

—— 'The United Nations Framework Convention on Climate Change: A Framework Approach to Climate Change' in Daniel A Farber and Marjan Peeters (eds), *Climate Change Law*, vol I (Edward Elgar 2016)

—— 'Guiding Principles and General Obligation (Article 2.2 and Article 3)' in Daniel Klein and others (eds), *The Paris Agreement on Climate Change: Analysis and Commentary* (Oxford University Press 2017)

—— 'Common but Differentiated Responsibilities' in Emanuela Orlando and Ludwig Krämer (eds), *Principles of Environmental Law*, vol VI (Edward Elgar 2018)

—— 'The Principle of Common but Differentiated Responsibilities and Respective Capabilities in the International Climate Change Regime' in Rosemary Lyster and Robert RM Verchick (eds), *Research Handbook on Climate Disaster Law: Barriers and Opportunities* (Edward Elgar 2018)

—— 'Integrating Human Rights in the Paris Climate Architecture: Contest, Context, and Consequence' (2019) 9 Climate Law 180

—— and Guérin E, 'Central Concepts in the Paris Agreement and How They Evolved' in Daniel Klein and others (eds), *The Paris Agreement on Climate Change: Analysis and Commentary* (Oxford University Press 2017)

Ramcharan BG, 'The Concept and Dimensions of the Right to Life' in BG Ramcharan (ed), *The Right to Life in International Law* (Martinus Nijhoff 1985)

Ramji-Nogales J, 'Migration Emergencies' (2017) 68 Hastings Law Journal 609

Rashid SF, Gani S, and Sarker M, 'Urban Poverty, Climate Change and Health Risks for Slum Dwellers in Bangladesh' in Rajib Shaw, Fuad Mallick, and Aminul Islam (eds), *Climate Change Adaptation Actions in Bangladesh* (Springer 2013)

Redgwell C, 'Multilateral Environmental Treaty-Making' in Vera Gowlland-Debbas (ed), *Multilateral Treaty-Making* (Springer 2000)

Rimi RH and others, 'Risks of Pre-Monsoon Extreme Rainfall Events of Bangladesh: Is Anthropogenic Climate Change Playing a Role?' (2019) 100 Bulletin of the American Meteorological Society S61

Robinson S and others, 'Vulnerability-Based Allocations in Loss and Damage Finance' (2023) 13 Nature Climate Change 1055

Rodríguez-Garavito C, 'Litigating the Climate Emergency: The Global Rise of Human Rights–Based Litigation for Climate Action' in César Rodríguez-Garavito (ed), *Litigating the Climate Emergency: How Human Rights, Courts, and Legal Mobilization Can Bolster Climate Action* (Cambridge University Press 2022)

Sakellari M, 'Media Coverage of Climate Change Induced Migration: Implications for Meaningful Media Discourse' (2022) 18 Global Media and Communication 67

Sands P, 'Treaty, Custom and the Cross-Fertilization of International Law' (1998) 1 Yale Human Rights and Development Journal 85

Savaresi A, 'Climate Change and Human Rights: Fragmentation, Interplay, and Institutional Linkages' in Sébastien Duyck, Sébastien Jodoin, and Alyssa Johl (eds), *Routledge Handbook of Human Rights and Climate Governance* (Routledge 2018)

Schipper ELF, 'Conceptual History of Adaptation in the UNFCCC Process' (2006) 15 RECIEL 82

—— and Burton I, 'Understanding Adaptation: Origins, Concepts, Practice and Policy' in E Lisa F Schipper and Ian Burton (eds), *The Earthscan Reader on Adaptation to Climate Change* (Earthscan 2009)

Schrijver NJ, 'The Future of the Charter of the United Nations' in Armin von Bogdandy and Rüdiger Wolfrum (eds), *Max Planck Yearbook of United Nations Law*, vol 10 (2006)

Schutter OD, *International Human Rights Law: Cases, Materials, Commentary* (3rd edn, Cambridge University Press 2019)

Scott M, *Climate Change, Disasters, and the Refugee Convention* (Cambridge University Press 2020)

Shamsudduha M and others, 'Multi-Hazard Groundwater Risks to Water Supply from Shallow Depths: Challenges to Achieving the Sustainable Development Goals in Bangladesh' (2020) 12 Exposure and Health 657

Shelton D, 'Law, Non-Law, and the Problem of "Soft Law"' in Dinah Shelton (ed), *Commitment and Compliance: The Role of Non-Binding Norms in the International Legal System* (Oxford University Press 2003)

—— 'Equitable Utilization of the Atmosphere: A Rights-Based Approach to Climate Change' in Stephen Humphreys (ed), *Human Rights and Climate Change* (Cambridge University Press 2010)

Simperingham E, 'State Responsibility to Prevent Climate Displacement: The Importance of Housing, Land and Property Rights' in Dimitra Manou and others (eds), *Climate Change, Migration and Human Rights: Law and Policy Perspectives* (Routledge 2017)

—— 'Climate Displacement and Community-Led Solutions: Perspectives from the Asia-Pacific Region' in Simon Behrman and Avidan Kent (eds), *Climate Refugees: Global, Local and Critical Approaches* (Cambridge University Press 2022)

Solomon MK and Warner K, 'Protection of Persons Displaced as a Result of Climate Change: Existing Tools and Emerging Frameworks' in Michael B Gerrard and Gregory E Wannier (eds), *Threatened Island Nations: Legal Implications of Rising Seas and a Changing Climate* (Cambridge University Press 2013)

Sommario E, 'Limitation and Derogation Provisions in International Human Rights Law Treaties and Their Use in Disaster Settings' in Flavia Zorzi Giustiniani and others (eds), *Routledge Handbook of Human Rights and Disasters* (Routledge 2018)

Stern N, *The Economics of Climate Change: The Stern Review* (Cambridge University Press 2007)

Sunstein CR, *Laws of Fear: Beyond the Precautionary Principle* (Cambridge University Press 2005)

Tabe T, 'Climate Change Migration and Displacement: Learning from Past Relocations in the Pacific' (2019) 8 Social Sciences 1

Tadgell A, Doberstein B, and Mortsch L, 'Principles for Climate-Related Resettlement of Informal Settlements in Less Developed Nations: A Review of Resettlement Literature and Institutional Guidelines' (2018) 10 Climate and Development 102

Tagliacozzo S, Ayeb-Karlsson S, and Ahmed B, 'An Integrated Governance Framework to Map Out and Act on the Interrelationships between Human Mobility and Disaster Risk' (2023) 50 Journal of Ethnic and Migration Studies 1277

Táíwò OO, *Reconsidering Reparations* (Oxford University Press 2022)
Tasioulas J, 'Custom, Jus Cogens, and Human Rights' in Curtis A Bradley (ed), *Custom's Future: International Law in a Changing World* (Cambridge University Press 2016)
Taylor CM and others, 'Frequency of Extreme Sahelian Storms Tripled since 1982 in Satellite Observations' (2017) 544 Nature 475
Thirlway H, 'The Law and Procedure of the International Court of Justice 1960–1989: Part Three' (1992) 82 British Yearbook of International Law 1
Thomas A, 'Protecting People Displaced by Weather-Related Disasters and Climate Change: Experience from the Field' (2013) 15 Vermont Journal of Environmental Law 803
Timmer A and others, 'The Potential and Pitfalls of the Vulnerability Concept for Human Rights' (2021) 39 Netherlands Quarterly of Human Rights 196
Tuck E, 'Suspending Damage: A Letter to Communities' (2009) 79 Harvard Educational Review 409
Uddin MS and others, '"Disasters Threaten Livelihoods, and People Cope, Adapt and Make Transformational Changes": Community Resilience and Livelihoods Reconstruction in Coastal Communities of Bangladesh' (2021) 63 International Journal of Disaster Risk Reduction 102444
van Asselt H, *The Fragmentation of Global Climate Governance: Consequences and Management of Regime Interactions* (Edward Elgar 2014)
van der Geest K and others, 'Climate Change, Ecosystem Services and Migration in the Marshall Islands: Are They Related?' (2020) 161 Climatic Change 109
van der Vliet J and Biermann F, 'Global Governance of Climate Migrants: A Critical Evaluation of the Global Compacts' in Avidan Kent and Simon Behrman (eds), Climate Refugees: Global, Local and Critical Approaches (Cambridge University Press 2022)
Verheyen R, 'The Legal Framework of Adaptation and Adaptive Capacity' in Joel B Smith, Richard JT Klein, and Saleemul Huq (eds), *Climate Change, Adaptive Capacity and Development* (Imperial College Press 2003)
Verschuuren J, 'Climate Change Adaptation under the United Nations Framework Convention on Climate Change and Related Documents' in Jonathan Verschuuren (ed), *Research Handbook on Climate Change Adaptation Law* (Edward Elgar 2022)
—— 'Introduction to Climate Change Adaptation' in Jonathan Verschuuren (ed), *Research Handbook on Climate Change Adaptation Law* (Edward Elgar 2022)
Vinke K and others, 'Migration as Adaptation?' (2020) 8 Migration Studies 626
Warner K and others, 'National Adaptation Plans and Human Mobility' (2015) 49 Forced Migration Review 8
Weiss EB, 'Intergenerational Equity', *Max Planck Encyclopedia of Public International Law* (Oxford University Press 2013)
Wewerinke-Singh M, State Responsibility, *Climate Change and Human Rights under International Law* (Hart 2019)
—— and Doebbler C, 'The Paris Agreement: Some Critical Reflections on Process and Substance' (2016) 39 University of New South Wales Law Journal 1486
—— and Van Geelen T, 'Protection of Climate Displaced Persons under International Law: A Case Study from Mataso Island, Vanuatu' (2018) 19 Melbourne Journal of International Law 666
Wiederkehr C, Beckmann M, and Hermans K, 'Environmental Change, Adaptation Strategies and the Relevance of Migration in Sub-Saharan Drylands' (2018) 13 Environmental Research Letters 113003
Wiener JB, 'Precaution and Climate Change' in Cinnamon P Carlarne, Kevin R Gray, and Richard Tarasofsky (eds), *The Oxford Handbook of International Climate Change Law* (Oxford University Press 2016)
—— 'Precautionary Principle' in Emanuela Orlando and Ludwig Krämer (eds), *Principles of Environmental Law*, vol VI (Edward Elgar 2018)
Wyman KM, 'Ethical Duties to Climate Migrants' in Benoît Mayer and François Crépeau (eds), *Research Handbook on Climate Change, Migration and the Law* (Edward Elgar 2017)

Xu L and Famiglietti JS, 'Global Patterns of Water-Driven Human Migration' (2023) 10 WIREs Water e1647

Yamada S, Burkett M, and Maskarinec GG, 'Sea-Level Rise and the Marshallese Diaspora' (2017) 10 Environmental Justice 93

Yamin F and Depledge J, *The International Climate Change Regime* (Cambridge University Press 2005)

Yamineva Y, 'A Legal Perspective on Climate Finance Debates: How Constructive Is the Current Norm Ambiguity?' in Alexander Zahar and Benoît Mayer (eds), *Debating Climate Law* (Cambridge University Press 2021)

Young IM, *Responsibility for Justice* (Oxford University Press 2011)

Young MA, 'The WTO's Use of Relevant Rules of International Law: An Analysis of the Biotech Case' (2007) 56 The International and Comparative Law Quarterly 907

—— (ed), *Regime Interaction in International Law: Facing Fragmentation* (Cambridge University Press 2012)

—— and Gehring MW, 'The Climate Regime and Other Areas of Law' (2023) 13 Climate Law 151

Zickgraf C, 'Immobility', in Robert McLeman and François Gemenne (eds), *Routledge Handbook of Environmental Displacement and Migration* (Routledge 2018)

—— 'Theorizing (Im)Mobility in the Face of Environmental Change' (2021) 21 Regional Environmental Change 126

Zodgekar K and others, 'A Dangerous Debt-Climate Nexus: In a Warming World, Technocratic Fixes Are Inadequate Responses to Vulnerability. Achieving Climate and Economic Justice in the Caribbean Calls for Reparative Solutions' (2023) 55 NACLA Report on the Americas 319

Intergovernmental decisions

African Commission on Human and Peoples' Rights (ACHPR)

'Resolution on Climate Change and Human Rights and the Need to Study Its Impact in Africa' (Banjul, The Gambia 25 November 2009), ACHPR/Res.153(XLVI)09

'Resolution on Climate Change in Africa' (Luanda, Angola 12 May 2014) ACHPR/ Res.271(LV)2014

'Resolution on Climate Change and Human Rights in Africa' (Banjul, The Gambia 20 April 2016) ACHPR/ Res.342(LVIII)2016

'Resolution on the Human Rights Impacts of Extreme Weather in Eastern and Southern Africa Due to Climate Change' (Sharm el Sheikh, Arab Republic of Egypt 14 May 2019), ACHPR/ Res.417(LXIV)2019

United Nations Convention to Combat Desertification (UNCCD)

Decision 7/COP.13, 'The UNCCD 2018–2030 Strategic Framework' (15 September 2017) ICCD/ COP(13)/21/Add.1

Decision 28/COP.13, 'The Positive Role that Measures Taken Under the Convention Can Play to Address Desertification/Land Degradation and Drought as One of the Drivers that Causes Migration' (23 October 2017) ICCD/COP(13)/21/Add.1

United Nations Framework Convention on Climate Change (UNFCCC)

UNFCCC Conference of the Parties (COP)

Decision 11/CP.1, 'Initial Guidance on Policies, Programme Priorities and Eligibility Criteria to the Operating Entity or Entities of the Financial Mechanism' (6 June 1995) FCCC/CP/1995/ 7/Add.1

Decision 5/CP.4, 'Implementation of Article 4.8 and 4.9 of the Convention (Decision 3/CP.3 and Articles 2.3 and 3.14 of the Kyoto Protocol)' (25 January 1999) FCCC/CP/1998/16/Add.1

Decision 2—24/CP.7, 'Marrakesh Accords' (21 January 2002) FCCC/CP/2001/13/Add.1

Decision 11/CP.8, 'New Delhi work programme on Article 6 of the Convention' (28 March 2003) FCCC/CP/2002/7/Add.1

Decision 17/CP.8, 'Guidelines for the preparation of national communications from parties not included in Annex I to the Convention' (28 March 2003) FCCC/CP/2002/7/Add.2

Decision 2/CP.11, 'Five-Year programme of work of the Subsidiary Body for Scientific and Technological Advice on impacts, vulnerability and adaptation to climate change' (30 March 2006) FCCC/CP/2005/5/Add.1

Decision 1/CP.13, 'Bali Action Plan' (14 March 2008) FCCC/CP/2007/6/Add.1

Decision 2/CP.15, 'Copenhagen Accord' (20 March 2010) FCCC/CP/2009/11/Add.1

Decision 1/CP.16, 'Cancun Agreements: Outcome of the work of the Ad Hoc Working Group on Long-term Cooperative Action under the Convention' (15 March 2011) FCCC/CP/2010/7/Add.1

Decision 2/CP.19, 'Warsaw International Mechanism for Loss and Damage Associated with Climate Change Impacts' (WIM) (31 January 2014) FCCC/CP/2013/10/Add.1

Decision 1/CP.21, 'Adoption of the Paris Agreement' (29 January 2016) FCCC/CP/2015/10/Add.1

Decision 10/CP.22, 'Report of the Green Climate Fund to the Conference of the Parties and guidance to the Green Climate Fund' (31 January 2017) FCCC/CP/2016/10/Add.1

Decision 2/CP.24, 'Local Communities and Indigenous Peoples Platform' (19 March 2019) FCCC/CP/2018/10/Add.1

Decision 17/CP.26, 'Warsaw International Mechanism for Loss and Damage Associated with Climate Change Impacts' (31 October 2021) FCCC/CP/2021/12/Add.2

Decision 2/CP.27, 'Funding Arrangements for Responding to Loss and Damage Associated with the Adverse Effects of Climate Change, Including a Focus on Addressing Loss and Damage' (20 November 2022) FCCC/CP/2022/10/Add.1

Kyoto Protocol COP Serving as the Meeting of the Parties (CMP)

Decision 1/CMP.3, 'Adaptation Fund' (14 March 2008) FCCC/KP/CMP/2007/9/Add.1

Decision 5/CMP.17, 'Fourth review of the Adaptation Fund (6 November 2022) FCCC/KP/CMP/2022/9/Add.1

Paris Agreement COP Serving as the Meeting of the Parties (CMA)

Decision 13/CMA.1, 'Matters relating to the Adaptation Fund' (2 December 2018) FCCC/PA/CMA/2018/3/Add.2

Decision 17/CMA.1, 'Ways of enhancing the implementation of education, training, public awareness, public participation and public access to information so as to enhance actions under the Paris Agreement' (19 March 2019) FCCC/PA/CMA/2018/3/Add.2

Decision 18/CMA.1, 'Modalities, procedures and guidelines for the transparency framework for action and support referred to in Article 13 of the Paris Agreement' (19 March 2019) FCCC/PA/CMA/2018/3/Add.2

Decision 19/CMA.1: 'Matters relating to Article 14 of the Paris Agreement and paragraphs 99–101 of Decision 1/CP.21' (19 March 2019) FCCC/PA/CMA/2018/3/Add.2

Decision 1/CMA.3, 'Glasgow Climate Pact' (8 March 2022) FCCC/PA/CMA/2021/10/Add.1

Decision 1/CMA.5, 'Outcome of the First Global Stocktake' (13 December 2023) FCCC/PA/CMA/2023/16/Add.1

Decision 2/CMA.5, 'Global Goal on Adaptation' (13 December 2023) FCCC/PA/CMA/2023/16/Add.1

Decision 5/CMA.5, 'Operationalization of the New Funding Arrangements, Including a Fund, for Responding to Loss and Damage Referred to in Paragraphs 2–3 of Decisions 2/CP.27 and 2/CMA.4' (6 December 2023) FCCC/PA/CMA/2023/16/Add.1

Decision 14/CMA.5, 'Enhancing climate technology development and transfer to support implementation of the Paris Agreement' (11 December 2023) FCCC/PA/CMA/2023/16/Add.2

Draft Decision -/CMA.5 'Glasgow–Sharm El-Sheikh work programme on the global goal on adaptation referred to in Decision 7/CMA.3' (2023) FCCC/PA/CMA/2023/L.18

United Nations General Assembly (UNGA)

Statute of the International Law Commission (adopted 21 November 1947, as amended by resolutions 485 (V) of 12 December 1950, 984 (X) of 3 December 1955, 985 (X) of 3 December 1955 and 36/39 of 18 November 1981) Resolution 174 (II)
'The human right to water and sanitation', Resolution 64/292 (28 July 2010) UN Doc A/RES/64/292
'Global Compact for Safe, Orderly and Regular Migration', Resolution 73/195 (19 December 2018) UN Doc A/RES/73/195

United Nations Commission on Human Rights (UNCHR)

'Forced Evictions', Resolution 1993/ 77 in 'Report on the Forty- Ninth Session of the UNCHR' (10 March 1993) UN Doc E/ CN.4/ 1993/ 122

United Nations Human Rights Council (UNHRC)

'Human Rights and Climate Change' (23 June 2014) UN Doc A/HRC/26/L.33
'Human Rights and Climate Change' (19 June 2017) A/HRC/35/L.32

Intergovernmental Panel on Climate Change (IPCC) reports

Abram N and others, 'Framing and Context of the Report' in HO Pörtner and others (eds), *IPCC Special Report on the Ocean and Cryosphere in a Changing Climate* (Cambridge University Press 2019)
Adger WN and others, 'Human Security' in Christopher B Field and others (eds), *Climate Change 2014: Impacts, Adaptation, and Vulnerability. Part A: Global and Sectoral Aspects. Contribution of Working Group II to the Fifth Assessment Report of the Intergovernmental Panel on Climate Change* (Cambridge University Press 2014)
Baede APM, van der Linden P and Verbruggen A (eds), 'Annex II: Glossary', *Climate Change 2007: Synthesis Report.* (Cambridge University Press 2007)
Bezner Kerr R and others, 'Food, Fibre and Other Ecosystem Products' in HO Pörtner and others (eds), *Climate Change 2022: Impacts, Adaptation and Vulnerability. Contribution of Working Group II to the Sixth Assessment Report of the Intergovernmental Panel on Climate Change* (Cambridge University Press 2022)
Birkmann J and others, 'Poverty, Livelihoods and Sustainable Development' in HO Pörtner and others (eds), *Climate Change 2022: Impacts, Adaptation and Vulnerability. Contribution of Working Group II to the Sixth Assessment Report of the Intergovernmental Panel on Climate Change* (Cambridge University Press 2022)
Caretta MA and others, 'Water' in HO Pörtner and others (eds), *Climate Change 2022: Impacts, Adaptation and Vulnerability. Contribution of Working Group II to the Sixth Assessment Report of the Intergovernmental Panel on Climate Change* (Cambridge University Press 2022)
Cissé G and others, 'Health, Wellbeing and the Changing Structure of Communities' in HO Pörtner and others (eds), *Climate Change 2022: Impacts, Adaptation and Vulnerability. Contribution of Working Group II to the Sixth Assessment Report of the Intergovernmental Panel on Climate Change* (Cambridge University Press 2022)
Hoegh-Guldberg O and others, 'Impacts of 1.5°C Global Warming on Natural and Human Systems', *Global Warming of 1.5°C. An IPCC Special Report on the impacts of global warming of 1.5°C above pre-industrial levels and related global greenhouse gas emission pathways, in the context of strengthening the global response to the threat of climate change, sustainable development, and efforts to eradicate poverty* (Cambridge University Press 2018)
IPCC, 'Climate Change: The 1990 and 1992 IPCC Assessments, IPCC First Assessment Report Overview and Policymaker Summaries and 1992 IPCC Supplement' (IPCC 1992)
——, 'Annex B: Glossary of Terms' in RT Watson and the Core Writing Team (eds), *Climate Change 2001: Synthesis Report. Contribution of Working Groups I, II, and III to the Third Assessment of the Intergovernmental Panel on Climate Change* (Cambridge University Press 2001)

——, 'Annex I: Glossary', *Climate Change 2007: Impacts, Adaptation and Vulnerability. Contribution of Working Group II to the Fourth Assessment Report of the Intergovernmental Panel on Climate Change* (Cambridge University Press 2007)

—— *Managing the Risks of Extreme Events and Disasters to Advance Climate Change Adaptation*, A Special Report of Working Groups I and II of the IPCC (Christopher B Field and others (eds), Cambridge University Press 2012)

——, 'Annex II: Glossary', *Climate Change 2014: Synthesis Report. Contribution of Working Groups I, II and III to the Fifth Assessment Report of the Intergovernmental Panel on Climate Change* (IPCC 2014)

—— 'Summary for Policymakers' in Christopher B Field and others (eds), *Climate Change 2014: Impacts, Adaptation, and Vulnerability. Part A: Global and Sectoral Aspects. Contribution of Working Group II to the Fifth Assessment Report of the IPCC* (Cambridge University Press 2014)

—— *Global Warming of 1.5°C. An IPCC Special Report on the impacts of global warming of 1.5°C above pre-industrial levels and related global greenhouse gas emission pathways, in the context of strengthening the global response to the threat of climate change, sustainable development, and efforts to eradicate poverty* (V Masson-Delmotte and others (eds), Cambridge University Press 2018)

—— 'Summary for Policymakers' in HO Pörtner and others (eds), *IPCC Special Report on the Ocean and Cryosphere in a Changing Climate* (Cambridge University Press 2019)

—— *Climate Change 2021: The Physical Science Basis. Contribution of Working Group I to the Sixth Assessment Report of the Intergovernmental Panel on Climate Change* (V Masson-Delmotte and others (eds), Cambridge University Press 2021)

—— 'Summary for Policymakers' in V Masson-Delmotte and others (eds), *Climate Change 2021: The Physical Science Basis. Contribution of Working Group I to the Sixth Assessment Report of the Intergovernmental Panel on Climate Change* (Cambridge University Press 2021)

—— *Climate Change 2022: Impacts, Adaptation and Vulnerability. Contribution of Working Group II to the Sixth Assessment Report of the Intergovernmental Panel on Climate Change* (HO Pörtner and others eds, Cambridge University Press 2022)

—— 'Annex II: Glossary' in HO Pörtner and others (eds), *Climate Change 2022: Impacts, Adaptation, and Vulnerability. Contribution of Working Group II to the Sixth Assessment Report of the Intergovernmental Panel on Climate Change* (Cambridge University Press 2022)

—— 'Summary for Policymakers' in HO Pörtner and others (eds), *Climate Change 2022: Impacts, Adaptation and Vulnerability. Contribution of Working Group II to the Sixth Assessment Report of the Intergovernmental Panel on Climate Change* (Cambridge University Press 2022)

—— 'Summary for Policymakers' in Core Writing Team, H Lee and J Romero (eds), *Climate Change 2023: Synthesis Report. Contribution of Working Groups I, II and III to the Sixth Assessment Report of the Intergovernmental Panel on Climate Change* (2023)

Mirzabaev A and others, 'Desertification' in PR Shukla and others (eds), *Climate Change and Land: Special Report on Climate Change, Desertification, Land Degradation, Sustainable Land Management, Food Security, and Greenhouse gas fluxes in Terrestrial Ecosystems* (2019)

—— 'Deserts, Semiarid Areas and Desertification' in HO Pörtner and others (eds), *Climate Change 2022: Impacts, Adaptation and Vulnerability. Contribution of Working Group II to the Sixth Assessment Report of the Intergovernmental Panel on Climate Change* (Cambridge University Press 2022)

Mycoo M and others, 'Small Islands' in HO Pörtner and others (eds), *Climate Change 2022: Impacts, Adaptation and Vulnerability. Contribution of Working Group II to the Sixth Assessment Report of the Intergovernmental Panel on Climate Change* (Cambridge University Press 2022)

New and others, 'Decision-Making Options for Managing Risk' in HO Pörtner and others (eds), *Climate Change 2022: Impacts, Adaptation and Vulnerability. Contribution of Working Group II to the Sixth Assessment Report of the Intergovernmental Panel on Climate Change* (Cambridge University Press 2022)

Niang I and others, 'Africa' in Vincente R. Barros and others (eds), *Climate Change 2014: Impacts, Adaptation, and Vulnerability. Part B: Regional Aspects. Contribution of Working Group II to the Fifth Assessment Report of the Intergovernmental Panel on Climate Change* (Cambridge University Press 2014)

Noble IR and others, 'Adaptation Needs and Options' in Christopher B Field others (eds), *Climate Change 2014: Impacts, Adaptation, and Vulnerability. Part A: Global and Sectoral Aspects. Contribution of Working Group II to the Fifth Assessment Report of the Intergovernmental Panel on Climate Change* (Cambridge University Press 2014)

Nurse LA and others, 'Small Islands' in Vincente R. Barros and others (eds), *Climate Change 2014: Impacts, Adaptation, and Vulnerability. Part B: Regional Aspects. Contribution of Working Group II to the Fifth Assessment Report of the Intergovernmental Panel on Climate Change* (Cambridge University Press 2014)

Parmesan C and others, 'Terrestrial and Freshwater Ecosystems and Their Services' in HO Pörtner and others (eds), *Climate Change 2022: Impacts, Adaptation and Vulnerability. Contribution of Working Group II to the Sixth Assessment Report of the Intergovernmental Panel on Climate Change* (Cambridge University Press 2022)

Pörtner H-O and others, 'Technical Summary' in HO Pörtner and others (eds), *Climate Change 2022: Impacts, Adaptation and Vulnerability. Contribution of Working Group II to the Sixth Assessment Report of the Intergovernmental Panel on Climate Change* (Cambridge University Press 2022)

Shaw R, Luo Y and Cheong TS, 'Asia' in HO Pörtner and others (eds), *Climate Change 2022: Impacts, Adaptation and Vulnerability. Contribution of Working Group II to the Sixth Assessment Report of the Intergovernmental Panel on Climate Change* (Cambridge University Press)

Trisos CH and others, 'Africa' in HO Pörtner and others (eds), *Climate Change 2022: Impacts, Adaptation and Vulnerability. Contribution of Working Group II to the Sixth Assessment Report of the Intergovernmental Panel on Climate Change* (Cambridge University Press 2022)

United Nations and treaty body guidance

Committee on Economic, Social and Cultural Rights (CESCR)

'General Comment No 3: The Nature of States Parties' Obligations (Art 2, Para 1, of the Covenant)' (14 December 1990) UN Doc E/1991/23

'General Comment No 4: The Right to Adequate Housing (Art 11 (1) of the Covenant)' (13 December 1991) UN Doc E/1992/23

'General Comment No 7: The Right to Adequate Housing (Art 11.1): Forced Evictions', (20 May 1997) UN Doc E/1998/22

'General Comment No 8: The Relationship between Economic Sanctions and Respect for Economic, Social and Cultural Rights' (12 December 1997) UN Doc E/C.12/1997/8

'General Comment No 12: The Right to Adequate Food (Art 11 of the Covenant)' 12 (May 1999) UN Doc E/C.12/1999/5

'General Comment No 14: The Right to the Highest Attainable Standard of Health (Art 12)' (11 August 2000) UN Doc E/C.12/2000/4

'General Comment No 15: The Right to Water (Arts 11 and 12 of the Covenant)' (20 January 2003) UN Doc E/C.12/2002/11

'Concluding Observations on the Initial Report of Bangladesh' (18 April 2018) UN Doc E/C.12/BGD/CO/1

'Climate Change and the International Covenant on Economic, Social and Cultural Rights: Statement of the Committee on Economic, Social and Cultural Rights' (8 October 2018) <www.ohchr.org/en/NewsEvents/Pages/DisplayNews.aspx?NewsID=23691&LangID=E> accessed 20 September 2019

Committee on the Elimination of Discrimination against Women

'General Recommendation No 37 (2018) on the Gender-Related Dimensions of Disaster Risk Reduction in the Context of Climate Change' (18 March 2018) UN Doc CEDAW/C/GC/37

Human Rights Committee (HRC)

'General Comment No 15: The Position of Aliens Under the Covenant' (11 April 1986) UN Doc HRI/GEN/1/Rev.1 18 (1994)

'General Comment No 18: Non-discrimination' (10 November 1989) UN Doc HRI/GEN/1/Rev.1 26 (1994)

'General Comment No 20: Article 7 (Prohibition of Torture, or Other Cruel, Inhuman or Degrading Treatment or Punishment)' (10 March 1992) UN Doc HRI/GEN/1/Rev.9 (Vol. I) 200 (2008)

'General Comment No 23: Article 27 (Rights of Minorities)' (8 April 1994) UN Doc CCPR/C/21/Rev.1/Add.5

'General Comment No 27: Article 12 (Freedom of Movement)' (2 November 1999) UN Doc CCPR/C/21/Rev.1/Add.9

'General Comment No 36: Article 6 (Right to Life)' (3 September 2019) UN Doc CCPR/C/GC/36

UNCHR

'Report of the Commission on Human Rights on the Protection of Human Rights in Chile' (4 November 1982) UN Doc A/ 37/ 564

'Guiding Principles on Internal Displacement' (11 February 1998) UN Doc E/CN.4/1998/53/Add.2

UNHRC

'Report of the Special Rapporteur on the Right to Food, Jean Ziegler' (10 January 2008) UN Doc A/HRC/7/5

'The Right to Adequate Housing: Report of the Special Rapporteur on Adequate Housing as a Component of the Right to an Adequate Standard of Living, and on the Right to Non-Discrimination in this Context, Raquel Rolnik' (9 August 2009) UN Doc A/64/255

'The Right to Adequate Housing: Report of the Special Rapporteur on Adequate Housing as a Component of the Right to an Adequate Standard of Living, and on the Right to Non-Discrimination in this Context, Raquel Rolnik' (20 December 2010) UN Doc A/HRC/16/42

'Protection of and Assistance to Internally Displaced Persons, Report of the Special Rapporteur on the human rights of internally displaced persons, Chaloka Beyani' (9 August 2011) UN Doc A/66/285

'Report of the Independent Expert on the Issue of Human Rights Obligations Relating to the Enjoyment of a Safe, Clean, Healthy and Sustainable Environment, John H. Knox: Preliminary Report' (24 December 2012) UN Doc A/HRC/22/43

'Climate Change and Poverty: Report of the Special Rapporteur on Extreme Poverty and Human Rights' (17 July 2019) UN Doc A/HRC/41/39

'Mandate of the Special Rapporteur on the Promotion and Protection of Human Rights in the Context of Climate Change' (4 October 2021) UN Doc A/HRC/48/L.27

'The Human Right to a Safe, Clean, Healthy and Sustainable Environment' (5 October 2021), UN Doc A/HRC/48/L.23 Rev.1

'Report of the Special Rapporteur on the Situation of Human Rights in Myanmar Annex II (Bangladesh mission), Thomas H. Andrews' (13 June 2022) UN Doc A/HRC/49/76

'Report of the Special Rapporteur on the Human Rights of Migrants, Felipe González Morales' (19 July 2022) UN Doc A/77/189

'Providing Legal Options to Protect the Human Rights of Persons Displaced across International Borders Due to Climate Change: Report of the Special Rapporteur on the Promotion and Protection of Human Rights in the Context of Climate Change, Ian Fry' (18 April 2023) UN Doc A/HRC/53/34

'Report of the Special Rapporteur on Adequate Housing as a Component of the Right to an Adequate Standard of Living, and on the Right to Non-Discrimination in this Context, Balakrishnan Rajagopal. A Place to Live in Dignity for All: Make Housing Affordable' (15 August 2023) UN Doc A/78/192

Submissions of the parties to the UNFCCC

Ad- hoc Working Group on the LCA, 'Ideas and Proposals on the Elements Contained in Paragraph 1 of the Bali Action Plan: Submissions from Parties (Part I)' (10 December 2008) FCCC/AWGLCA/ 2008/ MISC.5/ Add.2

—— 'Ideas and Proposals on the Elements Contained in Paragraph 1 of the Bali Action Plan: Submissions from Parties (Part I)' (19 May 2009) FCCC/AWGLCA/ 2009/ MISC.4 (Part I)

—— 'Additional Views on Which the Chair May Draw in Preparing Text to Facilitate Negotiations among Parties: Submissions

from Parties' (30 April 2010) FCCC/ AWGLCA/ 2010/ MISC.2

Climate Change Directorate, Republic of the Marshall Islands, 'Adaptation Communication' (2020)

—— 'National Adaptation Plan (Pāpjelmae)' (2023)

Eritrea: Ministry of Land, Water and Environment, 'NAPA' (2007)

Federal Republic of Nigeria, 'Nigeria's First National Communication' (2003)

—— 'Adaptation Communication to the United Nations Framework Convention on Climate Change' (2021)

Government of India, 'Submission of India for Consideration by the Standing Committee on Elements in the Development of the Further Guidelines for the Fifth Review of the Financial Mechanism' (2013) Submission to the COP

Government of the Fiji Islands, 'Climate Change-The Fiji Islands Response: Fiji's First National Communication' (2005)

Government of the People's Republic of Bangladesh, 'Initial National Communication under the UNFCCC' (2002)

—— 'Second National Communication of Bangladesh' (2012)

Government of the Republic of the Marshall Islands, 'Initial Communication under the UNFCCC' (2000)

Islamic Republic of Mauritania, 'NAPA' (2004)

Kiribati Government, 'Initial Communication under the UNFCCC' (1999)

Ministry of Environment and Fishery Resources, 'Burkina Faso NAP' (2015)

Ministry of Environment and Forest Government of the People's Republic of Bangladesh, 'National Adaptation Programme of Action (NAPA)' (2005)

Ministry of Environment and Forests, Bangladesh, 'NAPA: Updated Version' (2009)

Ministry of Environment, Conservation and Meteorology, 'Solomon Islands NAPA' (2008)

Ministry of Environment, Forest and Climate Change (MoEFCC), Government of the People's Republic of Bangladesh, 'Third National Communication of Bangladesh' (2018)

—— Government of the People's Republic of Bangladesh, 'National Adaptation Plan of Bangladesh (2023–2050)' (2022)

Ministry of Natural Resources, Environment, Agriculture and Lands, 'Tuvalu's NAPA' (2007)

Republic of Sudan, 'Sudan's First National Communication under the UNFCCC' (2003) Volume I: Main Communications

—— 'First Nationally Determined Contribution under the Paris Agreement' (2021)

Republic of Sudan: Ministry of Environment and Physical Development, 'NAPA' (2007)

—— 'NAP' (2016)

Republic of Vanuatu, 'Vanuatu National Communication to the COP to the UNFCCC' (1999)

Other

Abeysinghe A, Craft B, and Tenzing J, 'The Paris Agreement and the LDCs: Analysing COP21 Outcomes from LDC Positions' (International Institute for Environment and Development (IIED) 2016) Issue Paper

'About C40' (C40 Cities) <www.c40.org/about-c40/> accessed 25 October 2024

Action Aid, 'The Vicious Cycle: Connections Between the Debt Crisis and Climate Crisis' (2023) Policy Brief <https://actionaid.org/publications/2023/vicious-cycle> accessed 9 September 2024

Adaptation Committee, '25 Years of Adaptation under the UNFCCC: Report of the Adaptation Committee' (UNFCCC 2019)

—— 'Flexible Workplan of the Adaptation Committee for 2019–2021' (2020) Version 3, Update following COP 25/CMA 2

—— 'Synthesis Report for the Technical Assessment Component of the First Global Stocktake' (2022)

Adaptation Fund, 'Climate Adaptation Finance: Direct Access' (2020)

Adaptation Fund Board, 'Environmental and Social Policy' (2016)

—— 'Operational Policies and Guidelines for Parties to Access Resources from the Adaptation Fund (OPG)' (2022)

African Union, 'Africa Regional Strategy for Disaster Risk Reduction' (ARSDRR) (July 2004)

—— 'Programme of Action for the Implementation of the Sendai Framework for Disaster Risk Reduction 2015–2030 in Africa: In Line with the ARSDRR' (2016)

Arrojo-Agudo P, 'Special Thematic Report on Climate Change and the Human Rights to Water and Sanitation by the Special Rapporteur on the Human Rights to Safe Drinking Water and Sanitation: Outlining the Impacts of Climate Change on the Human Rights to Water and Sanitation around the World' (Office of the High Commissioner for Human Rights (OHCHR) 2022)

'Bangladesh' (IDMC) <www.internal-displacement.org/countries/bangladesh/> accessed 30 May 2024

Barkat A, Osman A, and Ahmed SA, 'Bangladesh Migration Governance Framework' (MiGOF) (International Organization for Migration (IOM) 2020)

Barnett J and Webber M, 'Accommodating Migration to Promote Adaptation to Climate Change' (The World Bank 2010) WPS5270

Bower E and Weerasinghe S, 'Planned Relocation in the Pacific: A Regional Snapshot' (PDD 2021)

Brookings Institution and University of Bern, 'Protecting Internally Displaced Persons: A Manual for Law and Policymakers' (2008)

Brookings Institution, Georgetown University, and UNHCR, 'Guidance on Protecting People from Disasters and Environmental Change through Planned Relocation' (2015)

Buontempo C, 'Sahelian Climate: Past, Current, Projections' (Met Office Hadley Centre 2010)

'Call for Comments to the "Study on the Impact of Climate Change on Human and Peoples" Rights in Africa" (African Commission on Human and Peoples' Rights, 23 October 2023) <https://achpr.au.int/en/news/press-releases/2023-10-28/call-comments-study-impact-climate-change-human-and-peo> accessed 19 August 2024

Campbell J and Warrick O, 'Climate Change and Migration Issues in the Pacific' (United Nations Economic and Social Commission for Asia and the Pacific 2014)

CARE, 'Gender Equality and Women's Empowerment in the Context of Food Security and Nutrition' (2020)

Center for International Earth Science and Information Network (CIESIN), Columbia University, 'Low Elevation Coastal Zone (LECZ) Urban-Rural Population and Land Area Estimates, Version 3' (2021) <https://doi.org/10.7927/d1x1-d702> accessed 19 August 2024

Centre de Suivi Écologique, Senegal, 'Reducing Vulnerability and Increasing Resilience of Coastal Communities in the Saloum Islands (Dionewar)' (2017) Project/Programme Proposal to the Adaptation Fund <www.adaptation-fund.org/project/reducing-vulnerability-increasing-resilience-coastal-communities-saloum-islands-dionewar/> accessed 22 August 2024

Cincotta R and Smith S, 'What Future for the Western Sahel? The Region's Demography and Its Implications by 2045' (Atlantic Council 2021)

Clement V and others, 'Groundswell Part 2: Acting on Internal Climate Migration' (World Bank 2021)

'Climate-Affected Marshall Islands Don't Intend to Relocate' *RNZ* (23 April 2021) <www.rnz.co.nz/international/pacific-news/441080/climate-affected-marshall-islands-don-t-intend-to-relocate> accessed 20 June 2024

'Climate Change-Induced Sea-Level Rise Direct Threat to Millions around World, Secretary-General Tells Security Council' *UN Meetings Coverage and Press Releases* (14 February 2023) <https://press.un.org/en/2023/sc15199.doc.htm> accessed 17 April 2023

'Climate Finance Delivery Plan: Meeting the US$100 Billion Goal' (2021) <https://ukcop26.org/wp-content/uploads/2021/10/Climate-Finance-Delivery-Plan-1.pdf> accessed 23 September 2022

'The Climate Refugee Crisis Is Landing on Europe's Shores—and We Are Far from Ready' *POLITICO* (20 February 2023) <www.politico.eu/article/climate-refugee-crisis-europe-policy/> accessed 25 October 2023

Commission of Small Island States on Climate Change and International Law (COSIS), 'Request for an Advisory Opinion' (2022) <www.itlos.org/fileadmin/itlos/documents/cases/31/Request_for_Advisory_Opinion_COSIS_12.12.22.pdf> accessed 1 February 2024

Conference of the Parties serving as the meeting of the Parties to the Paris Agreement (CMA), 'Nationally Determined Contributions under the Paris Agreement: Synthesis Report by the Secretariat' (2022) FCCC/ PA/ CMA/ 2022/ 4

Congo Y and Nienta I, 'Final Evaluation of the Programme Support for Climate Change Adaptation in the Vulnerable Regions of Mopti and Timbuktu (PACV-MT)' (2020) Final Report <www.adaptation-fund.org/project/programme-support-for-climate-change-adaptation-in-the-vulnerable-regions-of-mopti-and-timbuktu-2/> accessed 22 August 2024

de Albuquerque C, 'Climate Change and the Human Rights to Water and Sanitation' (2010) Position Paper

Dempster H and Ober K, 'New Zealand's "Climate Refugee" Visas: Lessons for the Rest of the World' (Center For Global Development, 10 January 2020) <www.cgdev.org/blog/new-zealands-climate-refugee-visas-lessons-rest-world> accessed 15 August 2020

Deutsche Gesellschaft für Internationale Zusammenarbeit (GIZ) and International Labour Organization (ILO), 'Labour Market Trends Analysis and Labour Migration from South Asia to Gulf Cooperation Council Countries, India and Malaysia' (GIZ and ILO 2015)

Displacement Solutions, 'Climate Change Displaced Persons and Housing, Land and Property Rights: Preliminary Strategies for Rights-Based Planning and Programming to Resolve Climate-Induced Displacement' (2010) <https://globalprotectioncluster.org/sites/default/files/2023-04/Climate_Change_Displaced_Persons_2010_EN.pdf> accessed 30 May 2024

'Draft Paris Outcome, Proposal by the President, Version 1 of 9 December 2015 (15:00)' <https://unfccc.int/resource/docs/2015/cop21/eng/da01.pdf> accessed 2 September 2023

Dreiling K and others, 'People on the Move: Strengthening Adaptation Responses to Support Human Movement in a Changing Climate' (United States Agency for International Development (USAID) 2021) Technical Report

Earth Negotiations Bulletin, 'Summary of the Paris Climate Change Conference: 29 November–13 December 2015' (2015) 12 International Institute for Sustainable Development

Eckstein D, Künzel V, and Schäfer L, 'Global Climate Risk Index 2021: Who Suffers Most From Extreme Weather Events? Weather-Related Loss Events in 2019 and 2000 to 2019' (Germanwatch eV 2021)

Fobissie K, Saleh BK, and Haile E, 'Terminal Evaluation Report: Climate Change Adaptation Programme in Water and Agriculture in Anseba Region, Eritrea' (2019) Final Project Report <www.adaptation-fund.org/project/climate-change-adaptation-programme-in-water-and-agriculture-in-anseba-region-eritrea/> accessed 22 August 2024

Food and Agriculture Organization (FAO) and IOM, 'Agriculture and Migration in the Context of Climate Change' (2017) <www.fao.org/3/a-i7541e.pdf> accessed 20 September 2019

Fatima R, Wadud AJ, and Coelho S, 'Human Rights, Climate Change, Environmental Degradation and Migration: A New Paradigm' (2014) IOM and MPI Issue in Brief

Foresight, 'Migration and Global Environmental Change' (UK Government Office for Science 2011) Final Project Report

Foster M and others, 'The Future of Nationality in the Pacific: Preventing Statelessness and Nationality Loss in the Context of Climate Change' (2022)

Francis A, 'Free Movement Agreements & Climate-Induced Migration: A Caribbean Case Study' (Sabin Center for Climate Change Law, Columbia Law School 2019)

General Economics Division, Bangladesh Planning Commission, 'Making Vision 2041 a Reality: Perspective Plan of Bangladesh 2021–2041' (February 2020)

Georgetown University, IOM and UNHCR, 'A Toolbox: Planning Relocations to Protect People from Disasters and Environmental Change' (2017)

Global Center on Adaptation and Climate Policy Initiative, 'State and Trends in Climate Adaptation Finance 2024' (2024)

Global Environment Facility (GEF), 'Operational Guidelines for the Application of the Incremental Cost Principle' (2007) GEF Policy Paper <www.thegef.org/sites/default/files/documents/OPERATIONAL.GUIDELINES.FOR_.THE_.APPLICATION.OF_.THE_.INCREMENTAL.COST_.PRINCIPLE_0_0_0_0.pdf> accessed 17 May 2024

Government of Kiribati, 'Kiribati Climate Change Policy' (2019)

—— 'Kiribati Joint Implementation Plan for Climate Change and Disaster Risk Management (KJIP): 2019–2028' (2019)

Government of the People's Republic of Bangladesh, MoEFCC, 'Bangladesh Climate Change Strategy and Action Plan (BCCSAP)' (2009)

Government of the Republic of Fiji, 'Republic of Fiji NAP: A Pathway towards Climate Resilience' (2018)

Government of Tuvalu, Climate Change Department, 'Te Vaka Fenua o Tuvalu: National Climate Change Policy 2021–2030' (2021)

Government of Vanuatu, 'National Policy on Climate Change and Disaster-Induced Displacement' (2018)

Green Climate Fund, 'Report of the Thirteenth Meeting of the Board, 28-30 June 2016' (2016) GCF/B.13/33

Green Climate Fund Board, 'Environmental and Social Policy' (2018) Decision B.19/10, paragraph (b), Annex X

Hett H, 'As the Planet Warms, Canada Faces an Influx of Climate Refugees' *Wired* (2022) <www.wired.com/story/as-the-planet-warms-canada-faces-an-influx-of-climate-refugees/> accessed 25 October 2023

Human Rights Watch, 'Bangladesh: New Restrictions on Rohingya Camps' (4 April 2022) <www.hrw.org/news/2022/04/04/bangladesh-new-restrictions-rohingya-camps> accessed 30 May 2024

Huq S and Ayers J, 'Climate Change Impacts and Responses in Bangladesh' (2008) IP/A/CLIM/NT/2007-09

Internal Displacement Monitoring Centre (IDMC), 'Global Estimates 2015: People Displaced by Disasters' (2015) <https://api.internal-displacement.org/sites/default/files/publications/documents/20150713-global estimates-2015-en-v1.pdf> accessed 20 November 2023

—— 'Project Brief: Pacific Response to Disaster Displacement' (2019)

—— '2020 Global Report on Internal Displacement' (2020)

—— and Norwegian Refugee Council, 'Global Report on Internal Displacement' (IDMC 2019)

International Federation of Red Cross and Red Crescent Societies (IFRC), 'Acting Early to Protect Bangladesh's Vulnerable Population from Cyclone Amphan' (*Forecast-based Financing*) <www.forecast-based-financing.org/2020/05/19/cyclone-amphan-is-approaching-bangladesh/> accessed 6 June 2024

—— 'Planned Relocation in the Context of Disasters and Climate Change: A Guide for Asia Pacific National Societies' (2021)

International Law Association (ILA), 'Declaration of Legal Principles Relating to Climate Change' (2014) Resolution 2/2014

—— 'Sydney Declaration of Principles on the Protection of Persons Displaced in the Context of Sea-Level Rise' (2018) Resolution 6/2018 <https://disasterdisplacement.org/wp-content/uploads/2019/06/ILA-Sydney-Declaration.pdf> accessed 14 September 2019

International Law Commission (ILC), 'Draft Articles on the Law of Treaties with Commentaries' (1966) II Yearbook of the ILC 187

—— 'Draft Articles on Responsibility of States for Internationally Wrongful Acts' (November 2001) Supplement No 10 UN Doc A/56/10, chp.IV.E.1

—— 'Fragmentation of International Law: Difficulties Arising from the Diversification and Expansion of International Law: Report of the Study Group of the International Law Commission, Finalized by Martti Koskenniemi' (13 April 2006) UN Doc A/CN.4/L.682

—— 'Fragmentation of International Law: Difficulties Arising from the Diversification and Expansion of International Law; Conclusions of the Work of the Study Group' (18 July 2006) UN Doc A/CN.4/L.702

—— 'Draft Articles on the Protection of Persons in the Event of Disasters' (2016) UN Doc A/71/10

International Union for Conservation of Nature (IUCN), 'Written Statement of the International Union for Conservation of Nature and Natural Resources—World Commission on Environmental Law, Ocean Law Specialist Group' (2023)

IOM, 'Migration, Environment and Climate Change: Assessing the Evidence' (2009)

—— 'IOM Contributions to Global Climate Negotiations: 22nd Conference of Parties to the United Nations Framework Convention on Climate Change (UNFCCC)' (2016)

—— and UNCCD, 'Addressing the Land Degradation-Migration Nexus: The Role of the United Nations Convention to Combat Desertification' (IOM 2019)

Kälin W, 'Guiding Principles on Internal Displacement—Annotations' (The American Society of International Law 2008)

—— and Schrepfer N, 'Protecting People Crossing Borders in the Context of Climate Change Normative Gaps and Possible Approaches' (UNHCR 2012)

Khan S and Islam MA, 'World Leaders Must Prepare for the Climate Migration Challenge' (Mayors Migration Council, 17 December 2021) <https://mayorsmigrationcouncil.org/news/climate-migration-oped-migrants-day/> accessed 31 May 2024

Khan SMMH, Huq S, and Shamsuddoha M, 'The Bangladesh National Climate Funds: A Brief History and Description of the Bangladesh Climate Change Trust Fund and the Bangladesh Climate Change Resilience Fund' (IIED 2011) LDC paper series

'Kiribati National Labour Migration Policy' (2015)

LDC Expert Group (LEG), 'National Adaptation Plans: Technical Guidelines for the National Adaptation Plan Process' (2012)

—— 'Synthesis Report for the Technical Assessment Component of the First Global Stocktake' (2022)

Madgwick FJ and others, 'Wetlands and Human Migration in the Sahel' (Wetlands International 2017)

Mahajan D, 'No Land's Man: Migration in a Changing Climate', *On the Move: Migration Challenges in the Indian Ocean Littoral* (Stimson Center 2010)

Marsh J and Aziz M, 'Impact of Climate Change on the Migration and Displacement Dynamics of Rohingya Refugees' (Danish Refugee Council 2024)

'Marshall Islands Rules out Relocation in Climate Fight' *RNZ* (4 March 2019) <www.rnz.co.nz/international/pacific-news/383854/marshall-islands-rules-out-relocation-in-climate-fight> accessed 20 June 2024

Marshall R and Rahman S, 'Internal Migration in Bangladesh: Character, Drivers and Policy Issues' (UNDP 2013)

McAdam J, 'Climate Change Displacement and International Law: Complementary Protection Standards' (UNHCR 2011)

—— and others, 'International Law and Sea-Level Rise: Forced Migration and Human Rights' (Fridtjof Nansen Institute 2016)
'Membership' (CVFV20.Org) <https://cvfv20.org/membership/> accessed 24 October 2024
Ministry of Disaster Management and Relief (MoDMR), Government of Bangladesh, 'National Strategy on Internal Displacement Management' (2021)
—— 'National Action Plan (2022–2042) to Implement the National Strategy on Internal Displacement' (2022)
Ministry of Economy, Republic of Fiji, 'Planned Relocation Guidelines: A Framework to Undertake Climate Change Related Relocation' (2018)
Ministry of Environment and Forest, 'Bangladesh: Climate Change and Gender Action Plan' (2013)
Ministry of Finance, Government of India, 'Climate Summit for Enhanced Action: A Financial Perspective from India' (2019)
MoEFCC, 'Mujib Climate Prosperity Plan 2022–2041' (2022)
'Nansen Conference on Climate Change and Displacement in the 21st Century: Chairperson's Summary and Nansen Principles' (2011)
The Nansen Initiative, 'Human Mobility, Natural Disasters and Climate Change in the Pacific' (2013) Background Paper
—— 'Human Mobility, Natural Disasters and Climate Change in the Pacific' (2013) Outcome Report
—— 'Climate Change, Disasters, and Human Mobility in South Asia and Indian Ocean' (2015) Nansen Initiative Background Paper
—— 'Climate Change, Disasters, and Human Mobility in South Asia and Indian Ocean' (2015) Outcome Report
—— 'Agenda for the Protection of Cross- Border Displaced Persons in the Context of Disasters and Climate Change' vol 1 (2015)
National Oceanic and Atmospheric Administration (NOAA) National Centers for Environmental Information, 'Monthly Global Climate Report for July 2023' (2023) <www.ncei.noaa.gov/access/monitoring/monthly-report/global/202307/supplemental/page-1> accessed 29 August 2023
Nellemann, C, Verma, R, and Hislop, L (eds) 'Women at the Frontline of Climate Change: Gender Risks and Hopes' (UN Environment Programme (UNEP) 2011)
'Niger, Mali and Burkina Faso Military Leaders Sign New Pact, Rebuff ECOWAS' *Al Jazeera* (6 July 2024) <www.aljazeera.com/news/2024/7/6/niger-mali-and-burkina-faso-military-leaders-sign-new-pact-rebuff-ecowas> accessed 19 August 2024
Nishimura L, 'Climate Change and International Refugee Law: A Predicament Approach' (*RefLaw*, 18 October 2018) <https://michiganreflaw.wordpress.com/2018/10/19/climate-change-and-international-refugee-law/#_ftn11> accessed 3 September 2019
Ober K, Huckstep S, and Miller S, '"It's Time for Us to Be Included": An Assessment of Refugee and Displaced People's Participation in National Adaptation Planning' (Refugees International 2023)
Office of the Prime Minister, 'Standard Operating Procedures for Planned Relocation in the Republic of Fiji' (2023)
OHCHR, 'Report of the Office of the United Nations High Commissioner for Human Rights on the Relationship between Climate Change and Human Rights' (UN Human Rights Council 15 January 2009) A/HRC/10/61
—— 'Human Rights and Indicators: A Guide to Measurement and Implementation' (2012) HR/PUB/12/5
—— 'Situation of Migrants in Transit' (2016) A/HRC/31/35
—— 'Analytical Study on the Impacts of Climate Change and the Human Right of Everyone to the Enjoyment of the Highest Attainable Standard of Physical and Mental Health, Report of OHCHR' (6 May 2016) UN Doc A/HRC/32/23

—— 'Analytical Study on the Relationship between Climate Change and the Full and Effective Enjoyment of the Rights of the Child' (4 May 2017) A/HRC/35/13

—— 'Key Messages on Human Rights and Climate Change' <www.ohchr.org/Documents/Issues/ClimateChange/KeyMessages_on_HR_CC.pdf> accessed 17 May 2017

—— 'Key Messages on Human Rights, Climate Change, and Migration' <www.ohchr.org/Documents/Issues/ClimateChange/Key_Messages_HR_CC_Migration.pdf> accessed 17 May 2017

—— 'Principles and Practical Guidance on the Protection of the Human Rights of Migrants in Vulnerable Situations, Report of the United Nations High Commissioner for Human Rights' (3 January 2018) UN Doc A/HRC/37/34

—— Slow Onset Effects of Climate Change and Human Rights Protection For Cross-Border Migrants (22 March 2018) A/HRC/37/CRP.4

—— 'Five UN Human Rights Treaty Bodies Issue a Joint Statement on Human Rights and Climate Change.' (19 September 2019) <www.ohchr.org/EN/NewsEvents/Pages/DisplayNews.aspx?NewsID=24998> accessed 20 September 2019

—— 'Human Rights, Climate Change, and Migration in the Sahel' (2021)

—— and FAO, 'The Right to Adequate Food' (2010) Fact Sheet No 34

Omar-Matsumi K, Barnett M, and Schalatek L, 'Broken Connections and Systemic Barriers: Overcoming the Challenge of the "Missing Middle" in Adaptation Finance' Global Commission on Adaptation Background Paper

Organisation for Economic Co-operation and Development (OECD), 'Aggregate Trends of Climate Finance Provided and Mobilised by Developed Countries in 2013–2020' (2022) <www.oecd.org/content/dam/oecd/en/publications/reports/2022/07/aggregate-trends-of-climate-finance-provided-and-mobilised-by-developed-countries-in-2013-2020_d1d90868/d28f963c-en.pdf/> accessed 22 September 2022

—— 'Scaling Up Adaptation Finance in Developing Countries: Challenges and Opportunities for International Providers' (2023)

Pacific Island Forum, 'Forty-Eighth Pacific Islands Forum: Forum Communiqué' (8 September 2017)

—— 'Boe Declaration on Regional Security' (September 2018)

—— 'Forty-Ninth Pacific Islands Forum: Forum Communiqué' (6 September 2018)

—— '2050 Strategy for the Blue Pacific Continent' (2022)

—— 'Pacific Regional Framework on Climate Mobility' (2023)

Park S, 'Climate Change and the Risk of Statelessness: The Situation of Low-Lying Island States' (UNHCR 2011)

'The Peninsula Principles on Climate Displacement within States' (Displacement Solutions 2013)

Potsdam Institute for Climate Impact Research and Climate Analytics, 'Turn Down the Heat: Climate Extremes, Regional Impacts, and the Case for Resilience' (World Bank 2013)

Rahman S, 'How Bangladesh Is Supporting Climate Refugees' *BBC* (8 December 2023) <www.bbc.com/future/article/20231206-how-bangladesh-is-supporting-climate-refugees> accessed 28 May 2024

Raman S, 'Progress Is Slow on Africa's Great Green Wall, but Some Bright Spots Bloom' *Mongabay Environmental News* (3 August 2023) <https://news.mongabay.com/2023/08/progress-is-slow-on-africas-great-green-wall-but-some-bright-spots-bloom/> accessed 20 August 2024

Rayfuse R, 'International Law and Disappearing States: Utilising Maritime Entitlements to Overcome the Statehood Dilemma' [2010] UNSW Law Research Paper No 2010-52

Republic of Fiji, '5-year & 20-year National Development Plan: Transforming Fiji' (2016) <https://www.adb.org/sites/default/files/linked-documents/LD4%205yr%20and%2020yr%20DP%20Transforming%20Fiji.pdf> accessed 20 June 2024

'Request for an Advisory Opinion on the Climate Emergency and Human Rights Submitted to the Inter-American Court of Human Rights by the Republic of Colombia and the Republic of Chile' (2023) <www.corteidh.or.cr/docs/opiniones/soc_1_2023_en.pdf> accessed 16 November 2023

'Request for an Advisory Opinion Submitted by the Commission of Small Island States on Climate Change and International Law: Amicus Brief Submitted to the International Tribunal for the Law of the Sea by the UN Special Rapporteurs on Human Rights & Climate Change (Ian Fry), Toxics & Human Rights (Marcos Orellana), and Human Rights & the Environment (David Boyd)' (International Tribunal for the Law of the Sea 2023) Case No 31 <https://www.itlos.org/fileadmin/itlos/documents/cases/31/written_statements/4/C31-WS-4-1_Amicus_Brief_UN_Special_Rapporteurs.pdf> accessed 29 September 2023

Rigaud KK and others, 'Groundswell: Preparing for Internal Climate Migration' (World Bank 2018)

Sahara and Sahel Observatory, 'Integration of Climate Change Adaptation Measures in the Concerted Management of the WAP Transboundary Complex: ADAPT-WAP' (2019) Regional Project Proposal <https://www.adaptation-fund.org/project/integration-climate-change-adaptation-measures-concerted-management-wap-transboundary-complex-adapt-wap-benin-burkina-faso-niger/> accessed 21 August 2024

Siddiqui T, Islam ME, and Billah T, 'Policy Architecture to Address Disaster and Climate Change Induced Displacement in Bangladesh' (Researching Internal Displacement 2023) <https://researchinginternaldisplacement.org/short_pieces/policy-architecture-to-address-disaster-and-climate-change-induced-displacement-in-bangladesh/> accessed 6 December 2024

Siddiqui T, Islam MT, and Akhter Z, 'National Strategy on the Management of Disaster and Climate Induced Internal Displacement (NSMDCIID)' (Bangladesh Ministry of Disaster Management and Relief 2015)

Solomon Islands Government, 'Planned Relocation Guidelines 2022' (IOM 2022)

'Statement of United Nations Special Rapporteur David R. Boyd on the Conclusion of His Mission to Fiji' (OHCHR 2018) <www.ohchr.org/EN/NewsEvents/Pages/DisplayNews.aspx?NewsID=23964&LangID=E> accessed 31 January 2023

Stephens T and French D, 'ILA Study Group on Due Diligence in International Law: Second Report' (ILA 2016)

Strauss B and Kulp S, 'Bangladesh and the Surging Sea: A Vulnerability Assessment with Projections for Sea Level Rise and Coastal Flood Risk' (Climate Central 2017)

Táíwò OO and Bigger P, 'Debt Justice for Climate Reparations' (Climate and Community Project 2022) <www.climateandcommunity.org/debt-justice-for-climate-reparations> accessed 9 September 2024

Task Force on Displacement, 'Report of the Task Force on Displacement, Recommendations' (2018)

'There Could Be 1.2 Billion Climate Refugees by 2050. Here's What You Need to Know' <www.zurich.com/en/media/magazine/2022/there-could-be-1-2-billion-climate-refugees-by-2050-here-s-what-you-need-to-know> accessed 25 October 2023

Türkelli GE and others, 'Collective Input to the Inter-American Court of Human Rights Regarding the Requested Advisory Opinion Submitted by the Republic of Chile and the Republic of Colombia Regarding "Climate Emergency and Human Rights"' (19 December 2023)

UN Children's Fund (UNICEF), 'No Place to Call Home: Protecting Children's Rights When the Changing Climate Forces Them to Flee' (2017)

—— 'Children Displaced in a Changing Climate: Preparing for a Future Already Underway' (2023)

UN Department of Economic and Social Affairs, Population Division, 'World Population Prospects 2019: Highlights' (2019) ST/ESA/SER.A/423

UN Habitat, 'Forced Evictions, Global Crisis, Global Solutions' (2011)

UN Office for the Coordination of Humanitarian Affairs (OCHA), '2024 Humanitarian Needs & Requirements Overview: Sahel' (2024)

UN Treaty Collection, 'Status of the 1951 Convention Relating to the Status of Refugees' <https://treaties.un.org/pages/ViewDetailsII.aspx?src=TREATY&mtdsg_no=V-2&chapter=5&Temp=mtdsg2&clang=_en> accessed 3 November 2023

—— 'Status of the ICRMW' <https://treaties.un.org/Pages/ViewDetails.aspx?chapter=4&clang=_en&mtdsg_no=IV-13&src=TREATY> accessed 5 June 2024

UNCCD, 'The Great Green Wall Implementation Status and Way Ahead to 2030' (2020) <https://www.unccd.int/sites/default/files/2024-08/1551_GGW_Report_ENG_Final_040920.pdf> accessed 20 August 2024

UNCCD Secretariat, 'The positive role that measures taken under the Convention can play to address desertification/land degradation and drought as one of the drivers that causes migration' (24 June 2019) UN Doc ICCD/COP(14)/19

UNDP, 'Human Development Report 2006: Beyond Scarcity—Power, Poverty and the Global Water Crisis'

—— 'Sahel Human Development Report 2023: Executive Summary' (2024)

UNEP, 'Livelihood Security: Climate Change, Migration and Conflict in the Sahel' (UNEP 2011)

UNFCCC secretariat, 'Climate Change: Impacts, Vulnerabilities and Adaptation in Developing Countries' (UNFCCC 2007) <https://unfccc.int/resource/docs/publications/impacts.pdf> accessed 17 October 2023

—— Progress Tracker: Work Programme Resulting from the Relevant Requests Contained in Decision 1/CP.21' (2018) <https://unfccc.int/sites/default/files/resource/Progress%20Tracker%20PAWP.pdf> accessed 17 October 2023

—— 'Synthesis Report for the Technical Assessment Component of the First Global Stocktake: State of Adaptation Efforts, Experiences and Priorities' (2022)

UNHCR, 'Bangladesh' (*Global Focus*) <https://reporting.unhcr.org/operational/operations/bangladesh> accessed 30 May 2024

—— 'Planned Relocations, Disasters and Climate Change: Consolidating Good Practices and Preparing for the Future' (2014)

—— 'Guidelines for International Protection No 12: Claims for Refugee Status Related to Situations of Armed Conflict and Violence under Article 1A(2) of the 1951 Convention and/or 1967 Protocol Relating to the Status of Refugees and the Regional Refugee Definitions' (2016) HCR/GIP/16/12

—— 'Legal Considerations on Refugee Protection for People Fleeing Conflict and Famine Affected Countries' (2017)

—— 'Legal Considerations Regarding Claims for International Protection Made in the Context of the Adverse Effects of Climate Change and Disasters' (2020)

USAID, 'Lake Chad Basin—Complex Emergency' (26 June 2020) Fact Sheet #3 <https://www.ecoi.net/en/file/local/2023391/lake_chad_ce_fs01_01-03-2020.pdf> accessed 12 August 2024

Walsham M, 'Assessing the Evidence: Environment, Climate Change and Migration in Bangladesh' (IOM 2010)

Warner K and others, 'In Search of Shelter: Mapping the Effects of Climate Change on Human Migration and Displacement' (CARE 2009)

WIM ExCom, 'Synthesis Report for the Technical Assessment Component of the First Global Stocktake' (2022)

Wood T, 'The Role of Free Movement of Persons Agreements in Addressing Disaster Displacement: A Study of Africa' (Platform on Disaster Displacement 2019)

World Bank, 'World Development Report 2010: Development and Climate Change'

—— 'Turn Down the Heat: Confronting the New Climate Normal' (2014)

—— 'High and Dry: Climate Change, Water, and the Economy' (2016).

World Bank Group, 'Systematic Country Diagnostic for the Eight Small Pacific Island Countries: Priorities for Ending Poverty and Boosting Shared Prosperity' (2016)

—— 'G5 Sahel Country Climate Development Report' (The World Bank Group 2022)

World Food Programme, National Agency for Civil Aviation and Meteorology of Senegal and Columbia University's International Research Institute for Climate and Society, 'Climate Risk and Food Security in Senegal: Analysis of Climate Impacts on Food Security and Livelihoods' (2013)

World Health Organization (WHO), 'Sahel Region' (2006) Health Action in Crises

World Meteorological Organization (WMO), 'Past Eight Years Confirmed to Be the Eight Warmest on Record' (11 January 2023) <https://wmo.int/news/media-centre/past-eight-years-confirmed-be-eight-warmest-record> accessed 29 August 2023

—— 'Global Temperatures Set to Reach New Records in next Five Years' (15 May 2023) <https://wmo.int/news/media-centre/global-temperatures-set-reach-new-records-next-five-years> accessed 29 August 2023

Yonetani M, 'Mapping the Baseline: To What Extent Are Displacement and Other Forms of Human Mobility Integrated in National and Regional Disaster Risk Reduction Strategies?' (PDD 2018)

Zhou M and Braam D, 'Community Resilience and Disaster-Related Displacement in South Asia' (Norwegian Refugee Council (NRC) 2015) Nansen Initiative Research Paper

Index

For the benefit of digital users, indexed terms that span two pages (e.g., 52–53) may, on occasion, appear on only one of those pages.

D